# Oil and Gas: Crises and Controversies
# 1961 – 2000

## *Studies and Commentaries by*
## *Peter R. Odell*
*Professor Emeritus of International Energy Studies*
*Erasmus University, Rotterdam*

## Volume 1: Global Issues

Multi-Science Publishing Company Ltd.
Brentwood, England.
© 2001

**ISBN 0 906522 13 7**

*to Jean and*
*Nigel, Deborah, Mark and Susannah*

# Contents

## Section I – Global Resources, Reserves and Supply

## Section II – International Oil Markets

# List of Figures

## List of Tables

# Foreword

The debts I owe for this book extend way back over time and widely over space. My forebears over three generations from the mid-nineteenth century worked in the mining and associated transport industries in the Nottinghamshire and Leicestershire coalfields, and I was thus assured of the genes and/or the social influences which caused energy to run in my blood. Their arduous and ill-paid work did little more than keep the proverbial body and soul together, but they could well lay claim to an importance for their work in respect of the country's economic development and strength. But they also recognised that education was the only possible means whereby a later generation could escape from the constraints imposed by the social system within which their lives had been lived. This was to be my privilege.

By the time I had passed through the portal of a lengthy period of exposure to secondary and higher education from 1941 to 1958 there were already clear signs of the beginning of coal's terminal decline as a source of energy so that my choice of the oil industry as my work environment was "a natural successor". Though my career with Shell International was limited to only a little over three years, it was long enough for familiarisation with the language of the oil industry and for achieving some "feel" for the factors that made it tick. My mentors over that period – notably Geoffrey Chandler and Napier Collyns and my colleagues in the Company's London-based economic division – ensured my understanding of the oil industry's critical variables: to the extent that my earliest contributions on oil and gas in the early 1960s as an academic at the London School of Economics were soundly based on the realities of the global oil and gas industry at that time. Over that period I was also privileged to know and to work with the outstanding UK economist with a central interest in international oil developments, *viz*. Professor Edith Penrose, whose experience and knowledge of the Middle East and of the role of the major international oil corporations in that area was formidable (Penrose, 1968). The rigour and effectiveness of her work was a major stimulus to my own efforts to interpret the then rapidly evolving and expanding international oil industry. In particular, the fortnightly seminars on international oil which we jointly organised each year from the mid-60s (when it was, indeed, the only academically-orientated study programme on the subject), opened up many issues that had long remained undiscussed: at the very time, coincidentally, when the companies were turning their attention to the North Sea on which therefore I was well-placed to comment and advise.

An invitation in 1967 by Michael Posner, then the economic adviser to the Ministry of Fuel and Power, to join him to work on the government's policy on oil and gas did not materialise, because of opposition from the civil servants to the appointment of an informed outsider known to have views on the North Sea which were contrary to those previously evolved within the Ministry under the earlier Conservative administration. Instead, I thus decided to accept an invitation to take up a professorial appointment at the Netherlands School of Economics in Rotterdam: then the epi-centre of Europe's rapidly expanding downstream oil industry and of trading in crude oil and oil products in the so-called Rotterdam market. Moreover, the Dutch natural gas industry based on the giant Groningen gasfield, was just beginning to emerge as a massive new source of European energy.

Thus, the atmosphere in Rotterdam was highly conducive to oil and gas studies. Indeed, as events unfolded over the decade from 1968, it was the perfect place to be from which to examine, record and interpret the fundamental changes in the international oil and gas industry: from the re-establishment of an ordered oil market in the aftermath of the 1967 Arab/Israeli conflict, to the first oil price shock of 1973/4, to the subsequent wholesale nationalisation of the major upstream assets of the international oil corporations in the countries which had collectively formed the Organisation of Petroleum Exporting countries, to the second oil price shock of 1979/81 with its subsequent adverse ramifications for the international oil industry and to the re-structuring of the European energy sector through indigenous oil and gas production

Throughout this exhilarating period, marked by a series of fundamental changes in my field of academic interest, the challenges to keeping pace with emerging events and burgeoning information and of understanding and interpreting them were formidable. Yet the Rotterdam School of Economics (soon to become part of Erasmus University – the first 'new' university in The Netherlands for many decades) provided an environment in which such work could be done – on the basis of the financial resources which were made available in generous measure; in large part, appropriately, as a result of the state's huge revenue windfalls at the time arising from its 50% interest in the production of very low-cost Groningen gas. In institutional terms this eventually led to the establishment within the University's Faculty of Economics of a Centre for International Energy Studies (known as EURICES), to which I was appointed the first Director in 1982.

During the whole of this challenging period Dr. Ken Rosing was my "right-hand man" in the many studies which were undertaken. I would wish to extend my full acknowledgement of his massive contribution to that work, with particular reference to the statistical and mathematical inputs which he made to a number of major research projects. Some of our jointly-authorised papers are included in this volume – with his permission.

A succession of numerous other assistants, students and academic visitors helped to create the opportunities for expanding the range of international energy studies undertaken, not only on oil and gas, but also relating to coal and nuclear power. An expanding range of national, European and international contacts in both academia and in government and industry also helped to stimulate and broaden the range of topics put under examination. This diversity of interests is reflected in the range of papers etc., dating from the mid-1970s to the early-1990s, which are included in this volume.

Following my retirement from EURICES in 1992, I resumed my earlier academic links in London, by now the global centre for the international oil business, at the London School of Economics where I was asked to teach a course on the International Political Economy of Energy in the M.Sc. programme developed by Professor Susan Strange on the Politics of the World Economy. This not only attracted an annual flow of students from many parts of the world – whereby my need to keep up with the subject was required, but it also enabled me to continue to make contributions on international oil and gas issues. Indeed, five of the chapters in this volume emerge from the stimulus given by my latter-day academic activities, with the final contribution dating from 2000: so extending the range of publications constituting this book to a period of almost 40 years. Within each of the two sections the work is presented chronologically by date of publication – except in one or two cases of earlier work for which there was a delay in publication.

Section I of the book is concerned with the physical attributes of oil and gas resources, reserves and supply – with emphases on "how much" and "where". The world's leading oil petroleum economist for many years, Professor M. A. Adelman of the Massachusetts Institute of Technology has characterised the issue of the world's ultimate resources of oil (and gas) as being "unknown, unknowable and unimportant (Adelman, 1993). He is, of course, right in his view on the impossibility of "knowing" the world's prospective ultimate volumes of oil and gas and he is, moreover, also right in theory over the unimportance of estimating how much there will ultimately prove to be. But he was ill-advised not to recognise the need for economists and other social scientists to discuss the issue in a world in which fears of the near-future exhaustion of oil resources have been so widely – and so adamantly – expressed by a succession of earth scientists. The latter's' fears of oil scarcity date back even to the early part of the 20th century (Williamson, 1963) and often led to costly and inappropriate policies and expenditures by governments anxious to mitigate the perceived dangers.

Thus, much of my work has been devoted to the issue so that no fewer than eight of the twelve chapters in Section 1 of this book are centrally concerned with the resources question. In the early 1960s (when the study published as chapter I-1 was written), global oil consumption was little more than 1000 million tons per year and the great oil wealth of the Middle East was only just being revealed. Nevertheless, even then a coming scarcity of oil was already held out to be a danger!

By 1970 oil use had doubled so that, in spite of the massive discoveries of new oil wealth over the preceding decade, the voices indicating a world running out of oil were even more robust – including, for the first time ever, contributions in this vein by some of the major oil corporations themselves (Warman, 1972). At that time, therefore, I though it necessary carefully to rebut the pessimistic and over-simplistic view of the future of oil (in Chapter I-3). Subsequently, based on a thorough and integrated examination of the range of probabilities for volumes of ultimately recoverable resources, for annual additions to reserves and for annual rates of use it became possible to present the complexity of the world's oil prospects for a 100-year period (in Chapter I-7).

In the early 1980s, however, the price of oil in the market escalated to record high levels as a result of political factors (examined in Section II of this book). This development not only led to a decrease in oil use, but it also stimulated large volumes of production and the establishment of significant reserves in hitherto unexpected places, most notably beneath the offshore waters around many of the world's continents. As a result, expressions of fears for oil scarcity were more or less silenced, though not before those fears had severely damaged not only the industry, but also the global economy. My valedictory lecture at Erasmus University in 1991 drew attention to the range of energy sector fictions and fallacies that had been propagated and to the difficulties that these had posed for economies and societies around the world (Chapter I-9).

Yet it was not very long before the misconceptions and misunderstandings over the future of oil and gas re-surfaced – but this time with the added component of fears for global warming and climate change as a result of the increasing use of fossil fuels (Campbell, 1997). Chapters I-11 and I-12 show that this most recent attempt to justify the new claim for a near-future scarcity of oil and gas is as much as ever at variance with the real-world prospects. In these chapters, moreover, the important alternative hypotheses on the origin and nature of hydrocarbons – as abiogenically, rather than biogenically, derived (Porfir'yev, 1974) – are introduced as pointing to a prospect for a potentially even longer-term availability of more than adequate supplies to match even optimistic views on the development of oil and gas demand in the 21st century.

The other chapters in the first section of the book are devoted to the geography of world oil and gas supplies. Chapters I-2 and I-4 examine the importance of the location of markets to decisions on the exploitation of newly discovered and technically accessible reserves. Chapters I-5 and I-6 – dating from research done in the early 1980s – then examine the impact of the changed relations between OPEC and the rest of the world on supplies from outside the OPEC countries. Chapter I-5 deals with the OECD set of countries and Chapter I-6 with the Third World within which the major international oil corporations had hitherto either been reluctant, or unable, to explore for oil, let alone exploit known reserves.

Chapter I-10 revisits these aspects of global oil developments after more than a decade of high prices for oil and gas, compared with the previous long period of low prices for most of the first 70 years of the 20th century. This up-date not only confirmed the existence of much greater oil resources in these countries than previously suggested by the oil corporations, but it also indicated a rapid and extensive *rapprochement* between the governments and companies concerned, given evidence of a "revealed mutuality" of interests between the parties.

Section II of the book is concerned with the economic and political inputs to global oil and gas industry developments since the early 1960s. Chapter II-1 sets the mid-1960s' scene for the organisation and structure of the industry after almost a half-century of effective control over the system exercised by the major multi-national oil companies. Chapter II-2, published in 1968, is concerned specifically with oil developments in the Middle East following a decade-and-a-half of concentration of international upstream investment in that small part of the petroliferous world: as a result of which it became by far the single most important region for oil production and, even more so, for exports of the commodity. By contrast, Chapter II-3 analyses the relationships between the international oil corporations and the highly nationalistic countries of Latin America at the time of acrimony over state/foreign private oil company relationships. It shows in a number of case studies how the political confrontation between the parties inhibited the effective development of oil exploration and exploitation across the continent.

Chapters II-4 to II-8 inclusive present my contemporaneous interpretations of the revolution in the world of oil from 1971 to 1986. The revolution encompassed many aspects of the political economy of oil. First, a shift in power relationships between the producing countries, on the one hand, and the international oil corporations, on the other. Second, a consequential dramatic – eventually a ten-fold – increase in the real price of oil. Third, important spin-off impacts on the global economic situation and prospects. Fourth, misinterpretations of the realities of the changes under-way and of their long-term significance. Fifth, the eventual emergence of competitive markets in crude oil and oil products, compared with the previous "ordered" markets as organised by the international oil companies. And last, but by no means least, a diminution in the role of, and the prospects for, oil use in the emerging global economy. These five chapters present my efforts to analyse what was happening over that 15-year period: and to synthesise the conflicting strands in the set of dramatic events.

Between that set of papers and essays and a later set devoted to developments post the oil price collapse of 1986, there is one chapter – II-9 – which was prepared as a contribution to the book published to mark the opening of the Maritime Museum in Rotterdam in 1986 – on a century of maritime oil transport. This is a particularly important aspect of the political economy of oil throughout most of the industry's history, given the geographical separation of supply and demand on a very grand scale: in terms of both the volumes involved and the distances over which most of it had to be shipped, with a requirement for continuity of supply no matter what physical or political restraints arose.

The final four chapters – Chapters II-10 to II-13 – incorporate my attempts to interpret the evolution of the global oil system, in terms of both challenges and responses, over the 15 years since 1986. They suggest the evolution of a partial recovery of the international oil industry from the days of confusion and little hope in the aftermath of six years of declining demand from 1979–1985. At the end of that period there were not only fears for the well-being of the western system, given the pressures to which the oil shocks had subjected it, but also concerns for the security of oil supplies from both the physical and political standpoints. And there were also questions as to the possible impact that the re-distribution of power in the oil system could have had on relations between the power blocs of the US and the USSR (prior to the latter's demise in 1991), and between the oil exporters (notably in the Middle East) and the importers over most of the rest of the world.

The continuity and still-growing complexity of the issues necessarily means that the book does not conclude with a set of rationally argued and well-defined conclusions as to present status of the global oil and gas system, let alone with forecasts as to how the politics and economics of the industry will develop. I hope that it does, nevertheless, show, first, that the oil and gas system has a long-term future (despite the fears for global warming from the increasing use of fossil fuels); and, second, that it exposes the set of inter-related issues from which the future structure of this long-since globalised industry will emerge.

The thirty-four selected contributions which are presented in this book have been chosen from a much larger number which I published during my continuing forty years' study of the global oil industry. References have been given to other papers not included in this selection, should greater scope and/or depth of description and analysis be required. Those that are in this book are presented much as they were originally published. All the opinions and interpretations as originally expressed have been maintained so as to show how things appeared to be at the time – rather then re-interpreted with the benefit of hindsight.

The contributions have been edited, but only in respect of syntax and clarity of presentation; and, in a few cases, to contain the length of contributions. Graphics, diagrams, maps, tables and bibliographies have also been maintained as in the originals (except, to avoid duplication, when references back to previously presented illustrations or tables concerned are given). I hope that the end-result is not only readable, but that it will also be enlightening with respect to an understanding of the sometimes tumultuous and always interesting years of global oil and gas in the second half of the 20th century.

London, 4 February 2001

## References

M.A. Adelman, *The Economics of Petroleum Supply*, The MIT Press, Cambridge, Massachusetts, 1993.

C.J. Campbell, *The Coming Oil Crisis*, Multi-Science Publishing Company Ltd, Brentwood, 1997.

P.J. McCabe, "Energy Resources; Cornucopia or Empty Barrel", *Bulletin of the American Association of Petroleum Geologists*, Vol.82, No.11, 1998, pp.110–134.

E.P. Penrose, *The Large International Firm in Developing Countries: the International Petroleum Industry*, G. Allen and Unwin Ltd, London, 1968.

V.B. Porfir'yev, "Inorganic Origin of Petroleum", *Bulletin of the American Association of Petroleum Geologists*, Vol.58, No.1, 1974, pp.3–33.

H. R. Warman, "The Future of Oil", *The Geographical Journal*, Vol.138, Part 3, October, 1972, pp.287–297.

H.F. Williamson et al, *The American Petroleum Industry, 1899–1959: the Age of Energy*, Northwestern University Press, Evanston, 1963.

# Section I

# Global Resources, Reserves and Supply

# Chapter I – 1

# The Industry's Resource Base and the World Pattern of Production*

## Reserves and Resources

The generally accepted theory of the process of oil formation implies the existence of a finite quantity of oil available for mankind's use and the spectre of the exhaustion of reserves has often been raised. It is estimated that by the end of 1962 some 18,000 million tons of oil had been used since the industry started about a century earlier. Of this total amount, however, almost 50 per cent was consumed in the decade 1950–60 when the rate of consumption increased very rapidly in most parts of the world. The 1960 consumption of over 1000 million tons has been compared with the figure for proven reserves of oil which stand at little more than 40,000 million tons. The relationship between current production and proven reserves apparently seems to indicate a future for the industry of only forty years and much less than this if consumption continues to expand at the current rate of about 7 per cent per annum. It is largely on the basis of this comparison that the charge is laid that the world is running out of oil, which should not, therefore, be relied upon to provide energy for the future. For example, Mr. W. Wyatt, the Member of Parliament for a constituency which includes the greater part of Leicestershire coalfield, in a speech in the House of Commons on Britain's energy position in 1959, commented:

> *"How safe will it be to reduce our coal consumption.... It may be 30 years before atomic energy can be relied upon to provide*

* Reprinted from *An Economic Geography of Oil*, G. Bell and Sons. London, 1963. Pages 3-27.

> *any substantial load of power. Meanwhile, oil is running out fast.... The consumption of oil in the world is going up by 7% a year.... In 20 years there will almost certainly be a major crisis in the world. The proved reserves today are only 40,000 million tons and even the wildest estimates for future discoveries do not put the total amount of oil in the world at more than 200,000 million tons. A good deal of that is in areas which it is extremely difficult to drill and exploit. In 20 years time the oil companies will be down to about the last 30 years of their supply at the rate of consumption which will then have been reached in the world as a whole."*[1]

In March, 1961, the economic adviser to the National Coal Board, Dr. E.F. Schumacher, expressed the belief that it is 'completely uncertain' whether sufficient oil will be available to supply Britain's energy requirements in the 1980s.[2]

Is the resource base of the oil industry as narrow as these observations suggest? If so, then it seems likely that the general world-wide preference for liquid over solid fuels will be foiled by a physical shortage of oil supplies in the later part of the twentieth century. The answer is not, however, to be found in a comparison of production trends with the figures for proven reserves, for the latter represent only the 'working inventory' of the industry and do not give any indication of the quantity of oil available for future use. Proven reserves is a concept concerned only with those pools of oil in the ground which are known to exist as a result of actual drilling operations and from which a certain proportion of the oil in place (varying from field to field depending on conditions) can be recovered by known production systems at the contemporary level of costs and prices.

Such reserves are proven by exploration drilling and by drilling to delimit the extent of the oil-bearing formations in areas in which oil has been discovered. This exploration and development is not carried out by companies for the sake of proving the underground existence of oil, but in order that they can locate potential supplies of oil sufficient to meet their future requirements,* the volumes of which are assessed by reference to current and expected levels of consumption. Thus, exploration – and hence the level of proven reserves – are a function of demand.

This is clearly brought out in an examination of the relationship

---

* Although in Venezuela it seems likely that legislation may be enacted shortly to require oil companies to 'prove' the whole of their concession areas, only one-eighth of which at the moment have been drilled.

between production (which may be roughly equated with demand) and proven reserves. World production of oil in 1938 was less than 250 million tons and proven reserves were then only a little more than 4000 million tons. By 1960 production had increased four times to over 1000 million tons, but proven reserves at that date stood ten times higher at over 40,000 million tons, even though in the intervening twenty-two years no less than 12,000 million tons of oil had been produced – three times as much as the proven reserves of the earlier date. Even over the shorter period 1950–60, production doubled, while proven reserves increased threefold. Over these ten years the cumulative production was equivalent to about 40 percent of the proven reserves of 1950. In fact, in every year in the post-war period the annual production of oil has been exceeded by the increase in the amount of proven reserves. Between 1958 and 1959, for example, production increased by about 70 million tons while proven reserves in the year increased by 1570 million tons.

**TABLE I-1.1**
**Distribution of world proven crude oil reserves, 1961**
(in millions of metric tons)

| | |
|---|---:|
| Middle East | 25,600 |
| North America | 6,100 |
| Soviet Sphere | 4.575 |
| South America | 3,175 |
| Far East | 1,400 |
| North Africa | 700 |
| Rest of World | 825 |
| *Total* | *42,375* |

*Source: Based on B.P. Statistical Review of the World Oil Industry 1961*

Figures of proven reserves – perhaps given unfortunate and misleading emphasis even by the oil industry – can thus be discounted in an assessment of the resource base. The distribution of these proven reserves at the end of 1961, shown in Table I-1.1, is little more than an indication of the near-future productive capacity in various parts of the world. It has been observed that 'their location and quantity varies by and large in accordance with the direction, intensity and skill of discovery work',[3] and does not indicate the distribution of oil resources. The distribution of proven reserves is, of course, constantly changing as new discoveries are made, resources are withdrawn from established fields and new assessments are made of the oil in place and the likely recovery possibilities. The magnitude of the changes is seen in the growth of Middle East proven reserves from only 700 million tons in 1938 to over 25,000 million tons by 1961, during which period its share of the world total increased from 17 per cent to over 60 per cent.

Having thus discounted the value of proven reserves for indicating the magnitude of the industry's resource base, it is necessary to study the assessments that have been made of the amount of oil in place in the earth's surface and to consider the possibilities of extracting it. The attempt to put a figure on the world's ultimate recoverable reserves has been described as an 'essentially speculative subject'[4] and we are also warned by one of the world's leading authorities in this field that the figure should not be looked on 'with anything like the same confidence as proved reserves'.[5] Even in the United States, where the petroleum geologists have made more extensive and intensive studies than elsewhere in the world, there have been many estimates of the country's oil producing potential.[6] Thus, on a world wide basis the estimates can be of only the roughest kind, for the exercise involves the assessment of vast areas of sedimentary basins which have never been drilled and the geology of many of which is little known.

In spite of the uncertainty, however, the possibilities appear to be very large, for 18 million of the world's 22 million square miles of sedimentary basins – the distribution of which is illustrated in Figure I-1.1 – are considered to be effective basin areas from the point of view of oil potential. Only about one per cent of these areas have been effectively explored for oil.[7] The sedimentary areas have been examined 'basin by basin' by L.G. Weeks with 'a careful study of each basin's geology over a long period'. He has assessed the likelihood of oil occurrence by reference to the geological facts of deposition, potential reservoir porosity, relative location of potential source and potential reservoir rocks and the existence of 'timely and adequate traps'.[8] From his studies he has conservatively assessed the total world ultimate potential of liquid petroleum resources recoverable by conventional primary methods in terms of current economics to be of the order of 300,000 million tons – enough for almost 300 years at the current rate of consumption.

This estimate, moreover, ignores the possibilities of using secondary methods of recovery (which will be examined later) and of the prospects of winning oil from greater depths. To date 85 percent of all the oil produced has come from depths of 2000–8000 feet, but it is 'expected that the deeper picture will improve with time as more and more deeper

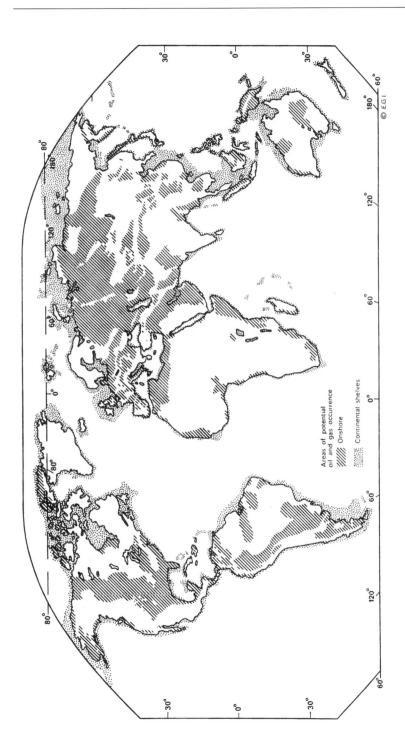

Figure I-1.1: The World's Sedimentary Basins

wildcats are drilled'.[9] This process of deeper drilling is now proceeding with the deepest production to date from a depth of 25,000 feet, compared with only 7500 feet in 1927. Little is known of the deepest possibilities, but it is generally believed that 'the condition of the potential reservoir rocks rather than the breakdown of oil may be the limiting factors. There may be no single figure applicable to all areas, for the critical depth probably depends on the original characteristics of the rock and the setting, including the nature of the cover and the temperature gradients in that cover'.[10] Theoretically, oil may occur in sandstones down to a depth of 65,620 feet and limestones to 51,300 feet.[11] Present exploration has thus hardly scratched the surface of the possibilities in depth. Technological progress in prospecting and in drilling seems certain to ensure that the theoretical limit will be more nearly approached in the future.

New areas of exploration are continually being sought and developed. The Sahara region provided a significant development in the late 1950s and major discoveries are still being made. At the present time there appear to be particularly good prospects in the U.S.S.R. where there are indications of an oil area in Siberia, covering about 400,000 square miles and probably containing larger reserves than those of the Volga-Urals fields which currently supply almost three-quarters of the country's total production. The Canadian Arctic Islands are also thought to offer favourable prospects, for exploration permits covering 43 million acres (about 67,000 square miles) have been taken up by twenty-two companies or groups of companies. An editorial in a Canadian oil journal described the prospects in the following terms:

> 'Sediments thousands of feet thick; structures rivalling in size those of the Middle East; reservoir rocks with porosity and permeability undreamed of in Western Canada; the hope of gushers like Spindletop. The dream of a 100,000 barrels per day (about 5 million tons per year) from a single well. The prospect of 50,000 million barrels (7000 million tons) of oil or more. More oil than the known reserves of Canada and the United States combined.'[12]

The Canadian Government commissioned an enquiry into the potentialities of the Arctic Islands and adjacent areas of the mainland. The report[13] confirms the physical possibilities of oil production on the large scale enthusiastically foreseen in the editorial quoted. There are also indications of new and important basins in parts of the Middle East from which oil is not currently produced. The Trucial Oman coast, for example, holds promise of becoming a major producing area, with particular and immediate attention on the Sheikhdom of Abu Dhabi, where the success of exploration efforts in 1961 and early 1962 have already led to the area being described as a second Kuwait.

If, however, the likelihood of the continued discovery of new fields is entirely discounted, the physical possibilities for oil production are still not limited to the proven reserves. On average, the figure for proven reserves assumes a recovery rate of the oil in place which can be commercially produced at present costs and prices of only 33 per cent. Thus, in the known fields, there is as much as 600,000 million tons of oil presently awaiting exploitation by means of improved recovery technology. A one per cent improvement in the overall recovery efficiency from these fields would produce the equivalent of almost two years' supply at current rates.[14] Through the development of methods of secondary recovery, in which water or gas is injected under pressure into the oil-pool or by means of underground combustion (known as fireflooding), major improvements in recovery rates are assured.

In the United States, where the average primary recovery efficiency is about 35 per cent, and where secondary methods are now being applied more extensively and systematically, it is estimated that the recovery rate will be increased to 60 per cent. In the rest of the world (outside the Soviet Union) much less progress has been made in the application of secondary methods and it is 'not yet possible to build up a world picture with any pretensions to accuracy'.[15] Projects in different parts of the world suggest, however, that the United States' experience may be repeated on a worldwide scale. The Pembina field in Canada, with an estimated 600 million tons of oil in place and a primary recovery of only 12.5 per cent, could be made to yield an additional 225 million tons of oil with the injection at high pressure of heavier gas fractions. Less ambitious secondary methods such as simple gas injection or waterflooding would at least double the recovery rate to give an additional minimum ultimate output from the field of 75 million tons. In Venezuela, in both the Eastern and the Maracaibo fields, the application of secondary recovery methods has increased the recovery rate by up to 14 per cent. In the Middle East, which is only now becoming an active area for secondary recovery operations, there have been comparable developments. In the Abquaiq field in Saudi Arabia it is estimated that secondary recovery methods will lift the percentage recovered from 35% to 70%. On the basis of the limited evidence accumulated to date, it is estimated that even the relatively small number of secondary recovery projects will lead to an additional 11,000 million tons of oil becoming available.[16] The trend in recovery efficiency is now 'strongly upwards' and this alone would seem to ensure an adequate supply of oil for the world for much of the remainder of the century. The application of secondary recovery methods to Week's estimate of total oil in place would raise the world's ultimately recoverable resources from 300,000 million to 525,000 million tons.

This, however, does not exhaust the physical possibilities of producing oil for there are other sources which have not yet been touched. These are the very widespread occurrences of oil shales and tar sands known to exist in many parts of the world. Given the present level of technology and the present price levels for crude oil, these are uneconomic to work, but, in many cases, only a little more than marginally so, such that a small improvement in technology or a slight rise in the price level will make oil extraction from the shales and sands possible. In the United States it is estimated that there are 220,000 million tons of oil in shales with an oil content of 10 gallons or more per ton of shale. Canada also has large shale reserves and, in addition, has some 80,000 million tons of oil contained in the Athabasca tar sands of Alberta. Weeks estimates that in the world as a whole the shales and sands contain energy resources 'as great or greater than those of liquid petroleum'.[17]

Thus, ultimately recoverable reserves of oil from the known and likely resource base seem likely to be in excess of 1 million, million tons and thus ensure the ability of the world oil industry to meet the demands placed on it for at least the rest of the twentieth century without an undue increase in the costs of crude oil production which, it has been estimated,[18] are under £1 per ton in parts of the Middle East and a little more than an average of £2 per ton in Venezuela.* Opinions, such as those quoted above, that the world is running out of oil would thus seem not to be well-founded.

It is in the light of this wide resource base that one must examine the possibility of physical limitations on the geographical distribution of the industry's productive capacity. In fact, it seems unlikely that any major oil producing area or country has reached its physical limits of production. Until recently, there has been a considerable body of opinion that production in the United States was approaching its

---

* It should, moreover, be emphasized that, as these production costs represent only a small part of the final prices of petroleum products to consumers, even a large proportional increase would not necessarily affect retail prices very significantly. For example, consumers in Britain pay about £70 per ton for gasoline, £30 for kerosene, £20 for home heating oils and £10 for fuel oil. Thus, the prices paid by consumers are much more a function of royalty and other payments to governments in producing countries and of refining and transport costs and of sales taxes than they are of the costs of lifting the oil out of the ground.

maximum after almost a century of increasingly intensive exploration and extraction.[19] Such prognostications have, in general, been based on a mechanical extrapolation of statistics of production, proven reserves and recovery rates from field to field. A recent comprehensive and authoritative survey of the energy situation, has, however, challenged both the general thesis and the methods by which such conclusions were reached. The survey has come to the conclusion that the total potential availability of crude oil in the United States in 1975 will be of the order of 850 million tons (compared with a current annual rate of production of 350 million tons) 'at no appreciable increase in constant dollar costs'.[20] It should be noted that this is not a judgement of what production will be in 1975, as this will depend on decisions concerning imports and the regulation of production – the main determinants of the degree of exploration and development efforts – in the whole of the intervening period. It is rather a 'judgement of what production could be, under constant costs, in the light of the resource position and foreseeable technological progress'.[21] This figure of availability for 1975 does not, moreover, include oil that would become available given a higher price level, thus permitting exploitation of higher cost reserves. A continuation or an intensification of the present United States policy of restricting imports of crude oil and products and thus effectively isolating United States' crude prices from those of the rest of the non-communist world, is likely to produce a situation in which more oil pools become commercially attractive. It would also produce a situation in which a start could be made in the use of the country's oil shale reserves which constitute 'an abundant potential domestic source of liquid fuels that is considerably larger than the estimated crude oil base'.[22]

Such an optimistic estimate of the physical possibilities for oil production in the United States is perhaps all the more remarkable in the light of the fact that nowhere else in the world has the potential for petroleum production been explored so thoroughly in the last 100 years, during which period the United States has produced more than 50 per cent of the total cumulative world output. As early as 1940 one well had been drilled to every twelve square miles of sedimentary basins compared with one well to every 1100 miles in the rest of the world.[23] In view of the past performance and the still remaining potential of the United States, which certainly does not have any unique geological advantages, it seems most unlikely that any other significant parts of the sedimentary areas of the world can be written off as having reached, or even as approaching their physical limits of oil production. Such hypothetical physical limits of production might thus be discarded as a determinant of the world pattern of petroleum extraction now and for much of the remainder of the century.

The pattern of production will depend rather on regional variations in the input of capital into more and deeper wells to achieve a higher rate of primary recovery; into research and equipment to improve the chances of secondary recovery; and into the technology and development of the extraction of liquid petroleum from the oil shales and the tar sands. Such regional variations in the application of capital will be in part a function of the physical factors of oil occurrence – such as the climate and associated disincentives to activity in certain regions; the size of reservoirs; and the depth of oil deposits – which obviously affect costs differentially. In greater part, however, they will depend on other considerations such as the location of the areas of potential production in relation to the main centres of consumption, the organization of the industry, the impact of the forces of political and economic nationalism and the internal and external effects of decisions based on a different set of priorities and motivations in the countries of the Soviet bloc.

## The World Pattern of Production

The total world output of crude oil exceeded 1000 million metric tons for the first time in 1960. Just one decade earlier production had reached 500 million tons for the first time so that in the 1950s the annual average increase in output was of the order of 7 per cent. In both 1960 and 1961 an additional 70 million tons was produced and by mid-1962 output was running at a level of over 1150 million tons. The world pattern of production in 1961 is illustrated in Figure I-1.2.

The United States stands out clearly as the major petroleum producer with an output over twice as great as that of any other country. However, production in the United States is now practically stable at about 350 million tons and the annual increases in production since 1958 have been very small. This reflects the impact of schemes, first introduced in the early 1930s, for restricting oil production. The major objective of these prorationing schemes has been the maintenance of crude oil prices.[24] Partly as a result, increasing imports of both crude oil and oil products have been attracted into the country and United States oil has, moreover, been priced out of the world market except for highly specialized products and for purchases of oil made by the United States military authorities for their forces overseas. Additionally, there has recently been a much slower rate of increase in the demand for petroleum and this has also adversely affected production. Thus, United

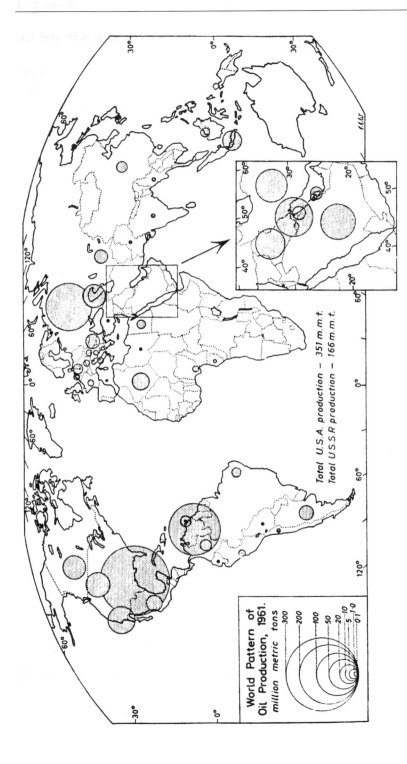

Figure I-1.2: World Pattern of Oil Production, 1961

States production is well below capacity in all of the major producing regions. For example, in Texas, which accounts for about one-third of total output, there is an effective system of restricting output in force. In 1961, only the equivalent of 104 days' output was allowed from each well. This was the lowest figure since the system of prorationing began and much less than the 'daily allowables' for recent years (126 for 1959, 122 in 1958 and 161 in 1957). An indication of the declining relative importance of the United States in the world picture of petroleum production is shown by the fact that it now produces little more than one-third of the world total; in 1950 its share was more than 50 per cent; and in 1940 it accounted for almost two-thirds.

The development of Canadian resources contrasts significantly with the long standing importance of the oil industry in the United States. Canadian production is essentially a phenomenon of the last decade. Production in 1945 was only 1.2 million tons. This increased to 4 million tons by 1950, following the major discovery of the Leduc field in 1947 and there then ensued an oil boom of large proportions which raised output to 18 million tons in 1955 and to 25 million in 1957. Thereafter, the boom conditions faded and it was not until 1961 that there was another significant increase in output (to 31 million tons). It is estimated that about half of western Canada's oil producing capacity was then idle under the impact of unfavourable marketing conditions, arising from a deceleration in the rate of increase in Canadian demand and from competition in eastern Canada from imported oil. This can now be landed at highly competitive prices because of the combination of weak world crude oil price levels and low tanker freight rates. The under-utilization of the available oil resources led to detailed government enquiries into the situation and, as a result of these, the government has set targets for the industry such that production is to be raised to 40 million tons in 1963. The 6 million ton jump in output in 1961 was partly a result of this state intervention in the affairs of the industry, but further significant developments are needed if the target for 1963 is to be met. It will necessitate, for example, increased exports to the United States which, as we have just seen, has under-utilized capacity of its own. Indeed, some sectors of the industry are seeking a curb on imports from Canada to match those on imports from every other country (except Mexico). It will also involve Canadian oil replacing imported oil from the markets of Eastern Ontario and Quebec – a development which will necessitate the construction of a pipeline from Alberta to Montreal. If there is continued weakness in world oil prices, the replacement of foreign oil by domestic supplies may well be achieved only by the imposition of mandatory controls on imports by the Canadian government.

The U.S.S.R. is now the world's second largest producer and is certain to remain so for the rest of the decade, when its output will be approaching that of the United States. The Soviet Union moved into second place only in 1961, when its production of 166 million tons exceeded production in Venezuela for the first time. Production almost doubled in the period 1950–55, and then more than doubled again between 1955 and 1960. This was largely as a result of the expansion of output from the new fields of the Volga-Urals region. Production from the old established fields of the Caucasus has risen little above its pre-war level of 30 million tons per annum in spite of the development of off-shore facilities in the Caspian Sea.[25]

Venezuela is the world's third producer and, in contrast to both the United States and the Soviet Union, it produces primarily for export. Of the total production of 151 million tons in 1961, only 6 million were consumed domestically. Production in Venezuela started in 1921 and by the outbreak of the Second World War had risen to almost 30 million tons a year. Production in the early years of the war varied as a result of German submarine activity but, once this menace had been overcome, output rose rapidly under the impact of greatly increased investment by United States companies. These, by 1945, accounted for over 70 per cent of total output (50.5 percent in 1939).[26] The rapid rate of growth persisted until 1957, when, as a result of the Suez crisis which reduced supplies to Europe from the Middle East, total output exceeded 145 million tons. Production fell back in 1958 as Europe reverted to restored supplies from the Middle East. Since then Venezuelan production has grown at a much slower rate under the impact of generally unfavourable political and economic conditions.

Venezuela has until recently dominated oil production in Latin America. Outside Venezuela, production in Latin America was less than 20 million tons until 1950. At that time Mexico accounted for about 50 per cent and Colombia for another 25 per cent of this amount. Since 1950, however, and more particularly since 1953, Latin American production has increased rapidly and by 1961 had reached a level of more than 50 million tons, of which Mexico contributed 15 million, Argentina 12.5 million, Colombia 7.5, Trinidad 6.6 and Brazil 4.7 million tons. Until 1957, Venezuela's production increased sufficiently rapidly to increase its share of the Latin American total to 84 per cent (compared with 78 per cent in 1950) By 1961, however, its share had fallen to a little over 75 per cent and this trend may be expected to continue over at least the next few years.

The Middle East as a whole is the world's major producing region outside the United States but, as shown in Figure I-1.2, output is divided among several different states, four of which produced more than 50 million tons each in 1961. Kuwait, a state smaller than Wales, produced about 83 million tons and was followed in order of importance by Saudi Arabia (68.5 million tons), Iran (58.8 million tons) and Iraq (50 million tons). The changing significance of the various producing countries in the Middle East is brought out in Figure I-1.3, which shows changes in the pattern of production since 1946. This shows the major changes that have taken place (within the overall pattern of a rapidly increasing output from the area as a whole), under the impact of, first, the contrasting geographical development of new facilities for producing and transporting oil and, second, of the various political crises that have affected the area differentially in the period.

In 1946 production was largely concentrated in Iran, where oil industry development dates from the early years of the century. Iran, however, lost its position as a result of the crisis following the nationalization of the Iranian oil industry in 1951. The giant Abadan refinery all but completely closed down and there was a cessation of exports. This situation persisted until 1954, when the political crisis was resolved, and since then production has risen steadily with new developments designed to off-set the declining production from the older fields to the north-east of Abadan. The consortium of companies that now produces oil on behalf of the National Iranian Oil Company★ has developed the Gach-Saran field, 150 miles east of Abadan. With the construction of a thirty-inch pipeline, with a capacity of over 20 million tons of oil a year, to a new export terminal on Kharg Island, twenty-two miles off the mainland, where tankers of up to 100,000 tons can be loaded, the export potential of the field has been substantially increased. Off-shore exploration in the vicinity of Kharg Island has produced evidence of a new field whose output may also be channelled through the export facilities on the island. Discoveries of oil only eighty miles south of Teheran seem likely to produce sufficient for local consumption and thus eliminate the long and difficult back-haul of products from the Abadan refinery. The Iranian Government's need for increasing oil revenues seems likely to ensure a continuation of the present situation in which increased production is encouraged.

Saudi Arabia was second in importance to Iran in 1946 and is now

★ This arrangement was made under the terms of the 1954 agreement. Ownership of the oil and the fixed assets of the industry are vested in the State.

second to Kuwait. Production has increased steadily over the whole period. New fields have been discovered with great regularity and the productive areas of the established ones constantly extended. The Ghawar field, for example, first discovered in 1948, is now one of the largest in the world. It is already known to be more than 150 miles long and is still being extended. The area of the Safaniya field, brought into production in 1957, was doubled in 1960 and is now the largest off-shore field in the world. Its output increased from 4 million tons in 1959 to over 9 million tons in 1961. Saudi Arabia's proven reserves, thought to be about 7000 million tons, are second only to those of Kuwait and are 50 per cent greater than the proven reserves of the United States. These vast reserves provide the basis for a continuing rise in output, particularly as the ownership of production has now been diversified with the development of the Neutral Zone off-shore facilities by a Japanese company which started to ship oil in 1961.

The small production in Iraq in 1946 came entirely from the north-east fields from which output has since increased more then ten times. The southern Iraqi fields did not start to produce until 1952, but now provide about 25 per cent of the total from the country. Production continued to increase during the early part of the period of disagreement between the government and the Iraq Petroleum Company, but failure to resolve the difficulties eventually led to stagnation in the level of output in 1962. This is likely to persist until the government arranges new organizational means of further exploration and development in the unused areas.

Kuwait was of small importance in 1946 with an output of under 5 million tons. This arose from the fact that little development had been possible during the war, when the need to expand output had generally been concentrated on fields elsewhere which were already producing significant quantities in 1939. Since 1946, however, the average annual rate of increase in Kuwaiti production has been about 12 per cent. The Burgan and Ahmadi-Magwa fields, which have been responsible for most of the oil production to date, continue to be developed, but the North Kuwait field, brought into production only in 1957, is already producing about 10 million tons a year. Moreover, exploration in new areas continues, both in the concession areas of the Kuwait Oil Company (jointly owned by B.P. and Gulf Oil) and in the Kuwait/Saudi Arabian Neutral Zone, where American independent companies are the concessionaires. The off-shore area of Kuwait has recently been awarded to Shell which hopes to produce at least 5 million tons from it in 1965 and, thereafter, to increase production rapidly to 25 million tons per year.

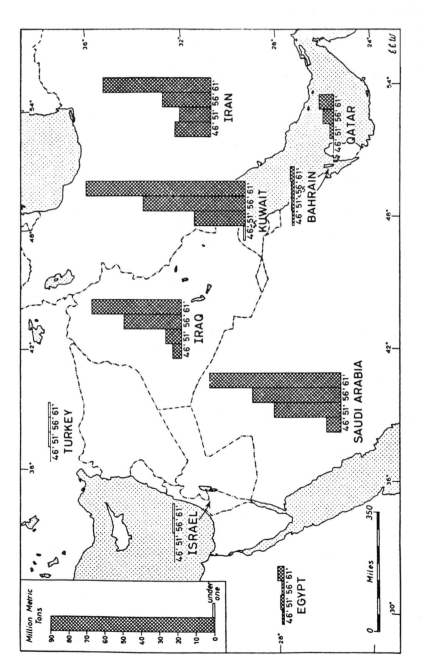

Figure I-1.3: The Changing Pattern of Middle East Oil Production, 1946–61

By that time Kuwait seems likely to be producing over 150 million tons a year and may then be rivalling Venezuela for the position as the world's third largest producing country.

Of the other nine sheikhdoms of the Persian Gulf only two, Bahrain and Qatar, are established producers. The first discovery of oil on the western side of the Persian Gulf was made in Bahrain in 1932, but there have been no further discoveries there since then so that production has stabilized at about 2.5 million tons a year. In contrast, production in Qatar commenced only in 1949 and has since risen to over 8 million tons a year. This is all produced from the Dukham field, now thought to be approaching its maximum output, but an off-shore field has recently been discovered which seems likely to raise Qatar's production to significantly higher levels.

The seven independent sheikhdoms collectively known as the Trucial States are all under concession to the Petroleum Development Company (an associated company of the Iraq Petroleum Company), but discoveries have so far been made only in Abu Dhabi. Although this field – the Murban field – seems to have very considerable potential (some reports have, in fact, indicated that Abu Dhabi is likely to become a second Kuwait) and is located only ten miles inland, the growth of production has been delayed as a result of the difficulty in locating an export terminal on the rocky and hitherto uncharted coast. In the meantime, however, Abu Dhabi joined the ranks of the producing countries in 1962 when an off-shore field, jointly developed by B.P. and C.F.P., was brought into production, following the completion of a submarine pipeline to the operating centre and tanker terminal of Das Island, about forty miles from the mainland.

Oil production in the rest of the world – Europe, Africa, the Far East and Australasia – amounts to less than 7 per cent of the global total. The only producing countries of any international significance are Indonesia, Algeria and Rumania. Production in Indonesia is about 20 million tons a year (with another 4 million tons from nearby areas in British Borneo). This early oil producing area (production first started in 1890) suffered severe damage during the Second World War, when output was reduced to very small amounts. As a result of immediate post-war rehabilitation, there has been a threefold increase in output since 1950, but, in the absence of any major discoveries and in the face of political difficulties between the major producers and the strongly nationalistic Indonesian government which has not granted any new concessions since 1949, the Indonesian contribution to total world oil production has steadily declined. It is as yet impossible to assess the likely impact of the state controlled development efforts, which are being supported by help from Japan and from Rumania.*

Rumania is the only producer of significance, apart from the Soviet Union itself, in the Sino-Soviet group of countries. Production from the old established fields, mainly near Ploesti, reached a peak of 8.7 million tons as long ago as 1936. Partly as a result of wartime damage, this fell to less than 4 million tons by 1947, since when there has been some increase. However, the doubling of Rumanian production from 5 million tons in 1950 to 10.5 million in 1955 was mainly accounted for by the discovery and development of new areas in the south (Oltenia and Pilesti) and the north-east (Moldavia). The second five-year plan envisaged an increase in production to 13.5 million tons by 1960, but this level was not achieved (1960 production was only 11.4 million tons) and the present plan calls for an increase only to a little over 12 million tons by 1965, indicating that no important new discoveries are anticipated.

In contrast with Indonesia and Rumania, the development of large scale oil production in Algeria is a recent phenomenon. This is in part a reflection of the difficulties of the physical environment in the Sahara, but in larger part it indicates the general lack of interest in the area among the major international oil companies, whose attention was riveted on the Middle East in spite of North Africa's greater proximity to the Western European markets. However, French state enterprise persisted in its efforts to secure a supply of oil that would involve France in no foreign exchange expenditure.

This perseverance was regarded in the oil industry generally as 'the pursuit of a mirage', but in 1955 the mirage became a reality with the first major discoveries of oil, and, thereafter, various commercial companies willingly went into partnership with the French for further exploration and development. As the result of intensive efforts, stimulated by the oil shortage in Western Europe at the time of the Suez crisis and by the prospects of Saharan oil securing a privileged position in the French and, possibly, the European Economic Community markets, production was initiated in 1958. By 1961 it reached a level of 16 million tons. Although there may be temporary setbacks to the growth in output as a result of the difficulties accompanying the transfer of political power from France, the new Algerian government is committed to maintain a favourable atmosphere for the further development of the industry. This, combined with the advantage of the country's location in relation to the

---

* These efforts will now be supplemented as a result of the agreement in 1962 by several North American oil companies to explore for and to produce oil on behalf of the Indonesian state oil enterprise, Pertamin.

European market, seems likely to result in a larger Algerian contribution to world oil supplies by the mid-1960s.

Elsewhere in the world, petroleum production is only of local significance as, for example, in Germany, where an annual production of about 6 million tons meets about 15 per cent of domestic demand, and in Austria, where an output of under 2.5 million tons fills about 60 per cent of market requirements. However, there are three African countries which seem likely in the next few years to make a contribution to the world oil industry. These are Nigeria, Gabon and Libya.

Oil exploration began in Nigeria in 1937, but it had to be suspended during the war and was not resumed until 1947. Another nine years went by before oil was finally discovered in 1956. Since then five fields in the eastern deltaic region of the Niger have been brought into production and the country's total output increased rapidly from 300,000 tons in 1958 to about 2.2 million tons in 1961. Steady expansion of output from these fields is expected during the 1960s, as facilities for exporting and refining the oil are developed. By the end of the decade an annual production of about 10 million tons per annum is expected. However, there are also prospects for oil development to the west of the River Niger, where several American and other non-British companies were granted concession rights early in 1962. If these efforts are successful Nigerian production may be significantly higher than 10 million tons by 1970.

Oil production in former French Equatorial Africa (Gabon and the Congo Republic) started in 1957 and had risen to over 1 million tons by 1961. Production from the existing fields seems unlikely to go ahead very quickly, but new exploration by a French State company, in conjunction with Shell and Mobil, both on-shore and off-shore, could lead to significant increases in the output of the area by the mid-1960s.

Of much greater immediate significance is the prospect of rapidly increasing production from Libya, where the search for oil did not start until 1955. It quickly became a favoured exploration area as a result of the discoveries at that time of important fields just across the Algerian border and the introduction of a new Libyan oil law which, by its favourable terms, encouraged oil companies to seek concessions. By 1960 almost the whole of the country was either under concession or sought by prospective concessionaires, which, in their eagerness to secure the more favourable areas, offered terms better than the 1955 law demanded. Oil was first discovered in 1958 and since then no fewer than twenty-five oil producing structures have been located. The 150 producer wells at the end of 1961 have a potential yield of at least 12

**TABLE I-1.2**

**The changing distribution of world oil production, 1930–70**

|  | 1930 | 1940 | 1950 | 1960 | 1970 (estimated) |
|---|---|---|---|---|---|
| World production (million metric tons) | 206 | 300 | 545 | 1050 | 4850 |
| **Percentage in** | | | | | |
| 1 *Western Hemisphere* | *78* | *77* | *73* | *56* | *40* |
| of which: United States | 63 | 64 | 54 | 36 | 21 |
| Canada | – | – | 1 | 3 | 3 |
| Venezuela | 10 | 9 | 15 | 13 | 11 |
| Rest of Western Hemisphere | 5 | 4 | 3 | 4 | 5 |
| 2 *Eastern Hemisphere* | *7* | *10* | *19* | *28* | *38* |
| of which: Middle East | 3 | 5 | 17 | 24 | 28 |
| Far East | 4 | 4 | 2 | 3 | 2 |
| Western Europe | – | – | 1 | 1 | 2 |
| Africa | – | – | – | 1 | 6 |
| 3 *Sino-Soviet Bloc* | *15* | *12* | *8* | *15* | *22* |
| of which: U.S.S.R. | 7 | 9 | 6 | 13 | 20 |

*Sources: Oil Industry Trade Journals. Author's estimates for 1970*

million tons per year. Actual production has had to await the completion of export facilities, but the first development was complete by 1961. This was a thirty-inch pipeline over the 110 miles from the Zelten field to the export terminal at Marsa el Brega. In the last few months of 1961, 500,000 tons were exported, but in the first full year of operation, exports increased to over 6 million tons, and it is expected that by 1963 the pipeline will be used to its 8.25 million tons capacity. This, however, can easily be increased by the addition of more pumping stations when the need arises. Pipelines from two other fields are under construction and should be carrying another 7 million tons of oil a year to the coast by mid-1963. Thus, less than five years after the discovery of oil in Libya, the country's annual production will probably exceed 15 million tons. By the end of the 1960s, Libya seems certain to be ranked among the world's major petroleum exporting countries, although the extremely favourable petroleum law of 1955 has now been modified so that it offers much less attractive conditions for exploration and production. The modified law, however, together with the practice of inviting competitive bids for new and relinquished concessions, only brings Libya into line with other major producing countries. Libya should, therefore, continue to attract investment, with the oil companies anxious to take advantage of the fact that its territory is attractive for exploration and that Libya is in a favourable location in relation to the major European markets.

The changing distribution of world petroleum production over the last thirty years is summarized in Table I-1.2, in which an estimate of the likely pattern for 1970 is also included.

## References

1       *Hansard*, House of Commons. 23 November 1959. Col. 101.

2       E.F. Schumacher, 'The Economic 'Approach to Heating in the Future.' A paper presented to the C.U.C.-N.C.B. Conference, Torquay, 24th March 1962.

3       D.C. Ion, 'Oil Resources in the Next Half Century.' A Paper presented to the Institute of Petroleum Summer Meeting, Torquay, June 1956.

4       D.C. Ion, *Ibid*.

5       L.G. Weeks, 'Fuel Reserves of the Future.' *Bulletin of the American Association of Petroleum Geologists*. Vol. 42, No. 2, February 1958, p.434.

6       S.H. Schurr and B.C. Netschert, *Energy in the American Economy*, 1850–1975, pp. 347–389.

7       L.G. Weeks, *Op. Cit*.

8       L.G. Weeks, *Ibid*.

9       G.M. Knebel and G. Rodriquez-Eraso, 'Habitat of some Oil.' *Bulletin of the American Association of Petroleum Geologists*. Vol. 40, No. 4, April 1956, p.527.

10       G.D. Hobson, 'Why are oil reserves continually growing? Geological Considerations.' *Institute of Petroleum Review*. Vol 15, No. 177, September 1961, pl.273.

11      J.S. Cloninger, 'How Deep Oil and Gas may be expected.' *World Oil*. Vol. 130, No. 6, May 1950, p.60.

12      *Oilweek* Editorial, 26 June, 1961.

13      G. David Quirin, *Economics of Oil and Gas Development in Northern Canada*, 1962.

14    M. Stephenson, 'The Potential of Secondary Recovery in increasing Oil Reserves.' *Institute of Petroleum Review*. Vol. 15, No. 179, November 1961.

15    M. Stephenson, *Ibid*.

16    M. Stephenson, *Ibid*.

17    L.G. Weeks, *Op. Cit*.

18    International Bank for Reconstruction and Development, *The Economic Development of Venezuela*. 1961, p.128.

19    See, for example, E. Ayres and C.A. Scarlott, *Energy Sources; the Wealth of the World*. 1952, pp. 29–47.

20    S.H. Schurr and B.C. Netschert, *Op. Cit.*, p.386.

21    S.H. Schurr and B.C. Netschert, *Ibid.*, p.386.

22    S.H. Schurr and B.C. Netschert, *Ibid.*, p. 387.

23    W. E. Pratt, *Oil in the Earth*, 1942, p.65.

24    See M.G. de Chazeau and A.E. Kahn, *Integration and Competition in the Petroleum Industry* Petroleum Monograph Series, Vol.3, 1959, Chapter 7 for an analysis of the prorationing system in the United States.

25    A recent analysis of the distribution of Soviet oil production and potential can be found in J.A. Hodgkins, *Soviet Power*, 1961, pp.100–133.

26    L.M. Fanning, *Foreign Oil and the Free World*. 1954, p.356.

# Chapter I – 2

# The Geographic Location Component in Oil and Gas Reserves' Evaluation*

My paper begins with a controversial hypothesis: *viz* that the whole concept of *world energy supply/demand* as a relevant parameter in policy planning by companies and nations is a myth. Such consideration of world energy supplies in relation to world demand is in fact relevant only to the very long-term discussion of the world's ultimate energy resource base. That subject is not, I think, the concern of this Symposium, which aims to concentrate our attention on those factors which will influence oil industry or national energy policy decisions over the next 10 to 20 years. Currently, however, there appears to be something of a systematic oil industry attempt to perpetuate and strengthen the myth of near imminent reserves exhaustion, perhaps as a means of securing important short-term commercial advantages from the pressures which can then apparently be justified for significant oil product price increases in those parts of the world which are dependent on oil traded internationally. These parts of the world include West Europe and Japan and also many of the world's poorest nations, which have to import their required oil supplies. Thus we are exposed to 'crisis talk' such as the alleged need to discover more oil in the next decade than has been discovered in the previous century at a cost 'x' times greater than the total investment made in oil exploration/development over the history of the industry. And this, in the final analysis, is a lot of nonsense. Indeed, even assuming no development whatsoever in the world's present understated oil

---

* Reprinted from the *Selected Papers of the International Oil Symposium*, The Economist Intelligence Unit Limited, London, 1973, pp.25–41.

reserves of about 90,000 million tons, plus a continuation of the present annual rate of increase in oil consumption of about 7 percent, to give a cumulative 10-year consumption figure for the 1970s of just over 30,000 million tons and, further, unrealistically assuming away all the barriers to international trade and exchange in oil, then no more investment actually *needs* to be made in oil exploration and the development of new fields for at least a decade. In 1980 there would then still be a 15-year reserves/production ratio, given a 1980 use of oil of over 4,250 million tons. In that the size of existing fields' reserves trends up with increasing production, the reserves/production ratio in 1980 would be even better than 15 years. If, over this same 10-year period, we took the opportunity to spend the money saved on exploration in order to finance a comprehensive examination and investigation of the patterns and functions of oil and gas use with a view to finding more efficient ways and means of utilising them, then, when exploration and development finally had to start again, we might well discover that the problem of maintaining an adequate resource base had disappeared!

Few will need persuading of the very low probability indeed of such a strategy being followed; not because it is inherently impossible, but because the organisation and behaviour of the oil industry is such that exploration and development work cannot be put on ice for a decade. This is partly because the large and powerful exploration departments of oil companies have to continue to have something to do and, more seriously, because the continuation of exploration and development work is oriented towards particular geographical supply/demand patterns not simply in overall oil industry terms, but also within the framework of the behavioural patterns of different companies. Which brings us back to the hypothesis with which I opened this paper – *viz* the inappropriateness of using the world energy supply/demand situation as the basis for policy planning by companies and governments both now and for the foreseeable future.

The validity of this hypothesis may be demonstrated by looking back into the postwar history of the oil industry when, according to conventional wisdom, the internationalism of decision-taking in oil was paramount. However, this conventional wisdom conveniently ignores the fact that during this period the world's major oil-consuming nation (the USA) cut itself off quite deliberately from the international oil supply/demand position and followed the same kind of autarchy in attitudes and policies towards oil as those followed by the USSR. Likewise, though on a smaller scale, we find a similar attitude to oil resources and supply patterns in China – and similarly with Argentina, and with Brazil and so on. And who would now care to argue that countries and/or regions, which have followed policies of energy and/or oil self-sufficiency, have taken inherently bad or completely inappropriate decisions? Perhaps we have been blinded into thinking that such strategies are abnormal by paying too much attention to the contrasting situations that have arisen in the cases of Japan and Western Europe. As far as Japan is concerned, its increasing dependence on oil resources produced elsewhere in the world has arisen out of sheer necessity, given the obvious inadequacy of Japan's currently known energy-resource base to sustain the country's massive industrialisation and the affluent society. And, in the case of West Europe, it arose from the 'choice' of policy-makers as they agreed with energy consumers that the indigenous coal industries should be abandoned in the face of an availability of much cheaper energy in the form of oil from that unique occurrence – the massive and multiple oil basins of the Middle East; workable at low cost and conveniently situated for supplying West Europe and, moreover, located in a region over which European powers until very recently exercised, directly or indirectly, a great deal of political and economic control.

I would now suggest that the strongly import-oriented energy supply patterns of West Europe and Japan over the last two decades are both temporal and spatial aberrations in a more general tendency towards evolving policies that emphasise self-sufficiency in energy within national or regional boundaries. Moreover, this is a tendency that is not only going to continue, but is also going to become accentuated over the period with which we are concerned for current industry and governmental planning purposes. Thus, any analysis of the long-term (15–25 year) demand/supply position for oil and gas with any pretence at concern for real-world situations has to be made within the framework of particular geographical regions, rather than on a general worldwide basis.

Within this required framework for analysis I think I may safely leave the United States to itself – and to other papers in this Symposium. Here, I would merely observe that its developing attitudes to the world's political and economic affairs hardly seem likely to be accompanied by less autarchy than its attitude to energy supplies in recent years. Thus, in the light of currently growing difficulties over the adequacy of domestic energy supplies emerging out of existing policies, one can confidently expect – and probably sooner rather than later – a fundamental reappraisal of national energy resources and consumption patterns, accompanied by political action designed to adjust them much more closely to each other. Furthermore, US policy strategy then seems likely to be devoted to securing the bulk of the required balance of oil and gas supplies which cannot be met indigenously from the rapidly expanding Canadian resource base. The main constraint in the evolution of this policy will be Canada's concern for its ability to withstand, both politically and economically, the inflow of US capital into resource development which the strategy implies. The most serious issue that has to be faced in this respect is the problem of how to absorb so massive an inflow of US oil capital without too great inflationary effects in Canada and their consequential impact on the somewhat precarious relationship between the dollars of the two countries. But, having noted these as the main likely trends in the development of the United States' energy supply over the next 10–15 years, there will still no doubt remain a relatively small residual element in the US total energy supply that will have to be imported either as oil or gas from traditional exporting areas overseas. Conventionally, this residual element in the United States' energy supply is estimated eventually to involve up to another 300 million tons of oil and 100 billion ($10^9$) cubic metres of gas per year. It thus implies a tremendous additional burden on the expected level of resource development and exploitation in the traditional exporting areas, with all that this means not only in terms of the political, economic and strategic world power balance, but also in terms of the capital sums which the industry will somehow have to find to invest in these areas and in terms of the impact it will have upon the price level at which the producing countries are prepared to sell their oil. The question whether or not the USA will, in the final analysis, accept or need to accept the level of imports which the conventional hypothesis indicates, does, as shown above, remain open to considerable doubt in the light of the possibilities for expanding the indigenous and Canadian resource base. In any case, this element of my argument and its conclusions concerned with the impact of burgeoning US imports on the international oil producing and transportation system is undermined by the fact of increasing autarchy elsewhere in the energy-consuming world. This is particularly true of West Europe, where the requirements for imported oil by the end of the next decade now seem almost certain to be less in absolute amounts than the present levels of imports. The significance of this development in West Europe must be seen within the framework of the conventional wisdom about the development of the energy-supply pattern in this region. The industry and the oil-producing and exporting nations are counting on a 50–100 per cent increase in West Europe's oil imports to a total of up to 1200 million tons per year by the early 1980s.

On the basis of such forecasts, there are, in fact, already plans and projects to ensure that physical provisions are made to cope with this size of demand in terms of both producing and carrying capacity for the oil.

The possibility of an emerging West European autarchic attitude towards energy-supply possibilities constitutes such a traumatic and hitherto unconsidered effect on the whole postwar framework within which oil has been considered in the 'world energy context' that it necessitates our special attention in any evaluation of the future organisation and structure of the oil industry in the next two decades. But, at the same time, one should continue to bear in mind that virtually every other country and region in the world is very actively pursuing policies that aim to locate, discover and develop indigenous oil and gas resources. These, if successful, will almost inevitably lead to the utilisation of the indigenous hydrocarbons in preference to the continued importation of energy requirements. In these other cases, much the same arguments concerning the utilisation of any resources discovered – and the financing of their development – will apply, as in the case of West Europe, and thus accentuate the implications for so-called international oil of the continuing geographical proliferation of oil exploration and discovery. Of course, no one national success – with the one exception of Japan – which is considered below – would make much overall difference to the opportunities for the international oil industry and to the demand on Middle East/North African oil reserves, but an accumulation of national successes would, in the final analysis, lead to a situation in which the growth potential for international oil is further markedly curbed. If we assume no more than a modest level of probability that oil will be found in significant enough quantities to stimulate indigenous, as opposed to international, interest in initiating production from amongst the fifty or so nations in which exploration is currently under way, then one still ends up with the elimination of a significant part of the currently expected demand for international oil.

Japan, as noted above, is the one single country outside West Europe where a successful local search for hydrocarbons could in itself have a rapid and traumatic effect on international oil. This is so not only because of the quantities of energy that are involved, but also because of Japan's present almost complete dependence on supplies of international oil. To date, owing to a variety of political-cum-institutional factors, the search for indigenous hydrocarbons in the geologically not unattractive offshore areas of Japan has barely got under way. Now, however, within the framework of acceptable agreements reached with several foreign oil companies, the pace of development is certain to build up and who, looking back to the immense success of Japanese enterprise in virtually every other sector of economic activity in the postwar period, can doubt but that even modest success with the drill will quickly bring a rate of development of the resource base discovered will ahead of that to which the oil world is used. Moreover, the development would most certainly be stimulated within the framework of an energy policy that would given preference to indigenous resources.

The local oil and gas resource base from which such a Japanese policy could evolve remains, of course, to be discovered, but this is not so in the case of West Europe. The now familiar story of the rapid and almost wholly successful opening up of part of the immense potential hydrocarbon-bearing acreage of the North Sea basin is well known enough not to have to be retold in this paper. However, what is needed is the presentation of an alternative evaluation of the resources which takes us well away from the make-believe world of the official presentations by means of which the most interested parties in West Europe, *viz* the energy-consumers, are persuaded that North Sea oil (and gas) developments are going to make no *essential* difference either to the price at which oil can be made available, or to the presently very high degree of dependence of West Europe on oil brought into the continent by a colluding group of international oil companies from a colluding group of oil-exporting countries. Elsewhere, I have examined the development of, and the motivations for, this collusion between producing countries and international companies[1] and have also attempted to establish a rationale for the deliberate understating of the North Sea's potential. More recently, I have outlined the implications for West Europe of a rate of indigenous oil production of the order of 300 million tons per year and of an indigenous natural gas production offering roughly the same amount of energy.[2] The underlying basis for this appraisal lies in two different – but related – considerations. First, in the establishment of a reasonable hypothesis on the size of the indigenous resource base on which a production potential may be determined; and, second, on an evaluation of the melange of political and economic factors at work in determining how far and how quickly this potential will be developed.

Conventionally, the resources and potential resources of the North Sea are 'put into perspective' by comparing them with the reserves of the Middle East, which is much the same as belittling the new multi-millionaire by comparing him with the world's wealthiest man! In this way it is not too difficult to 'prove' that the North Sea, with its currently declared reserves of 12 billion barrels, or its ultimately recoverable reserves of 30–40 billion barrels is of no real significance in the world oil scene compared with the Middle East's reserves of 367 billion barrels: and similarly, though with not quite so emphatic a difference, in the case of natural gas – with the North Sea basin's resources estimated at less than 3 per cent of the world's total reserves and at only one-third of the gas reserves that have been 'accidentally' discovered in the search for oil in the Middle East.

In order further to convince European policy-makers and energy consumers that what has been – or even can be – found in the North Sea basin is little more than enough oil and gas to provide a small part of the increasingly affluent continent's rising demand for energy, the modest calculations of the area's reserves discovered to date are then translated into equally modest annual production potential figures.

The discrepancy between a realistic evaluation of the resources and what is 'known' publicly about the area is startlingly revealed in the gap between the number of fields declared to have been discovered – *viz* 39 or 40, plus several others not yet declared – and the number of fields about which any data at all about reserves have been given – *viz* only 11, of which five are gas fields in southern British waters discovered between four and seven years ago, and most of the rest are oil and gas fields in Norwegian waters about which the Norwegian government has published some information of an excessively modest kind. The absence of data on available resources from other fields is in some cases due to the lack of time since their discovery for making a first publicly declarable indication of their size. Even allowing for this, however, and for appropriate commercial caution in revealing what has been found, given the fact that allocation procedures for concession areas in different parts of the basin are continuing and that trading of concession and drilling rights is in full swing, the lack of information concerning the resources is more than sufficient to arouse suspicions that the oil companies find it advantageous to keep the facts of the rapidly developing situation from the public and the governmental policy-makers in Western Europe.

These suspicions are further confirmed by an equally startling contrast in the 'official' reticence about the prospects for the North Sea basin and the story about the basin which emerges from the already famous article on the *World's Giant Oil and Gas Fields*.[3] Though published in 1970 – that is, well before exploration in the more northerly parts of the North Sea got under way – this article already listed eight fields in West Europe, six of them in the North Sea basin, amongst the world's 266 giant oil and gas fields. By the present time, the six giant fields in the

North Sea basin have at least doubled in number and thus already equal
or exceed in number all the giant fields discovered in the whole of Latin
America over an exploration period dating back more than three-
quarters of a century! What is equally surprising is that the evaluation of
the likely ultimately recoverable reserves of the giant North Sea basin
oilfields already discovered ignores the warning given by Michel
Halbouty in his "Giant Oil Fields" article concerning the evaluation of
new fields; *viz*.

> *"We recognise, however, that… reserves estimates are subject
> to considerable upward revision, particularly in the case of
> newer fields that are not fully developed and thus lack
> adequate production history."*[4]

This warning of the dangers of viewing initial or early declarations
of oil and gas fields' reserves as a basis for long-term decisions on rates
of depletion is illustrated in the case of the first major North Sea basin
field – the Groningen gas field, which for years following its discovery
was said to be a phenomenon of only modest proportions and which,
even in the eighth year of its development, was still declared to the world
at large as a field of only 500 billion cubic metres, the depletion of which
only 'enabled' the operators to sell its limited production as a premium
fuel in the European energy market at, of course, high prices.[5] The fact
that the field is now evaluated as at least four time that size makes for an
extreme contrast between the information on which the policy was
accepted by the public and the currently known position. The latter has
opened up a whole new set of alternative prospects on which to make a
dynamic evaluation of the likely ultimate recoverable reserves of oil and
gas in the North Sea basin. In this respect, too, one notes the initial low
recovery rate figure of 16% used to establish the Ekofisk field as having
only 1,140 million barrels of recoverable oil. An ultimate three to
fourfold increase in the estimate of the reserves recoverable from the 7
billion barrels of oil in place in the field would seem to tie in with
experience from elsewhere in the world for the very light oil in the
Ekofisk field.

In the light of the outside observer's experience to date, therefore,
concerning the contrast between initial and later information on the
development of the amount of oil and gas recoverable from discovered
fields, it would not appear unreasonable to apply an average
multiplication factor of at least 3.0 to the information initially provided
by the companies on the reserves situation of their North Sea discoveries
so far. Thus, instead of the 12 billion barrels of declared recoverable oil[6]

we ought instead to be looking at the policy option implications arising from an ultimate oil production potential available from some 36 billion barrels of recoverable reserves. And this, moreover, is from only ten of the total of 16 oilfields that have been discovered up to now. If we further assume that the *average* size of the other six discovered fields is no more than half the average size of the first ten fields, we then reach the figure of an ultimately recoverable oil potential from the fields already discovered of almost 50 billion barrels! Furthermore, we also know that only about one-fifth of the major structures revealed by the seismic work on the North Sea basin have yet been drilled and so, if we only very conservatively assume that drilling the remaining 80 per cent will discover only as much oil again as from the first 20 per cent of the structures which were drilled up (implying, of course, a very significant fall in both the success rate and the average size of field discovered), then we still end up with ultimately recoverable reserves from the whole basin of the order of magnitude of 100 billion barrels[7] – so ultimately giving the North Sea basin roughly a third of the reserves of *those currently known* to exist in the Middle East (though not, of course, one-third of the ultimately recoverable reserves in that region, in that there are numerous structures in the Middle East that remain to be explored, as well as fields whose ultimate recoverable reserves will turn out to be much greater than those currently declared).

This admittedly tentative but, nevertheless, not unreasonable evaluation of the North Sea oil province adds up to a resource base which will be capable – when fully developed – of sustaining a rate of production of, say, 12 million b/d (ie 600 million tons per year). Such a rate of production will, of course, make a huge dent in the oil import requirements of Western Europe, particularly when, as I have argued elsewhere,[8] the rising availability of indigenous oil will be automatically accompanied by a rising availability of natural gas, in part from the associated gas production of the oilfields. Potential gas availabilities will serve inevitably to take away much of the growth in currently anticipated oil demand. This will occur in a Western European energy demand situation, moreover, in which oil has already lost most of its traditional (25-year-old) growth potential by having largely completed the substitution of coal – given the fact that both the British and the West German governments seem likely to continue their policies of sustaining the remaining demand for coal by subsidies of various kinds. Thus, the present expectation of a doubling of European demand for oil in the 1970s and, thereafter, a further doubling in the 1980s now seems likely to be little more than half-achieved, such that the hitherto forecast 1980 demand for 1200 million tons may only be achieved by 1990. Yet it is over this same 20-year period that West Europe's indigenous oil production potential could rise to a level of 600 million tons per year, sustainable, thereafter, on the basis of a 100 billion barrel resource base, throughout the remaining part of this century and at least through the first quarter of the next. Beyond that, of course, it remains an open question as to whether our societies and economies will still be oil and gas-based at all and so puts us beyond the time-scale of the problems under consideration in this paper.

Moreover, the development of the oil production potential of West Europe is from reserves which, for only the second time in the oil industry's history, lie in a geographical location and in the context of special social, political and economic circumstances that together combine to establish the motivation for, and the commercial possibility of, their being exploited to the technically maximum possible extent as quickly as possible. The previous occasion in which this happened was, of course, in the case of the oil industry in the USA, where the oil and gas resource potential has similarly had every incentive to be thoroughly worked over in the light of a favourable combination of economic, technological and political factors. Thus, in spite of bad production practices in an earlier period of the industry's history (some of which, in fact, have persisted until very recently indeed) and some inappropriate legal provisions, both of which gave rise to the inefficient production of some of the fields, US resources have, over a period of more than a century, almost completely sustained the rising demand for oil – and for gas – in the world's most energy-thirsty region. West Europe seems unlikely to develop either the same voracious appetite for energy or the same propensity to allow wasteful systems of oil and gas production. Thus the reserves of gas and oil of the magnitude that we here envisage are, within a decade, going to place West Europe to a large degree outside the area of operations of the international oil industry, one of the prime functions of which has hitherto been to determine which of the more than adequate reserves of oil around the world should be utilised in meeting the distantly located demands of consumers in oil-deficient areas. Middle Eastern oil reserves, out of which these deficiencies have been met in the post-1945 period, have thus been consistently underutilised in relation to their maximum producible capacity. They seem likely so to remain, given West Europe's rapidly approaching ability to follow the examples of the USA and the USSR in pursuing autarchic energy policies and to eliminate the current and hitherto still expected major growth element in the demand for "international oil reserves"

The more the oil-producing and exporting countries attempt to force up the price of crude by the imposition of quite disproportionate tax rates on a low-cost commodity, so the exploitation of the potential oil and gas resources in all other parts of the world will move ahead even more quickly. Fears for the interruption of supplies traded internationally and of continued unilateral action on the part of oil-exporting countries over taxes and other charges, and concern for the impact of expensive oil imports on the balance of payments and suspicion of the activities and motivations of the international oil companies, together form a powerful set of reasons for favouring the rapid exploitation of whatever hydrocarbon resources can be found indigenously in a consuming country or region: and hence unequivocally spell out the need, both politically and economically, to evaluate resources of oil and gas quite specifically in terms of where they are in relation to demand centres, rather than simply in terms of a given world total (or a 'free world' total) which are described as providing so many years' supply at a current rate of consumption on a world basis.

Viewed in this way, quantitatively apparently quite limited resources – compared with the immense, but severely underutilised, reserves of the Middle East and North Africa – achieve an importance well in excess of that which is attached to them by traditional aggregated analysis and, as in the case of the currently developing situation in West Europe, provide a realistic base on which a new evaluation of the long-term supply/demand position for a region must be made. This alternative, geographically-based, approach to the evaluation of oil and gas resources does now seem to be particularly important, given the rapidly increasing capabilities for exploring for oil and gas in offshore waters. Such new areas for exploration in the case of many important consuming countries provide much more attractive prospects for oil and gas discoveries than do on-shore sedimentary areas in these energy-short regions.

In brief, following through the hypothesis presented in this paper, the immense importance that has traditionally been attached to Middle Eastern reserves is seen to become more than inappropriate. Indeed, it becomes positively dangerous in that the existence of these underutilised – and even under-utilisable – resources has somehow led not only oil consumers and governments of oil-importing countries, but also the oil industry itself, automatically to think in terms of the desirability of a discovered and proved resource base that needlessly provides, as at the moment, a 40-year availability of supply at current rates of consumption.

The fact that this 40-year reserves/production ratio is simply an accident of discovery – arising from the high success rate in Middle

Eastern exploration and the prolific nature of many of the fields – is ignored not only by casual observers of the oil scene, but also by others in both the industry and governments who should know better. They continue to interpret important discoveries in closer geographical relationships with areas of energy consumption in terms of their size relative to those of the Middle East; and argue, implicitly if not explicitly, for the maintenance of this ridiculously high reserves/production ratio. The maintenance of such a ratio implies a level of exploration and development expenditure which represents a wasteful use of scarce capital resources; given that a 15 to 20 year 'apparent' availability of oil and gas from resources indigenous to areas of consumption and from basins and fields undergoing continued development, is an adequate base on which to evolve secure and long-term energy policies. Let us, therefore, cease to present new consuming-region-oriented discoveries in terms of their insignificance as compared with the oil that is known to exist in the Middle East, for this is an illogical comparison of the utility of oil and gas which is *immediately and continuously* developable, with oil and gas whose utility will, if ever, only become fully realisable in the next century. In spite of the very low production costs attached to most Middle Eastern oil (though offset, of course, by high taxes on production and by transport costs), no energy-consuming nation or region is going to give it a second thought, as other than a supplementary source of energy, once significant enough indigenous discoveries of oil and gas have been made which offer a local availability adequate to cover anticipated rates of consumption over a period in excess of a decade. West Europe is currently entering this stage and will within a further eight years or so be in a position to spurn the ridiculous demands of the oil-exporting nations and the international oil companies for oil at $5 to $7 per barrel. This will reflect a situation in which there will be an availability of large quantities of local oil which can be profitably produced and delivered to refineries at no more than $3 per barrel and supplemented by equally large quantities of both associated and non-associated natural gas at even lower costs.

Japan's turn in this respect seems likely to come next, as its continental shelf provides the next attractive region for intensive exploration; and, in the longer term, so will the less developed oil-importing countries. These at the moment are forced to watch their increasing needs for oil imports with a great deal of political and economic trepidation. They, too, will in the future be able to insist on local exploration and, given success, on maximum possible indigenous production. Such attitudes to oil and gas resources are rapidly going to take the oil industry out of its traditional international phase into one in which a whole series of national interests will become paramount and within the framework of which the international companies will be obliged and, in most cases, pleased to work, as they have been in the case of US autarchy in oil. Their longer-term opportunity to do this, however, depends on their willingness in the immediate future not to throw in their lot, for the sake of short-term higher profits, with the traditional oil-exporting countries. The latter can for only a few years more continue to act as though they have the energy-consuming world at their mercy.

## References

1       P.R. Odell, "Europe and the Oil and Gas Industries in the 1970s", in *Petroleum Times*, Vol.76, No. 1929 and No. 1930 (January 1971)

2       P.R. Odell, "Europe's Oil", in *National Westminster Bank Quarterly Review*, (August 1972), pp.6–21.

3       M.T. Halbouty et al., "World's Giant Oil and Gas Fields: Geologic Factors Affecting Their formation and Basin Classification", in *American Association of Petroleum Geologists*, Memoir 14 (November 1970), pp.502–556

4       *ibid*. p.509.

5       P.R. Odell, *Natural Gas in Western Europe*, Ewen F Bohn, Haarlem, 1969.

6       "Oil and Gas Journal" estimate on basis of discoveries made by January 1, 1972.

7       Compare this figure with the figure of 50 billion barrels of recoverable oil strongly rumoured in American oil circles to have been estimated on the basis of the geological and seismic evidence etc as likely to be available from the North Sea basin lying between the 59th and 62nd lines of latitude – or roughly, the relatively small part of the North Sea between the Orkneys/Shetlands and southern Norway. Interestingly, the range of estimates for the North Sea oil province, *viz* 20 billion to 100 billion barrels of recoverable oil, is almost identical with the range of estimates made for Alaska's oil potential. The higher figures for the Alaskan potential are taken quite seriously in official evaluations of energy policy options open to the US over the medium to longer term. Europe would appear to need to do the same in respect of the higher figures of North Sea potential, especially as the technical and other (eg conservation) constraints on getting the oil out of the ground are probably less serious than in the case of Alaska whilst the economics and geography of the local supply/demand position are certainly more favourable for the North Sea.

8       P.R. Odell, *op cit.*, 1972

# Chapter I – 3

# The Future of Oil: a Rejoinder*

In his article, 'The Future of Oil' (*Geographical Journal*, Vol.138, No.3), H. R. Warman justifiably observed 'in our energy hungry society oil has assumed a dominant role with its future affecting economics, politics and the patterns of life' (p.295). It is because oil has come to fulfil this role that, as with the question of war which is too important to be left to the generals, the future of oil is similarly too important to be left only to the oil companies. Hence this lengthy rejoinder on a subject which needs to be opened up to more general debate.

During the past two decades of very high annual rates of increase in oil consumption in Western Europe, Japan and elsewhere, the spectre of the imminent danger of the exhaustion of conventional oil resources has often been raised – particularly by spokesmen for coal industry lobbies anxious to find arguments in favour of the protection of indigenous, but high cost, coal (Odell, 1963). Until very recently, however, such arguments were constantly derided by the oil industry whose spokesmen were emphatic in their response that no restraint should be introduced on the growth of oil demand and who pointed out that the argument for limitation emerged from generally inadequate and incomplete evaluations of the nature of the oil resource base. In as far as the resource-exhaustion spectre was raised largely on the basis of a comparison of the current year's oil production with the so-called proven reserves of oil, the evaluations could certainly be described as inadequate and incomplete, for such proven reserves represent nothing

* Reprinted from *The Geographical Journal*, Vol.139, Part 3, October 1973, p. 436–454

more than the working stock of the oil industry. Thus, proven reserves can only be defined as a function of the demand for oil, given the fact that proving it is so expensive that no one – neither companies nor governments – has any incentive to undertake the task merely to show that the world, or a particular part of it, is oil rich. Oil is sought – and found – because oil companies and state entities have to make long-term investment and infrastructure plans to produce, transport, refine and market the oil they expect to sell and thus must have a firm idea as to where their crude oil supplies are to come from over the next ten years. Moreover, given their expectation, based on their experience over the period since 1945, that the demand for oil will build up year by year they know that the reserves available to them on which to justify such long-term investment must be equal to about 15 times the current annual level of consumption. This, in theory, sets what might be termed the overall optimum level for proven reserves. In practice, in a world of less-than-perfect mobility for the trading of commodities and one divided, moreover, into hostile or potentially hostile groups, the search for security of oil supply will necessitate the discounting of the apparent availability of some resources and the discovery of others in areas on which greater strategic and political reliance can be placed.

This is one of the two main reasons why the optimal reserves/production (R/P) ratio has been greatly exceeded throughout the post-war period (Fig. I.3.1). It should be noted that it was not generally exceeded before 1940 when the proven reserves of the world were concentrated in the United States whose continuing willingness to supply oil was never really in doubt, so removing the incentive for exploration elsewhere. The second main reason for the high post-war R/P ratios is quite simply an accident: the accident of very little exploration effort producing 'too much' by way of reserves.

This was a consequence of the easily discoverable nature, the huge size and the high productivity of the oil-fields of the Middle East. Given specific oil company and oil-consuming-country demand patterns, depletion of these fields can only be achieved over scores of years rather than the 'normal' 15 to 20. This accident of Middle East oil occurrence, discovery and development has thus lifted the R/P ratio well above the optimal. Unfortunately, the post-war build-up of the proven reserves situation has created something of a psychological feeling that the position which was achieved in the late 1950s and early 1960s must be maintained. This is shown in Warman's concern for the future long-term security of oil supplies on the basis of the apparently more recent difficulties involved in maintaining adequate 'remaining reserves' in the

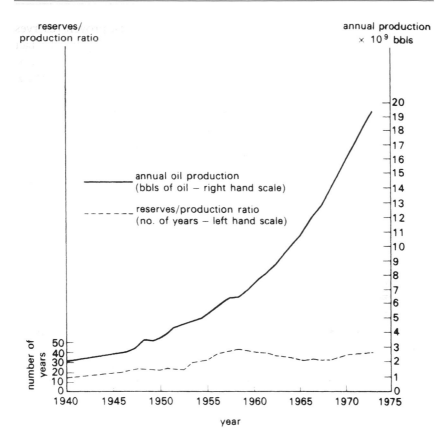

Figure I-3.1: Oil production and reserves/production ratios 1940–70

*In spite of the very rapid growth in oil production especially since the late 1950s, note that the R/P ratio is currently over twice as favourable as it was in 1940 and better than at any other time except for a few years in the late 1950s when it was pushed to more than 40 years by a series of spectacular discoveries in the Middle East.*

light of growing levels of consumption. Here we would wish to challenge the validity of both Warman's use of the concept of 'remaining reserves' and the hypothesis which he employs for calculating the future demand for oil.

## Remaining reserves

Warman's presentation of the concept of remaining reserves *implies* that remaining reserves constitute the world's total store of accumulated oil which is inevitably being used up at a faster and faster rate. From the article itself, however, it can be calculated that the so-called 'remaining reserves' constitute only one-third of the author's pessimistic estimate of the world's total recoverable oil. The relationship between the two totals and the gross of total recoverable oil both require further investigation.

Remaining proven reserves estimates are very much a matter of the timing of an evaluation, for reserves figures grow over time even without new discoveries being made and in spite of continuing oil production from existing fields. This paradox emerges from the initial difficulties and caution in the evaluation of newly discovered fields – both in respect of their size and of the oil-in-place that is likely to be recoverable. As a result, initial declarations of producible reserves err very much on the conservative side as made clear in a recent memoir of the American Association of Petroleum Geologists in which Dr M. T. Halbouty wrote as follows: 'We recognize, however, that... reserves estimates are subject of considerable upward revision particularly in the case of newer fields that are not fully developed and thus lack adequate production history' (1970). Over time, as experience is gained from producing a field and as knowledge of its characteristics increases, so both the estimates of the oil-in-place and of the percentage recovery rates possible within the parameters of unchanging economic conditions are normally enhanced. This process is described in an average appreciation curve as shown in Figure I-3.2 evolved from the study of a very large number of fields over a period of more than 20 years by the Alberta Resources Conservation Board. This is a public authority whose collection, collation and analysis of all the relevant date supplied by the oil companies participating in exploration and production, enables valid advice on resources to be given. A similar body is, in fact, required in Western Europe for advising on the development of North Sea oil and gas. Thus, in spite of an increasing rate of production from a set of fields, remaining reserves increase over a period of years after initial discovery – though how quickly for any given set will, of course, depend upon the relationship of the several variables involved. On average, however, as the diagram shows initial reserves declarations in Alberta have had to be multiplied by 8.89 to get a measure of the true ultimately recoverable reserves: reserves declared three years after discovery must be multiplied by 2.31: and even ten years after discovery the reserves figure must be appreciated by 1.10.

Unfortunately, oil companies are seldom under an obligation to reveal publicly, or even confidentially to governments, the information on discoveries and reserves that they have to give in Alberta. Until

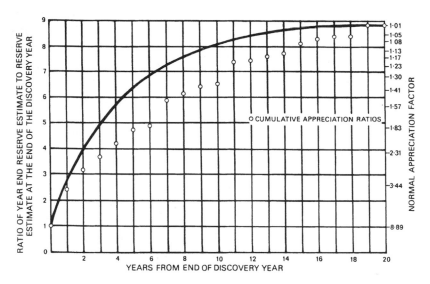

Figure I-3.2: Normal appreciation of initial recoverable crude oil reserves

*From the Alberta Energy Resources Conservation Board, Calgary, publication on Reserves of crude oil etc., December 31, 1971, pp. –31. The accompanying text reads as follows; 'Appreciation factors have been computed by aggregating, for all the reserves considered, the appreciation experience over successive years. The actual values have been used to construct a curve yielding smoothed appreciation factors, or estimates of average appreciation'. 128 oil pools were involved in the analysis which, as far as the author is aware, cannot be undertaken elsewhere given the usual high degree of oil company secrecy over their reserves position. However, the importance of the appreciation factor for any realistic interpretation of the reserves situation as an essential component in policies towards oil, suggests that such secrecy must be lifted and the public given access to the data.*

recently, in many parts of the world, they were under no obligation to give any information at all. Thus 'reserves' figures often reveal nothing more than what the companies think is appropriate for the public to know (a position which exists to quite a large degree in respect of the North Sea basin where the information which companies must give is limited and is, moreover, given on a confidential basis only to the government concerned). Thus, it is impossible to produce an average appreciation curve for the world-wide situation, but two important items of evidence appear to indicate that the Alberta situation is not unusual. Firstly, there is evidence from the United States that the average appreciation rate is of a similar order of magnitude. Secondly, one can compare the world's producible oil reserves for a certain year as they are

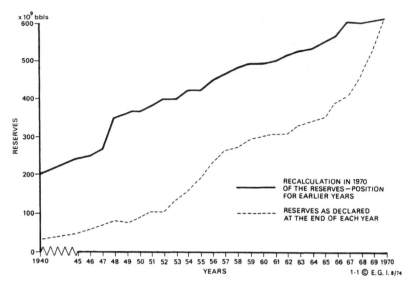

Figure I-3.3: Proven oil reserves 1940–70

*The contrast between proven oil reserves as they were thought to exist in a given year and as they are now known to have existed given 'normal appreciation' over time, clearly reveals the danger in using the presently declared figures of reserves as giving a real indication of the amount of recoverable oil that has been discovered. Given normal appreciation over 20 years, the slope of the upper line will be much steeper to the right giving a 1970 proven reserves position of at least 800x10$^9$ bbls.*

now known to exist, with the declaration of the reserves as made in the year itself. This comparison is made in Figure I-3.3 showing, for example, that the 33 billion barrels (bbls) of recoverable oil as declared in 1940 are now known to have really been 200 billion barrels: while the 90 billion in 1950 were really 365 billion barrels. Given, from the Canadian evidence, the continued appreciation of reserves from discovered fields over a period as long as 20 years, all post-1950 discoveries need to have their reserves figures multiplied by the appropriate appreciation factor (from Figure I-3.2) in order to make them comparable with earlier annual figures and to give a realistic picture of the total quantity of the currently discovered reserves. This is done in Table I-3.1. Thus, for example, the 1955 discoveries now defined at 16 billion barrels, have to be multiplied by 1.03, the appreciation factor for 16 years old fields. Thus a new total of discovered reserves for the year, *viz*. 16.5 billion barrels, is established. Total recoverable reserves from fields discovered between 1951 and 1971 are thus likely to be

## TABLE I-3.1
## Recalculations of annual discoveries of oil 1951–71
## applying the appropriate appreciation factor

| Year | Oil discovered (x 10⁹ bbls) | Appreciation factor for no. of years since discovery year | Adjusted total of oil discovered |
|------|------|------|------|
| 1951 | 30 | 1.01 | 30.3 |
| 1952 | 4 | 1.01 | 4.0 |
| 1953 | 28 | 1.02 | 28.6 |
| 1954 | 12 | 1.02 | 12.2 |
| 1955 | 16 | 1.03 | 16.5 |
| 1956 | 20 | 1.03 | 20.6 |
| 1957 | 24 | 1.04 | 25.0 |
| 1958 | 18 | 1.05 | 19.0 |
| 1959 | 12 | 1.06 | 12.7 |
| 1960 | 8 | 1.08 | 8.6 |
| 1961 | 28 | 1.10 | 30.8 |
| 1962 | 20 | 1.13 | 22.6 |
| 1963 | 45 | 1.17 | 52.7 |
| 1964 | 21 | 1.23 | 25.8 |
| 1965 | 22 | 1.30 | 28.6 |
| 1966 | 35 | 1.41 | 49.4 |
| 1967 | 25 | 1.57 | 39.3 |
| 1968 | 15 | 1.83 | 27.5 |
| 1969 | 16 | 2.31 | 37.0 |
| 1970 | 10 | 3.44 | 34.4 |
| 1971 | 8 | 8.89 | 71.1 |
| Total 1951–71 | 417 | | 596.7 |

almost 600 billion barrels instead of just over 400 billion. The difference of 180 billion barrels between the two figures must be added to the current estimate of 611 billion barrels of remaining proven reserves. These thus become almost 800 billion which, viewed in relation to the 1971 oil consumption of just over 17 billion barrels, indicates a R/P ratio of nearly 50 years. This clearly shows that the present situation is certainly no more unfavourable than it has been over most of the period since 1950. One can conclude, therefore, that this is hardly the moment in time for pessimism and scaremongering over the global proven reserves position.

The *known* position, in other words, concerning proven remaining oil reserves remains good. So good, in fact, that given a world without barriers to trade, one could even argue for a moratorium on oil exploration and the development of new fields for the duration of the present decade. In this period one can look for a cumulative production of oil of about 200 billion barrels and a 1980 rate of annual demand of about 30 billion barrels (assuming a 7 per cent per annum growth rate throughout the period). This means that, in 1980, one would still have a R/P ratio of about 20 years. However, in a world with trading barriers and with highly nationalistic policies towards resources such a moratorium is, of course, unlikely. This is partly because the large and powerful exploration departments of oil companies have to continue to have something to do over this period! Unfortunately there is no possibility of their being converted into well-financed departments responsible for undertaking comprehensive investigations into the patterns and functions of oil and gas use with a view to increasing the efficiency of energy utilisation. Were such financing of consumption problems possible and appropriate action taken, then, by the end of the decade, we might well find that any apparent problem of maintaining an adequate resource base would have disappeared, as each barrel of known reserves would in the meantime have had its ability to sustain economic development greatly increased.

## Ultimately recoverable reserves

The continuation and even intensification of oil exploration and development work may be expected. It is thus necessary to add to what we have presented above (on known remaining reserves) other elements which allow for (a) the fact that much oil remains to be discovered and (b) the probability of continued improvements in the methods of oil recovery and hence in the percentage of the oil-in-place actually recoverable. In both respects the presentation by Warman is excessively pessimistic – and, to some degree, his arguments are internally inconsistent. For example, on recovery rates, he asserts 'it is my contention that… no additional recovery techniques are either known or forecast that can in the next two or three decades seriously alter the recoveries used as the basis for the reserve estimates'. However, as he has also pointed out, recovery rates are, in part, a function of 'economics' – or, more specifically, of the relationship between the additional coasts of improving recovery rates and the price obtainable for the marginal barrel of oil. The oil industry generally expects very significant increases in oil prices over the next decade – from the present average of about $3 per barrel to $6, or even $10 (Chandler, 1973). Such price increases would lead to an enormous increase in the value of all oil-in-place that it would most certainly stimulate investment in secondary and tertiary recovery techniques well within the timespan of Warman's 'two or three' decades. This leads one to the conclusion that Warman's assertion is inattentive to the economics of the oil industry. This conclusion is certainly strengthened when one contrasts his assertion with an apparently general consensus of geological opinion in the United States. There, it is argued, given the right kind of economic incentive, recovery rates could go as high as an overall average of 60 per cent – the figure used as the basis for estimates of the country's future recoverable resources of oil (Cram, 1971; Kastrop, 1972). An explanation as to why the rest of the world should be left so far behind the United States in respect of probable significant upward adjustments in recovery rates is surely required before the validity of Warman's highly pessimistic assertion can be accepted.

Similarly, one also looks for further description and explanation of the 'independent work by Warman and his colleagues in BP' for their calculation that '1800 billion barrels is a reasonable maximum' of the ultimately recoverable reserves of all crude oil from traditional sources. The evidence advanced in the paper itself for such a fundamental break in the trend of estimates made by all other experts is slight indeed. The significantly 'off-trend' nature of the estimate is shown in Figure I-3.4. It is certainly not in line with the 'consensus of opinion', as Warman claims, unless one ignores the significance of the sequential timing of the estimates by simply averaging the set of forecasts as they have been made over the last 20 years. But the temporal element is critical given that the sequence of figures emerges from increasing information on hydrocarbon resources as a result of geographically more extensive, and scientifically more accurate, knowledge on the nature of the oil resource base in the earth's crust. The mid-point of Warman's estimate of 1200–2000 billion barrels of ultimately recoverable reserves takes us back over a decade in time to 1960 in terms of the forecast trend concerning the size of the resource base: but no evidence of any kind is presented as to why the immense developments both in the known resource base and in the expectations of producible oil being found at greater depths and in increasingly large offshore locations should be ignored (Weeks, 1973).

One notes, however, the importance which Warman attaches to the decline in the 'annual discoveries of oil in the recent past' as a basis

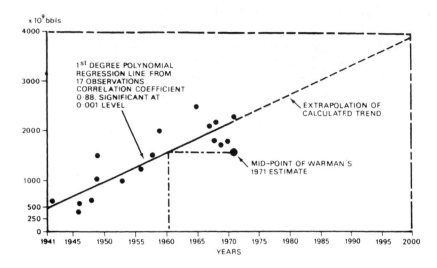

Figure I-3.4: Estimates of world ultimate reserves of crude oil from conventional sources (showing extrapolation to the year 2000)

*A statistically significant trend line has been fitted to all the estimates of ultimately recoverable oil made between 1941 and 1971 – except that of Warman, the mid-point of whose estimate of 1200–2000 x 10⁹ bbls takes us back in time to 1960–61 in terms of the trend line fitted to the data. Extrapolation of the trend line shows a year 2000 situation of almost 4000 x 10⁹ bbls of ultimately recoverable reserves: a situation which would enable the existing relationship between consumption and reserves to be more than maintained.*

for his pessimism. These data are interpreted by Warman as indicating a deterioration of the position. This, however, as shown in Figure I-3.5, does not appear to be a reasonable interpretation of the data set. The only statistically significant trend lines which can, in fact, be fitted to it are, as shown in Figure I-3.5a, impossible to extrapolate in any meaningful way – as no one would argue that discoveries will cease immediately or within a few years. This statistical analysis serves to confirm that the time series data as presented are not a valid set of observations in that the discovery figures for more recent years need adjustment by the Reserves Appreciation Factor previously described. However, statistically significant trend lines based on Warman's data for all years except the last five (the ones most seriously affected by the lack of an applied appreciation factor) can be established and, as shown in Figure I-3.5b, are strongly positive. This indicates how exploration by the industry has responded more than adequately to increasing oil demand over time.

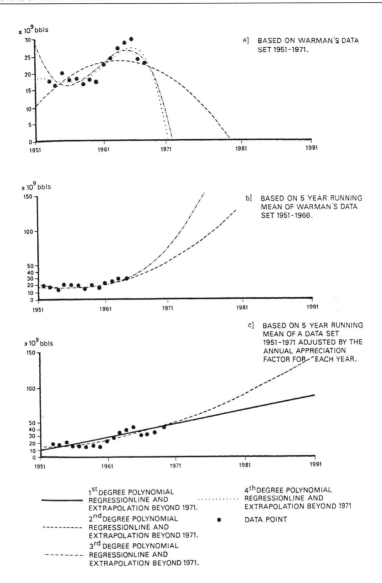

Figure I-3.5: Trends in oil discoveries since 1951

*These exercises try to pick out statistically significant trends in sets of data on oil discoveries. (a) indicates that there is something wrong with the data set as all statistically significant trend lines indicate an almost immediate cessation of oil discoveries – which is obviously nonsense. What is wrong is the non-adjusted nature of the figures for discoveries in recent years – discoveries, that is, that will appreciate strongly over time with increasing information and less secrecy. (b) therefore, eliminates the last five years data from the unadjusted data set and the significant trend lines are now very strongly upwards – also to give somewhat nonsensical forecasts of future annual discovery rates. (c) is based on the data set as adjusted by the appropriate appreciation factor for each year since 1951 and gives a very good fit straight line trend (correlation coefficient 0.93, significant at 0.001 level), suggesting that discoveries to date have emerged from the requirements of rapidly increasing oil consumption.*

However, it would seem even more reasonable to utilize the calculated range of appreciation factors for each year's oil discoveries since 1951 (see Table I-3.1) before trying to establish the trend in the situation. This is done in Figure I-3.5c in which the adjusted figures for annual discoveries have been used to calculate a set of data giving the five year running mean. The statistically most highly significant curves have been fitted to this data set. Both the first degree and the second degree curves demonstrate a highly positive conclusion as to the way in which new oil resources have continued to be found, whilst extrapolation of the fitted trends to 1991 offers evidence which appears to invalidate the argument that we may be 'at or just over the peak in the finding rate'.

## New Reserves from New Exploration

Thus, the basis for Warman's pessimism over the future availability of oil reserves can now lie only in his belief that the world in running out of opportunities for discovering new oil. This belief is backed by two arguments – firstly, by his feeling that Middle Eastern discoveries will decline and secondly, from his assertion that technology has so improved that 'major fields do not go undetected for long'.

The first argument implies that the Middle East has been thoroughly searched for oil. This is really too much to accept given the still relatively limited period of only 25 years in which exploration in the Middle East has been constantly under way (compared with twice as long in Venezuela, three times as long in Mexico and four times as long in the United States). In Alberta, where the oil resources have been developed over roughly the same time – but very much more intensively with 12,758 exploration wells in a total sedimentary area only about 50 per cent that of the Middle East where, by contrast, fewer than one-tenth as many such wells have been drilled – the State Conservation Board still calculates that less than half the required work for a full investigation of the oil resources has yet been made (Alberta Energy Resources Conservation Board, 1971).

Exploration in the Middle East is much less well developed, with one main reason being the sterilization until very recently of very large potential oil bearing areas by the original system of concessions. This 'awarded' whole countries to a single company for exploration and, given phenomenal success by the company concerned in its search for oil simply by looking in very limited parts of the total concession area, there was no incentive at all for its exploration efforts to expand geographically. Iran eliminated this situation in 1954 when it limited the Iranian Oil Consortium's exploration rights to cover only a small part of the country and thus opened up opportunities for new companies to move in to look for oil. The oil search was thus geographically and organizationally diversified and this has led to new fields and new production which would not otherwise have been developed – and has consequently helped to make Iran the world's largest oil exporting country in the last few years.

Much more recently the other three main oil producing countries in the Middle East – Saudi Arabia, Kuwait and Iraq – have similarly withdrawn the monopoly or near-monopoly concession rights previously granted to single companies and are in the process of reallocating very large acreages to other oil entities. In Iraq, for example, the Iraq Petroleum Company has had to give up over 90 per cent of its acreage, most of which it had had no incentive to explore during the concession period. Thus, in these countries the rate of oil search may now be expected to build up beyond the level which one would previously have anticipated from the monopoly operators whose long-term expectations of their crude oil requirements from the Middle East were already more than fully met by the reserves they had discovered in small parts of their total concession areas.

In other words, one would argue that the probability of future finds of oil in the Middle East is being enhanced by the bigger total exploration effort which now has to be made by the larger number of companies with concession and/or contractual interests emerging from the new policies of the oil producing countries. The scope for increased exploration activity is demonstrated in the contrast between the 5251 geophysical crew-months of exploration activity in the United States in 1971 with the less than 1500 crew-months in the whole of the Middle East (*International Petroleum Encyclopaedia*, 1971). If the already intensively explored United States offers scope for so much new exploration – with obvious expectations of sufficient success to justify the cost – then how much more scope remains in the Middle East with its as yet still largely unexplored basin areas.

The second argument used to sustain the view that major discoveries are coming to an end is that modern technology has ensured that what exists has been found. This view can be challenged by two counter-arguments. First, by the fact that the discovery of major oil fields by whatever technology is available depends on active exploration being undertaken in the sedimentary areas concerned. The major North Sea basin fields remained undiscovered until now because the oil province was not explored on any meaningful scale until the last decade. Elsewhere in the world outside North America, there remain many times more potential oil bearing basins to explore than have already been explored. Exploration will, however, be initiated successively in the new areas as economic and political conditions establish the appropriate incentives – today in the case of the North Sea, tomorrow in offshore Japan and the day-after-tomorrow in, for example, the Amazon basin or offshore south-east Asia. Then, and only then, given the appropriate technology, will the major fields be quickly found – but their discovery is a function of it being worthwhile for some company or state-entity to spend money on the effort involved!

Secondly, there is evidence to suggest that even following exploration with modern technology, major fields can and do remain undiscovered over long periods of time. The United States has led the way in advances in oil exploration and development technologies – both onshore and offshore – and such advances have naturally been applied in the continuing domestic search for new oil and under the stimulus of attractive economic, political and legal conditions. In spite of this historical combination of highly favourable circumstances, the current expectation in the United States is that the presently known 45 giant oil fields (i.e. fields with over 500 million barrels of recoverable oil) will increase to 51 as 'recent discoveries are eventually recognized as giants' and that, in addition, another '25 to 28 more giant fields remain to be found between now and 2060' when the last one – based on the extrapolation of current trends – will be discovered (Moody, Mooney, and Spivak, 1970). With this kind of result of the analysis of the potential for further major field's discovery and development in the United States, with its long and detailed exploration experience involving the use of a gradually evolving and innovating technology, it seems surprising, to say the least, that the rest of the oil world is presented as being so much more efficient in its near immediate ability to discover all major fields so soon after initial exploration. What makes the view even more surprising is that almost all the companies involved in the rest of the world are also involved in exploration in North America – why do they perform so much less efficiently in the latter area than they are apparently able to do in the former?

## Oil reserves and demand

Thus the validity of the bases on which Warman apparently made his pessimistic evaluation of potential oil supply can be questioned to such an extent that we must be forced to conclude that his case remains unproven. Alternative evidence, as presented in this rejoinder, appears to point towards the existence of a much larger reserves potential for the sustenance of future demand – in both the medium and longer term. One indication of the size of the discrepancy lies in the difference between Warman's 600 billion barrels figure for discovered oil and the 800 billion barrels that this becomes with the application of appropriate appreciation factors to the declared figures. Another indication is in the contrast between Warman's estimate of some 230 billion barrels of producible oil left in the United States and other estimates which put the figure from 20–70 per cent higher at 275 to 390 billion barrels (Kastrop, 1972). In light of the continuing American experience in finding new oil, we would, therefore, reassert our argument that growing demands do and can continue to provide the motivation for the expansion of the known resource base over long periods of time and well beyond the earlier expectations on reserve availability. The growth of demand and future supply are thus symbiotically related (Odell, 1963) in a way which makes it impossible for consumption to run away from the supply, as suggested by Warman when he compares an exponential growth rate in demand with his pessimistic evaluations concerning potential supply. If, in fact, the future production potential of oil is as bad as is argued, then demand – even in the medium term – will be automatically adjusted to the scarcity situation through the pricing mechanism. This will price oil out of those markets which cannot afford to pay higher prices for it, so altering the shape of the demand curve and making societies more conscious of energy conservation – in, for example, their insistence that mass transit facilities replace the use of private cars for commuting, whilst long distance rail passenger and freight transport replace the use of air and road facilities. Societies will, in other word, have to become less 'energy hungry' and oil will lose part of the dominance that it has secured in energizing development. The response will, it should be emphasized, be automatic so that the crisis at which Warman hints will never arrive.

Quite apart from this in-built kind of constraint on the growth of oil demand, however, one also has to question the validity of the use of 'the hypothetical case of demand increasing at 7.5% per annum to the end of the century'. This gives a cumulative use of oil from 1971–2000 of nearly 1750 billion barrels and a year 2000 use of about 130 billion barrels (compared with 17 billion in 1971). This is a frightening forward estimate of the emerging situation. Its use also involves, according to Warman, a 'need' to discover another 4000 billion barrels of oil over this period in order to sustain a minimum 15 years reserves/production ratio at all times. Compared with total discoveries of oil to date of less than 25 per cent of this amount, the apparent 'needs' are astronomical, not only

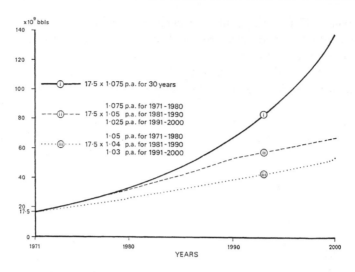

Figure I-3.6: Annual oil consumption 1971–2000 at different growth rates

*This compares the result of Warman's assumed 7.5 per cent annum growth rate in oil consumption with the results of other possible sets of growth rates. It demonstrates how quite small changes in the assumptions about growth rates can make enormous differences to future annual levels of oil consumption.*

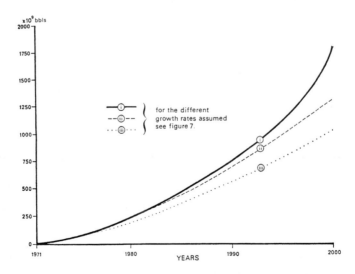

Figure I-3.7: Cumulative oil consumption 1971–2000 at different growth rates

*As in Figure I-3.6 above, different assumptions about future growth rates in oil consumption make enormous long-term differences to the cumulative use of oil and thus indicate both the need for very careful analysis in attempts to predict the future, and the desirability of making sure that each barrel of oil is effectively used.*

in relation to Warman's pessimistic calculations about the size of the oil resource base, but also compared with the alternative, larger resources' potential set out in this rejoinder.

## A reasonable demand hypothesis 1970–2000

Warman's 7.5 per cent annum cumulative-rate-of-increase-in-demand hypothesis does have the merit of simplicity: but simplicity is far from appropriate, as forecasting demand is even more difficult than forecasting the potential development of supplies. Simply to dismiss the whole demand problem in a few lines of an article entitled *The Future of Oil* is not good enough. There is even worse than this, moreover, in that the author, having recognized that extrapolation of recent trends is inappropriate, given the likelihood that the 7.5 per cent per annum rate 'will probably drop to 4–5 per cent per year in the ensuing decade', then ignores the lower figure in presenting his comparison of 'needs' with 'availabilities'. Yet, as Figures I-3.6 and I-3.7 show, the choice of growth rates makes a world of difference to the evolving situation. The substitution of lower than 7.5 per cent growth rates over all or some of the period reduces the cumulative use of oil from nearly 1750 billion barrels to 1250 and 975 billion barrels respectively: and give a year 2000 consumption of 61 and under 50 billion barrels instead of the 130 billion of Warman's hypothesis. Finally, the impact of the lower trend rates in 1970–2000 consumption on the need for new oil reserves by 2000 (additional that is to the 600 billion barrels already known in 1970), implies only 1500 billion barrels in the first case and 1150 billion in the second, instead of the 4000 billion barrels hypothesized by Warman. In other words, the problem is only 37.5 per cent or 27.5 per cent as great as Warman would have us face in making our energy policy evaluations.

Quite apart from the automatic pricing mechanism adjustment of demand to supply as discussed above, there are very good and immediately recognizable reasons for forecasting a reduced rate of increase in the demand for oil in the last quarter of the twentieth century as compared with the third quarter. Three main components can be recognized in this reduced-rate-of-increase-of-oil-demand hypothesis – *viz*. the relationship of coal and oil; the relationship of oil and natural gas; and the relationship between economic development and energy use.

*Coal and oil* – the last twenty-five years have been marked by an intensive substitution of coal by oil in the world's traditional regions – North America, Western Europe and Japan. In these regions coal has been abandoned or almost abandoned in many end uses (as, for example, on the railways, for steam raising in factories and for home heating) so that annual coal consumption has diminished by as much as 500 million tons from the levels of peak demands in the different areas at different times (eg. by 1943 in the US and by 1950 in Western Europe). By and large the use of coal was replaced by the use of additional oil. Without this substitution of coal by oil the average annual rate of growth in oil use in the post 1945 period would have been reduced by about one-sixth to 6.4 per cent instead of the 7.7 per cent shown by the unadjusted figures. This reduction is significant in any present attempt to forecast the future of oil demand in that oil for coal substitution has either ended or is now virtually at an end in the regions mentioned above, where present levels of coal use will either be maintained (as in the United Kingdom and West Germany as a result of government help), or even increased (as in the United States where coal is fully competitive with other fuels for electric power generation over large parts of the country). In other words, there will be no further growth in oil use as a result of oil replacing the use of coal in the world's most intensive energy consuming regions. Moreover, in the rest of the world the substitution of coal by oil is seldom of any importance in that the use of coal as an energy source never developed to anything like the same degree as in the 'old' industrialized countries and there is, therefore, little coal use to substitute. This factor thus provides an automatic check on the high oil demand growth rates to which we have become accustomed. For this reason alone extrapolation of the 7.5 per cent annual rate of increase in oil consumption is an unsubstantiated and unrealistic procedure.

*Oil and gas* – This second factor is also concerned with substitution – but this time the substitution of oil by natural gas. This phenomenon developed in the United States in the late 1950s and the 1960s as the pipeline infrastructure for distributing the gas from points of production in Texas, Louisiana etc was expanded to cover the main energy consuming regions of the country. The inherent advantages of natural gas (*viz.* cleanliness, ease of handling and automatic delivery) have enabled it to replace oil fuels in many uses – often in spite of premium prices – in most industrialised countries. More recently there has been similar replacement of oil by natural gas in Mexico, Venezuela and the Soviet Union. To date, however, the opportunities for such replacement have been limited by the technical and economic constraints on transporting natural gas. Very large quantities produced in the Middle East, North Africa and South America as a 'by-product' of oil production have not been able to find markets and have thus simply had to be flared. Recently, however, the technology of transporting natural gas in a liquefied state has made very significant advances and large contracts have been signed for the long-distance delivery of LNG (liquefied natural gas) from its main areas of production to Western Europe, Japan and the United States. These developments represent the beginning of the use of a hitherto unutilized large energy potential, the supply of which is automatic (as a consequence of oil production) and at virtually zero cost at source (except for collection and liquefaction costs). Thus it will increasingly be burnt in the end-uses for which oil would otherwise have been used – especially when and where there are firm environmental constraints on the use of oil products which cause atmospheric pollution.

Gas for oil substitution, however, is also developing from another set of circumstances with probably even greater short to medium-term importance affecting oil demand growth rates. This is the recent discovery of major natural gas resources in the heart of energy using regions – including relatively small energy using countries like Australia and New Zealand, but also including the world's most energy intensive using region of Western Europe. Here the rapidly expanding production of the Dutch Groningen field (the largest gas field in the world outside the Soviet Union) has virtually 'killed' growth in overall oil use in the Netherlands, where the consumption of fuel oil is already 30 per cent less than it was in 1969. Exports of Dutch gas, plus the rapid increase in locally produced supplies, are also having a major impact on the hitherto expected rates of growth in oil demand in Germany and Belgium.

North Sea gas has similarly provided Britain with a new energy source. By 1975 this will be providing energy which had previously been expected to come from some 300 million barrels of oil. Even more recently, more northerly parts of the North Sea between Scotland, the Orkneys and Shetlands and the coast of Norway have been proved to have a series of giant oil and gas fields with an annual natural gas production capability which is the equivalent of at least 750 million barrels of oil. Given the intensive region of energy demand surrounding the North Sea at relatively short distances from the points of potential production, the rapid incorporation of this new energy source in the European energy economy is assured. The first contracts for the delivery of over 26 Bcm of natural gas per year for a minimum of 20 years from these fields have already been signed. Gas from the Ekofisk field in Norwegian waters will go to customers in Germany, the Netherlands, Belgium and France, whilst the British Gas Corporation has negotiated for 14 Bcm of gas per year from the Frigg field, which lies partly in Norwegian and partly in British waters. As the quantities involved are so large, there can be no question of it merely filling Europe's incremental energy markets. It must replace some oil use rather than coal in that the two countries – Britain and West Germany – where most of it will be delivered, have decided to protect and guarantee coal production and use at about present levels. Thus, in a decade's time, as a result of the build-up of the availability of gas, the annual use of oil will be nearly 3 billion barrels less than has hitherto been expected and so undermine the basis on which the continuation of the post-war rate of increase in oil demand has previously been predicted.

The possibilities of a similar, though certainly a not so immediately dramatic development as far as Japan is concerned also exists – given the recently initiated intensive search for natural gas on the extensive continental shelf of that country, the first successful results of which have already been announced. Thus, from the overall point of view of the global trends in energy use one can evaluate that the substitution of natural gas for oil is set to play an increasingly important role in limiting the future rate of growth of oil demand, most notably in those regions of the world where energy is used most intensively and where oil and energy growth rates have been most formidable over the last 25 years.

*Oil and energy demand* – The third main factor contributing to a future moderation in the rate of expansion of oil demand is concerned with the role and function of energy in economic development (see Figure I-3.8) and, in particular, with the phenomenon of a less rapid rate of growth in energy use as economies become more complex and tend to expand mainly in activities which are not energy intensive (*viz.* in the tertiary and quaternary economic sectors). The United States entered this phase more than a decade ago so that, in spite of continued growth in the gross national product at a rate of 4 per cent per annum on average, its energy use has recently been expanding at only 3 per cent per annum. By contrast post-war economic expansion in Western Europe and Japan has been more concentrated in the industrial sector with higher energy inputs per unit of output. As a result energy use has been expanding at annual rates of up to 12 per cent. Nevertheless, in these areas, too, the energy intensive phase of development is now approaching its end, to such an extent that required increases in energy use can be expected to moderate significantly. This already appears to be happening in Western Europe, where energy growth rates over the last two years have been very modest indeed.

In that the USA, Western Europe and Japan account between them for no less than 60 per cent of total world demand for energy, the

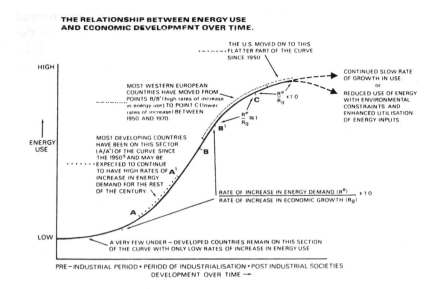

Figure I-3.8: The relationship between energy use and economic development over time

*The world's growing use of energy depends on the changing relationships over time between energy use and economic development in different economies. Note, at the moment, that the world's most important energy-using nations (the USA, Western Europe etc.) are moving off the steepest part of the curve whilst most developing countries are progressing along it. Very few countries are moving from low Re/Rg ratios to higher ones. Post-industrial societies which pay increasing attention to environmental considerations and to questions of efficiency in energy use could conceivably continue to expand economically whilst using less energy overall.*

rate of growth of their developing energy use will play a key role in establishing the overall trend for the foreseeable future. Furthermore, as far as the rest of the world is concerned we find already that most nations have already entered the period of economic expansion based on industrialization and the use of mechanized transport and, as shown in Figure I-3.8, are in that period in which there is a high rate of growth in energy use based, in almost all Third World countries, on the use of oil. Their high rates of increase in oil demand are therefore already written into the 7.5 per cent per annum growth rate of the last 20 years. Thus, whilst we certainly cannot expect any down-turn in their rate of oil use, their continuing industrialization will not, on the other hand, lead to an upward trend in the growth curve.

There is one final point to note in this brief examination of the relationship between energy growth rates and economic development:

*viz.* that there is no need for all economies to become as energy intensive as the United States in the process of continued development. In contrast with the United States, where a combination of cheap energy, very largely from indigenous sources, and the lack of any environmental constraints on energy uses have produced very high rates of energy consumption, Sweden has achieved an almost comparable per capita standard of living with an energy use per capita which is little more than half that of the United States. The rest of Europe (with the possible exception of the United Kingdom) is tending to follow the Swedish line – a trend which will be strongly reinforced by the increasing attention now being given by governments to the environmental questions of pollution and resource conservation. The extravagance of the USA in terms of energy use now seems likely to turn out to be the exception rather than the rule and, if so, then it provides yet another reason for predicating a reduced rate of energy growth in general – and of growth in the use of oil in particular – over the next 30 years.

## Conclusion

One must thus conclude that Warman's use of an exponential 7.5 per cent growth rate in oil demand for the rest of the century is unjustified and cannot reasonably be used to demonstrate the danger of a near imminent exhaustion of the world's conventional oil resources. Realistically one must prognosticate rates of growth below this level. The examples set out above (in Figs. I-3.6 and 3.7) illustrate the markedly reduced cumulative use of oil over the period 1971–2000 and the much lower annual rate of oil consumption emerging in 2000 from somewhat lower but, nevertheless, still quite significant rates of growth. The consequential 'requirements' for new discoveries of oil fall dramatically from the 4000 billion barrels estimated by Warman to only 1500 and just over 1000 billion barrels, respectively. Given, in addition, our earlier argument that the currently proven 600 billion barrels of proven reserves will, in fact, turn out to be at least 800 billion – on the basis of the appropriate appreciation factors for discovered oil – then the discoveries of oil which remain to be made to provide a minimum 15 years R/P ratio up to and including the year 2000 lie only between 800 and 1300 billion barrels. Averaging these data, we arrive at a figure of 1050 billion barrels for required additional oil reserves. This is, coincidentally, almost identical with the total oil discovered by 1970. This implies, on average, a future performance by the world-wide oil industry which is no better than that which it has achieved on average over the last 25 years or so. Given the new technologies of exploration and development work and given the virtual virginity of the considerable offshore areas for such work, as well as the possibilities for continued new discoveries in existing areas, then the task does not appear to be very formidable. Nor does the requirement threaten to exhaust even the resource base which Warman has estimated at 1600–1800 billion barrels, but which, as shown in Figure I-3.4, is already more generally believed to be one-quarter to one-third larger. Moreover, given the extrapolation of the calculated trend, it could reach almost 4000 billion barrels by the year 2000. In brief, the oil resource base in relation to reasonable expectations of demand gives very little apparent cause for concern, not only for the remainder of this century but also, thereafter, well into the twenty-first century. The level of consumption could then be five or more times its present level: and all this is without any exploitation whatsoever of the even more plentiful resources of potential oil from shales and tar-sands or from the distillation of coal.

By the middle of the twenty-first century, when our successors have finally mastered the economic and safe utilization of nuclear and hydrogen power and have, moreover, developed techniques for the direct utilization of the heat energy of the sun and the mechanical power of the tides and the winds, they will have to ponder on what all the late twentieth century Malthusian style fuss was about in respect of oil – a temporarily useful but, nevertheless, a still relatively inconvenient and dirty source of the world's energy needs.

## References

Alberta Energy Resources Conservation Board, *Annual Report*, 1971, Calgary.

Chandler, G. 1973, 'The changed and changing energy scene'. Paper presented at the Summer Meeting of the Institute of Petroleum. (To be published by the Institute in 1974.)

Cram, I.H. 1971, Future petroleum provinces of the United States: their geology and potential. *American Association of Petroleum Geologist, Memoir* No.15, pp.1–54.

Halbouty, M.T. et al. 1970, World's giant oil and gas fields: geological factors affecting their formation and basin classification. American *Association of Petroleum Geologists, Memoir* No.14, pp.509–540.

*International Petroleum Encyclopaedia* 1971, Tulsa: Petroleum Publishing Company.

Kastrop, J.E. 1972, Oil recovery technology: key to future US supply. *Petroleum Engineer* (December), pp.37–40.

Moody, J.D., Mooney, J.E. and Spivak, J. 1970, Giant oil fields of North America. *American Association of Petroleum Geology, Memoir* No.14, pp.8–16.

Odell, P.R. 1963, *An Economic geography of oil*. Bell, London.

Warman, H.R. 1972, The future of oil. *Geographical Journal*, Vol.138, No.3, pp.287–97.

Weeks, L.G. 1973, *Subsea petroleum resources*, UN report to Secretary General, New York.

# Chapter I – 4

# Political Economics of Offshore Oil Exploitation*

## 1. Introduction

Today, over much of the world, continental shelves and slopes are being increasingly explored for hydrocarbons. Exploration for offshore oil and gas and their development now seems likely to increase rapidly in the short to middle term as country after country determines that its potential should be investigated. But most countries of which the offshore areas are, or soon will be, the subject of such investigation, have little or no experience of this sort of resource development (eg. India, Ireland, Togo and Norway), while for other countries the offshore development may simply be seen as nothing more than a logical extension of the earlier exploration and production of oil and gas on land and in shallow coastal waters (eg. Venezuela, Indonesia, United States and the United Arab Emirates). In both cases, however, the technical and economic considerations involved in development decisions for deep water continental shelf and slope oil exploitation pose problems which are markedly different from all earlier experience for the countries concerned.

In most cases the countries involved have dealt, and will probably continue to deal, with one or more of the international oil companies. Such dealings have been and remain based on concessions, licensing, or some other permutations of government/company relationships which

* Reprinted from *Natural Resources Forum*, Vol.2, 1978, pp.227–239. (NB. The joint author of this paper, Dr. K.E. Rosing, has given his permission for this reprint. This permission is gratefully acknowledged.)

give the company the right to expend its own money in looking for oil and/or gas in return for a part of the profit which, if the company is successful, will be made from the production and sale of oil. The government involved will benefit from the activity in several ways. First, through the fees which they can levy on allocating the right to explore; secondly, through the establishment of support industries in their country, with a particular impact on employment; and thirdly – and especially – through the taxes which they impose on the production of any oil which is found. In addition, of course, national balances of payments will benefit either through the fact that imported oil can be substituted or because oil surplus to domestic requirements can be sold abroad.

## 2. Considerations in Establishing Exploration Legislation

The nature of the legal relationships between company and government are many and varied; but, in general, one can say that by legislation the government creates a local politico-economic environment within which the company or companies must operate if they wish to participate in the exploration of a particular national territory. Each of the various companies interested in such exploration will then judge the risks of operation within that nation's jurisdiction (ie. in that particular politico-economic environment) against its assessment of the likelihood of successful exploration and production. When a government sets its taxation and other policy instruments at too stringent a level the companies will respond by not participating in the exploration because the size of their share of the potential profit is viewed as insufficient to offset the risks which the exploration is likely to engender – compared with their opportunities elsewhere. Even then, once a transnational company has decided to participate in the investigation of a particular nation's possibilities, the magnitude of its investment in exploration will, at any given time, be a function of its total budget for exploration and its assessment of the profit potential/risk involvement in that area, relative to all other national areas within which the company has the possibility of exploring.

From the individual nation's point of view the harshness of the terms which it can dictate, regarding the ultimate division of the possible revenue from production, is governed by the assessment of the potential for their area which is made by the interested companies. The companies want to try to invest their limited exploration capital in regions where the after-tax-profit to risk ratio is the most favourable. This tendency is, however, moderated by the fact that each individual company is in competition for acreage with all other companies and must, as it were, try to 'keep a foot in the door' in the most highly prospective areas, even when contemporary governmental policies in respect of oil and gas exploration and development appear to them to be unsatisfactory. Governments, after all, can change – and so can conditions.

Thus, while the government responsible for a nationally designated area of continental shelf has the nominal freedom to set any standards which it wishes for the exploration and development of its sub-sea resources, the setting of the conditions at too stringent a level will result in no, or only a very limited, effort on the part of the companies to explore the acreage. Conversely, the establishment of the state's requirements at too generous a level will result in the government foregoing some or all of the economic rent which it could obtain during the productive life of the region. The choice of a policy is thus critical to the exploitation potential and to the returns which a country may expect.

## 3. Post-Discovery Considerations

After the discovery of oil the profit/risk environment is altered for both companies and governments. Governments have in the past responded to this alteration by modifying the terms under which initial concessions were granted in order to increase their share of the economic rent. Sometimes this has been done not just in respect of areas yet to be allocated, but also retroactively for areas previously allocated. This has the effect of forcing a new economic environment upon companies which may have already incurred considerable expense in the early exploration phases of their work. From the companies' view these new regulations necessitate a complete reassessment of the profit potential/risk involvement in continuing with the project. The company's only way of recouping its previous investment – or for that matter of turning in a profit – lies in its continuing to invest through the development phase to the production of oil. On the other hand, as the level of investment in exploration is trivial – in comparison to the level of investment required for production – a dissatisfied company could at this stage decide either to withdraw or postpone development once a find is made. Thus, post-discovery, when the risk of not producing oil is much reduced, the volume of investment which has to be placed at risk to secure development is so much greater that the government involved is still not in a position to impose excessive additional costs through its taxation; if it did so, then the companies involved would terminate or withdraw from their operations.

The governments' share of profits to be earned from oil/gas

developments, through taxes and royalties, is determined by nationally applied legislation. However, any company's reaction to the legislation is in terms of an individual – and unique – oil accumulation. The ultimate profit position of the company depends upon the share of the revenue which it can retain from production when set alongside the magnitude of the investment which it must make in order to produce some volume of oil from an accumulation. The company must then try to make an investment decision, within the politico-economic environment dictated by the government, in a way which maximises the share of the profit which will accrue to it following production and sale of oil from the field – after allowing for government take.

Various governments tax and control the development of oil by private companies in different ways. These, however, can be generalized to four methods, some or all of which can be applied in different 'mixes', depending on contrasting national conditions and propensities. *First*, there is a royalty provision whereby a particular – or even a varying – percentage of the oil produced or the monetary equivalent of that volume of oil must be delivered to the government without any allowances to the company for its expenses or investments needed to produce the oil. *Second*, there are special petroleum production taxes. This tax is applied to the value of the company's share of its proceeds from any oil produced. It may be applied on a field by field basis (rather than on a national basis) and allowances are made, when calculating the tax due, for capital expenses and other direct costs involved in the development of an accumulation. *Third*, there is Corporation Tax. This tax is applied on the profit earned by the company on its producing (and other) operations in accordance with the general laws governing the taxation of corporate profits as constituted by a particular nation. They generally include write-off provisions for capital investment and running costs. *Fourth*, a government may take an equity interest in the oil production either through a government oil company or by means of direct government involvement. This involvement is sometimes a general one involving a fixed percentage government participation in all enterprises or else a special arrangement made at the government's discretion in order to achieve participation at some desired level in oil activities. The government may purchase its participation rights by paying a proportion of the incurred exploration and development costs equal to the equity interest which it assumes – but this is not always the case. In its most extreme form the latter constitutes nationalization of the company's assets.

## 4. The Impact on Revenues of Increasing Production

Royalties to the government will, obviously, increase directly with any increase in production. Revenue from a special petroleum production tax will also rise, generally, as production increases, but the increase occurs in a more complex manner. Some proportion of capital costs, frequently more than 100%, are allowed to be written off against the value of the company's share of the oil before the calculation of tax due. Then, for the government, the larger the production per unit of investment, the greater the return to the government from this tax. As the level of investment required for production increases, the return per barrel of production to the government decreases. However, total government revenue from this tax will continue to rise, through increased production, until long after a company has ceased to invest additional capital because of the declining marginal return.[1] Government revenue from Corporation Tax accrues in a manner very similar to that of the preceding tax. Since it is a tax on profits the revenue generated will increase with any larger amount of production, if costs per unit of production do not rise. As the costs to the company rise (both capital and running costs), the return to the government will start to fall per unit of production. In absolute terms, however, the tax-take will continue to increase so long as the company continues to make profits from its investment.

The company or the consortium of companies which is developing a field is in quite a different position from the government. In order to obtain any return on the exploration costs, it must make an investment to install a production system. The magnitude of its minimum investment will be determined by the physical conditions of the reservoir and its location. Thereafter, however, the company will have some freedom of choice in determining its investment level. Capital costs – of platforms, wells and perhaps a pipeline – will be largely incurred over a two or three year period before first production begins, ie. before there is any cash flow to offset against the investment. The company must, of course, take into account the interest which it has to pay on money borrowed to finance the development and it also has to discount the future earnings by the international opportunity cost of the capital involved. Only by doing this can it assess its total financial position over the possible development of the field – and compare this with all other opportunities it has for using the same amount of money elsewhere in its system.

When the government is involved in the oil field through the holding of an equity interest in the development, such as through a government oil company, the interests of the government more closely parallel those of a private company or consortium. The closer to full participation the national oil company comes, the greater the similarity of interests. Indeed, when the national oil company participates fully (ie. pays its full appropriate share of the development costs at the time the costs are incurred in strict accordance with the proportion of the oil produced which it receives), then its interests will become identical with the private company or consortium – except that it may be working with a somewhat different opportunity cost of capital.

There are other benefits which accrue to the nation-state through the development of an off-shore oil field. First, there is a benefit to the national balance of payments either through the replacement of expensive imported energy or through the production which is surplus to domestic needs. This can be sold abroad to earn foreign exchange. In this respect there is no reservation at all as to what best suits a country's interests: the greater the production of oil, the greater the benefit to the national accounts, providing, of course, that all the oil can be sold at a price which exceeds total production costs. In balance of payments terms, certain early payments for overseas purchase of technology and equipment will certainly have to be made, but these will generally be relatively small compared with the total potential value of the oil to be produced. If this were not so, no commercial company or consortium would be interested in developing the reserves because their benefit only arises on an after-tax and a royalty-paid basis, whereas the national economy benefits on a full value added basis.

The mix of these considerations in the producing interests of companies, on the one hand, and governments, on the other, will, of course, vary from country to country depending on the nature of the legislation. On the whole, however, one can hypothesise that the closer the volume of reserves produced from a field approaches the technically recoverable reserves, the greater the net benefit of the development for the government. It should be noted, incidentally, that this maximising of production is also in the best interests of a world which has a limited volume of liquid and gaseous hydrocarbons for its future use – suggesting an overall desirability of constraining the rights of companies to determine for themselves the limits to which any field's reserves shall be exploited.

## 5. The Commercial Company's Interest in Development

The interest of a company or consortium of companies in respect of each individual potentially developable accumulation of liquid or gaseous hydrocarbon is to make a profit. The profit to be made must be *first*, sufficiently great to compensate for the difference between the early need to invest and the later flow of cash from production; *second*, sufficiently great to match the risk that the technical analysis of the reservoir and its producible reserves is in error, so giving some chance that insufficient production will be available to provide the profit; and *third*, sufficiently great to allow for the risk that the commercial analysis may have been conducted on the basis of cost and price assumptions which do not obtain when production is under way. A company which does not act in this way will not generate, overall, a sufficient cash flow to continue exploration and development work, including the cover required to meet losses on bad development decisions. Its decision takers must, therefore, seek the greatest possible profit from each development in which they participate.

In the development of an offshore, deep water field, present technology requires that large, expensive, platforms be set on the ocean floor. From these platforms wells are drilled, production is undertaken and controlled and various ancillary tasks are performed. Each platform supports a number of wells (maximum about 60), each of which taps a spatially different portion of the reservoir. In order to do this each well is deviated, from the vertical, at some angle and drilled to a precise position within the reservoir. The angle of the maximum deviation is technically limited so that the total area of the reservoir which can be reached from any one platform is also limited. Typically, in a reservoir 2500 metres sub-surface, a circle with a radius of about 3000 metres from the location of the platform currently marks the limit which can be reached. To produce a large field effectively, several platforms must thus be located, from each of which producing wells will be drilled. A single multi-well platform may easily cost $300 to $400 million (1977 dollars) and this investment must be made in order to produce any oil at all. In comparison to this large investment need, the cost of drilling the wells into the reservoir is relatively insignificant – one or two million dollars each.

An accumulation of oil invariably occupies an irregular volume of space such that each field has a unique configuration. A hypothetical field, of a not uncommon shape, is shown in section A of Figure I-4.1. The thickness of the oil-saturated sand of the reservoir is indicated by contour lines (compare the cross-section with the map). For this example we shall assume that all the sands of the reservoir have uniform characteristics and that its producibility is equal across the reservoir. The absolute quantity of oil in place can now be calculated as a function of the volume of oil-saturated sand, the porosity of the sand, and a factor for the oil's contraction when it is brought to the surface and the gas and water fractions removed from the fluid produced. This quantity is defined as the *stock tank oil in place* and is measured in barrels. Some fraction of this stock tank oil can be recovered using the technology which is available today. This fraction will vary depending on the particular set of technical parameters of the oil and the reservoir. The fraction can, however, be calculated from estimates of these parameters obtained by drilling into the reservoir and from laboratory tests on material recovered from the drill cores. By multiplying this fraction with the figure for the stock tank oil in place, one can calculate the *technically recoverable reserves* of the field. The value of this – as with stock tank oil in place – is, therefore, a function of the physical characteristics of the oil and the reservoir. *Economically recoverable reserves* on the other hand are much more complex to determine. Of course, the physical characteristics are again considered because the technically recoverable reserves set the upper limit, However, the economic environment within which the company has to operate, together with the company's determination of the magnitude of the investment which it wishes to devote to that accumulation in the special economic conditions, serve to define what oil can be economically recovered.

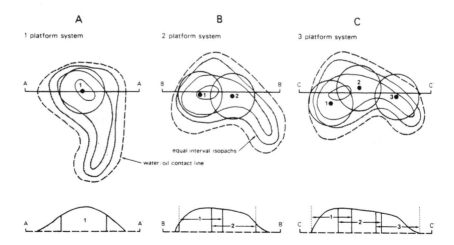

Figure I-4.1: A hypothetical reservoir shown in plan and cross-section

*The sands are assumed to be equally productive and hence the relative volumes of oil which can be recovered from various potions of the reservoir are a function of the sand thickness only. Sections A, B and C show the platform locations and nominal draw areas for three alternative production systems involving 1, 2 and 3 platforms, respectively.*

Figure I-4.1 shows, diagrammatically, three alternative development schemes which might be considered by a company in developing a field. As shown in part A of this Figure the one platform system consists of a single platform located in such a position that the largest volume of reservoir rock, and therefore of oil recoverable in an economically relevant time period, is inside the circle which can be directly tapped from the platform. Of course, in reality the circumference of this circle does not exist as a line. In a reservoir, oil, gas and water are mixed to form a hot fluid which is mobile under the great pressure of the reservoir. When the pressure is released at a point through drilling a well, fluid flows from other portions of the reservoir towards the point of release, *viz*. the well. There is a resistance to this movement from the matrix of rock in the reservoir. This resistance, the permeability, results in the pressure gradient around the well.[2] Such a gradient of recovery surrounds each of the wells drilled from the platform: collectively, they create the so-called platform draw area. To simplify this, for purposes of modelling the reservoir's producibility, all technically recoverable reserves inside the circle are considered as being produced, within some economically relevant time period (although this is not true), whilst none of the reserves outside the circle are considered to be produced (equally untrue). Thus, the one-platform's system has its likely productivity defined.

Part B of Figure I-4.1 shows that designing the two-platform system does not simply mean adding another platform to the one-platform system (as defined in part A). Instead, there is an alternative spatial pattern for the reservoir development in terms of two platforms and their wells' locations and, thus, of the area from which production is achieved. In order to maximize the recovery from the two platforms working together, the location of neither is the same as the location of the one platform in the one-platform system.

In this simple example, the investment for the two-platform system is now approximately double that of the one-platform system. However, the producible reserves of the two-platforms are less than twice the producible reserves of the one platform system. Thus, the investment per barrel of production from the two-platform system is higher than from the one-platform system. The same phenomenon – of progressively decreasing productivity of the investment – can be made with respect to the three-platform system illustrated in part C of Figure I-4.1.

## 6. The Economic Environment

A company assessing alternative investment policies for developing a field has to estimate a large number of unknowable factors in the external economic environment in order to do successfully. The most important of these factors are the magnitude of the capital costs, the future selling price of oil, the cost of money, the risk and, finally, the share of the revenues which the government owning the oil resources will allow the company to keep. The confidence with which each of these factors can be estimated will vary. The last can be estimated fairly accurately (baring changes in the law) from the terms of the taxation and licence regulations in force at the time, whilst others – such as the future price of oil – are little more than guesswork. However, the company has to choose a development plan which seems likely to produce the best economic result for itself. The declaration of economically recoverable reserves through one of the alternative levels as shown above (each with a different investment regime) will then be the result of this decision.

Most of the components of this economic environment are, of course, outside the control of either company or government, while the share of the total revenues from the oil production which goes to the company is determined by the fiscal legislation of the government. A government can thus, if it wishes, use the legislation as an instrument to modify the economic environment within which the company takes its decisions. This fact has been generally recognized in relation to the exploitation of smaller fields, the so-called 'marginal fields' in relation, for example, to Petroleum Revenue Tax in Britain and Progressive Incremental Royalty in Canada. In these special circumstance certain financial concessions are given in respect of the initial development of any field. These allow a company to retain a larger proportion of the revenue from production in the early years.

A marginal field can, however, be defined as such not because it is small, but as one which would be profitable under optimistic economic forecasts and unprofitable given pessimistic economic forecasts. In this case, the application of an option open to the government to change the government/company division of the revenues – for example by giving extra allowances against tax liability in the early years (when the forecasts are least uncertain and production is most valuable) – is able to change an unfavourable assessment and thus result in the development of an oil field which would otherwise have had to remain undeveloped. Such tax concessions or other devices are then in the best interest of both the company involved and of the government. Each benefits from an income which would not otherwise have been generated because the field would not have been developed.

Investment in the development of an offshore oil (or gas) reservoir is quite unlike investment in most other sorts of economic activities. It is, as shown in Figure I-4.2, characterised by a rising unit cost curve with an increasing scale of production; and there is also the extraordinary "lumpiness" of the unit of investment (platform by platform). Such high costs and the rising unit cost curve will frequently cause a less intensive development of a field than might otherwise be the case, since the development decision regarding a field is made to develop the field with the objective of *maximizing the financial benefit to the company* and not with a view to *maximizing the production of oil*.

The unit of investment with which the company must work, *viz.* the cost of a platform, is very large and while the return can also be very large, the marginal return per unit of investment *decreases* for each additional unit of investment (platform) placed on a field. Figure I-4.2 shows the shape of this curve for a particular development with four platforms. (Note, moreover, that this ignores running costs, the inclusion of which could further increase the differential in favour of the smaller system). Furthermore, as the development of the field becomes more intensive and complex, the risk element in the assessment must be increased, so increasing the effective cost of money. The overall result will thus be the limitation of the commercially declared reserves to some volume below that of the technically recoverable reserves, while the production system selected by the company will entail the fullest possible development (as many wells as required) from the minimum number of platforms which will cover the richest section of the field. In contrast the greatest royalty and tax return to the government would accrue from the most extensive development, while the country concerned would also derive the greatest benefit to its balance of payments from this development. In such circumstances the interests of the company and the interests of the nation and government are clearly divergent.

The governmental instruments which are available to modify the economic environment have generally been used to date to encourage the exploration for, and the initial production of, oil from small marginal fields. The reduction of the government's share of the revenues which it takes from such oil production certainly creates the opportunity for small fields to be developed, when otherwise they would not have been produced at all. However, such limited governmental intervention also *increases* the differential profitability between the first and successive units of investment on a larger field and could thus be a positive handicap to the full development of such larger fields.

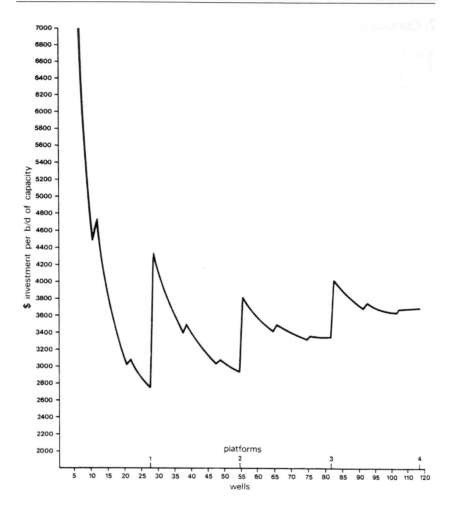

Figure I-4.2: Average Investment per Barrel per Day of Capacity – with a Four
           Platform System

*For the installation of this system on the field, four separate platforms, each with an
ultimate capacity for handling 27 wells are required. For each platform there is a falling unit
cost curve as average productivity increases with the increasing number of wells – except
that costs of expanded platform facilities at 12 and 20 wells per platform create the upward
kinks in the curves. The installation of each additional platform, however, reduces overall
productivity and there is a jump back to a higher unit investment cost – for example, from
the minimum $3350 when full productivity is achieved from the third platform to $4000 as
the first well on the fourth platform comes into production. Note the steadily increasing unit
cost from the most productive situations on Platform 1 to Platform 2... to Platform 4: that
is, from $2750; to $2950; to $3350; and to $3700 with the four platforms.*

## 7. Conclusion

In the development of almost all offshore oil fields there will be a *marginal system of development.* Governments have, as indicated, recognized that for a small field even the smallest possible development system may be marginal for a company and have responded by allowing tax concessions to encourage small fields' development. However, what has not so far been recognized is that a company in looking at any off-shore field is faced with a series of development alternatives each of which, generates potential production increases; but at the cost of more and more complexity in the development and higher unit costs of production. The marginal system is not necessarily the most productive. But the development of a large offshore field cannot, given present technology, be incremental over time. Instead, it depends essentially on the original investment decision. Thus, the company is obliged, in its own commercial interests, to limit the development to one which is most clearly economic.[3]

At the present time governments' policies on royalties and taxation reinforce this commercial limitation on economically recoverable reserves to a volume which is less than the technically recoverable reserves. But the best interests of the government – and, indeed, of an energy-short world – are for the fullest possible development of each accumulation of hydrocarbons. Companies responsible for developing fields must, and can, be encouraged to exploit a field's resources to its fullest possible extent by an appropriate set of governmental instruments able to adjust the economic environment within which the companies operate and so stimulate development and production to the benefit of both parties.

## References

1       These comparisons should be made in terms of Net Present Values. Expression in that form increases the difference between the two parties. See: P.R. Odell and K.E. Rosing, *Optimal Development of the North Sea's Oil Fields: A Study in Divergent Government and Company Interests and their Reconciliation,* London, Kogan Page, 1976, pp.41–49 and 65–76.

2       Obviously time is a variable here as well and over a sufficiently long time (perhaps hundreds of years) nearly as much oil would be recovered from one well as from 10 or 100. The number of wells to be drilled from a platform and the spacing of the wells to reach an

economically optimum rate of sustained production is therefore another element of the company's decision on its investment strategy.

3     P.R. Odell and K.E. Rosing, *Optimal Development of the North Sea's Oil fields: A Study in Divergent Government and Company Interests and their Reconciliation, op. cit.*, 1976; provides a full investigation of this and other points in three detailed case studies of development decisions in the context of the economic environment of British legislation.

# Chapter I – 5

# World Energy in the 1980s: the Significance of Non-OPEC Oil Supplies*

## I. Introduction

Global energy supply and demand considerations for the 1980s revolve essentially around oil. Except in limited and special circumstances – such as the opportunity for a continued rapid expansion in natural gas production and use in Western Europe – it is simply not possible, within the time span of a decade, either to expand the production of alternative forms of energy on a large scale; or to change patterns of energy consumption, so as to make really big differences to the shares of the various energy sources to overall demand.

Coal and nuclear power represent the only technically possible alternative energy sources for the medium term, but both are subject to severe constraints in their development. Coal is restrained in its potential contribution because traditional producing areas cannot help very much (indeed, the problem here is to prevent continued decline in output) and because new areas of coal production generally involve excessively long lead times for political, environmental and/or technico-economic reasons. And as far as nuclear power is concerned, environmental and safety considerations as well as surplus generating capacity in many electrical systems arising out of the post-1974 fall in the electricity growth rate, and the very high capital costs involved in nuclear developments, make much more than a token gesture in most countries

* Reprinted from the *Scottish Journal of Political Economy*, Vol.26, No.3, November 1979, pp.215–231 (the 14th Annual Lecture of the Scottish Society given at Heriot-Watt University, Edinburgh, 9 May, 1979).

well-nigh impossible.[1]

On the demand side, almost all nations in the non-communist world have, in the period since the early 1950s, developed systems – both technical and societal – which run on oil and have thus committed their futures to oil-based economies and societies. Neither individuals nor communities can afford to change other than slowly in respect of this dependence on oil – both because of the costs that would be involved and also because of the dangers of 'withdrawal symptoms' in the event of oil having to be replaced by alternatives incapable of sustaining the complete range of activities based on the use of oil products. Thus, the attractiveness of alternative energy sources is much reduced in the short term.

## II. Oil Supply and Demand: A Mistaken Evaluation with Dangerous Implications

Set against the background of a general recognition of this inevitable dependence of our systems on oil for the medium-term future, there is a very widely presented view that oil is inherently scarce so that some sort of a relatively near-future 'crunch' in the system is inevitable. The two essential components in this evaluation are, first, a belief that the demand for oil around the world is still evolving very rapidly and second, that there is a severe limit on the physical availability of oil reserves and resources such that production will necessarily have to be constrained. The usual form of presentation of this view of the future of oil is shown in Figure I-5.1.

Unfortunately, however, neither the demand nor the supply components in this analysis appear to be valid. On the one hand, worldwide oil demand outside the communist countries is now increasing only very slowly indeed. Over the two decades prior to 1973 the annual average rate of increase in the demand for oil was of the order of 7 per cent but since then, under the stimulus of the changed energy, economic and political conditions which arose out of the oil 'crisis' at that time, the rate of increase in oil use has fallen to about 1 per cent. Given the continuation, or even the exacerbation, of those conditions (rising real oil prices, economic recession and attempts by governments to reduce the use of oil), the outlook for the demand for oil seems unlikely to be very different from the recent experience. The prospect of an annual increase in demand approximating to that of the 1953–73 period is virtually non-existent, yet it is that experience which still appears to dominate contemporary thought on the future size of the oil industry.

On the other hand, there can be no overall situation of physical scarcity of oil in the 1980s – or even the 1990s – given only the world's remaining proven reserves of the commodity. These amount to some $750 \times 10^9$ barrels, some 50 per cent more than the oil 'needed' to meet total consumption between 1980 and the year 2000 – given the continuation of present growth rates. But remaining proven reserves constitute a minimum estimate of total future resources of conventional oil. Even conservative interpretations of the future potential for conventional oil put the ultimate availability about three times higher at

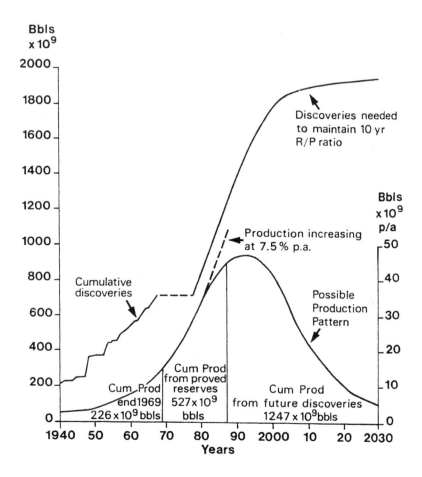

Figure I-5.1: A model of oil discovery production and depletion showing peak
year production about 1990 with a $2000 \times 10^9$ barrels resource base.

*Source: H. Warman, "Future Problems in Petroleum Exploration", Petroleum Review, March 1971*

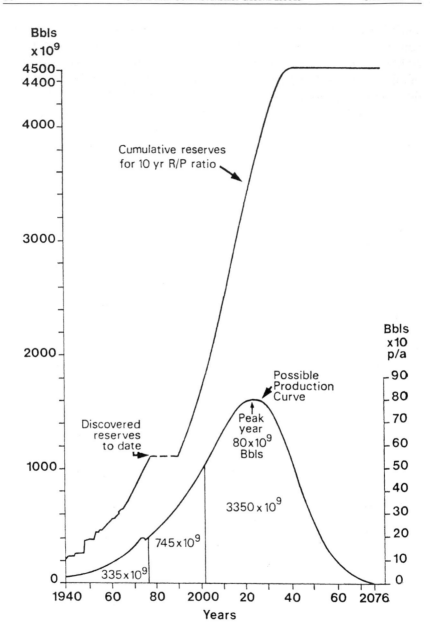

Figure I-5.2: An alternative model with a 4500x10⁹ barrels resource base and a
lower rate of growth in demand. Peak production is now delayed to
the 2020s.

some $2000 \times 10^9$ barrels whilst there are alternative estimates of total conventional oil resources which range up to $6,000 \times 10^9$ barrels. In addition there are, moreover, also so-called non-conventional resources of oil[2] which, it is generally agree, will eventually supply at least as much oil again as that from conventional oilfields.

Overall, if we take an essentially conservative view of the ultimate future availability of oil – and so use a figure of $4500 \times 10^9$ barrels to represent this – and couple this with the now rather generous view of the development of oil demand – which we assume will grow at 3 per cent per year for as long as the resource base makes possible – then the annual production of oil could continue to grow until the middle 2020s when it would reach a theoretical peak of about $80 \times 10^9$ barrels: more than three times higher, that is, than the present level. This exercise is shown in Figure I-5.2.

Viewed against this background, the concept of an imminent and inevitable scarcity of oil is grossly mistaken. This, in itself, would not matter if the engendered fear of scarcity had an effect only on the evolution of demand by assisting in the achievement of a more conservationist approach to the use of oil. But this is not the case. The presentation of the idea of relatively near-future oil scarcity also has a feedback effect on national and international decisions about future levels of oil production and of investment in production facilities. This dangerous reaction arises as follows. First, because of the fear of scarcity, the production of known resources is constrained by governments anxious about the future availability of oil for their own populations. As a consequence, scarcity is created. Second, because oil is said to be scarce, the motivation to invest to find more of it is reduced and as oil which is not sought cannot be found, so, again, the belief in scarcity becomes a self-justifying proposition.

These considerations are of much more than academic interest for, as we shall show below, they clearly influence the outlook for oil in the short to medium-term future: more specifically, they not only influence the price of the commodity in the immediate future (as even a reduced demand outstrips the willingness to supply), but also its supply over the somewhat longer term as the failure to invest inhibits the discovery of new resources.

## III. The Power of OPEC

It is in this context that the power of OPEC becomes apparent. Given the continuing dominance of OPEC oil in the world market outside the communist countries, the major oil exporters are firmly in a position to lever up the price of oil. This can be either when they choose to do so (as in December 1978 with their decision to raise prices by 14.5 per cent over the year 1979), or when an event affecting the level of production in one or more of the member countries of the Organization makes it inevitable (as in the early months of 1979, following the change of government in Iran). These countries are, moreover, able to argue, in light of accepted views on the "scarcity" of oil that such price developments are appropriate in order first, to curb the growth of oil demand on the grounds that too high a growth rate is not "good" for the system because oil is "known" to be scarce; and second, to protect their own resources from too rapid a rate of depletion so that some of their oil may be saved for their own longer term futures.

Such price and supply developments do, of course, affect the demand for oil. It will fail to grow as fast as it did previously or as fast as was expected under earlier price conditions. This, however, need have no adverse effect on OPEC's members as further price increases and supply limitations can be introduced, as appropriate, to protect their positions in terms of government revenues and foreign exchange earnings, etc. The essential solidarity of OPEC as an effective international organization makes this possible. Moreover, in periods of particular supply difficulties in the market (as in the first half of 1979), its members can not only take advantage of sales opportunities at prices standing well above official levels, but can then also use these to argue that their official oil prices are not high enough to clear the market. In this context their power is limited only by the potential collapse of the western system under the pressures so created, and/or by the fear of retaliation by the United States if it were pushed too far over the supply and price of oil.

## IV. The Policies of the OECD Nations

An effective answer to the challenge of OPEC power by the world's rich, industrialized nations has been made impossible by the combination of energy/oil policy elements. These are as follows. First, for some OECD countries higher oil prices and constrained OPEC supplies appear to offer national advantages in that considerable benefits can be secured in the short term by those countries, such as the US, the UK, the Netherlands, Norway and Canada, which have relatively low-cost oil and natural gas resources themselves. There is not much motivation on their part to inhibit developments in the international oil situation when the countries concerned view the options only from their short-term point of view as major oil producing countries.

Second, most OECD countries still appear to accept that policies which aim severely to restrict energy use as such, are economically and/or politically unacceptable. The short-term costs of such policies would, it is generally thought, exceed the short-term benefits. Thus, reductions in energy use can only be sought through exhortation and by policies which do not affect the accepted energy intensive way of doing things and going place. This often applies even in respect of the use of the pricing mechanism, let alone to the more interventionist, regulatory policies which are required to curb demand. As a consequence, energy savings at any given level of economic activities are likely to be modest, especially when the waste of energy engendered by the low and decreasing real prices for energy in the twenty year up to 1973 has been curtailed.

Third, many OECD countries have decided that their own oil and gas resources are so 'valuable' that they ought not to be produced in the short term to anything like the maximum technical, or even the optimum economic, level. Restraints both on production from discovered resources and on investment in potential resources emerge from the belief in scarcity, described in section II above, and so lead to a significant slow-down in the rate at which indigenous resources are developed and used.

Overall, too much concern for the long-term future availability of energy combined with too little concern for the much more serious shorter term politically inspired shortage of oil, jointly produce an inherent weakness in the western system: so much so, indeed, that the ability of the system effectively to survive the shorter term arises. In this context, the prospects for non-OPEC oil supplies in the 1980s is of high significance.

## V. Alternative Oil Supplies in the 1980s

In qualitative terms additional non-OPEC oil supplies in the 1980s are highly significant because they will become – or rather could become – available in a stagnant or near-stagnant western oil market. One must be careful to judge them in this context, rather than in the context of a rapidly expanding oil industry – as hitherto – in the framework of which it was correct to argue that a new North Sea oil province, for example, had to be discovered each year in order simply to keep up with the incremental growth in demand. Now, on the contrary, the combination of the elimination of waste as a result of much higher oil prices, some response to exhortation, the impact of government policies designed to stimulate the use of energy alternatives and the high probability of modest economic growth rates overall in the western economic system, seems likely to produce no more than a 1–1.5 per cent per annum average rate of increase in oil demand over the decade. If so, then by 1990 demand will be less than 12 per cent higher than its current level – by a maximum, that is, of only 6 million barrels per day (300 million tons per annum) more than the approximately 50 million b/d (2500 million tons per annum) of oil which the non-communist world will use this year.

Such an increase in the use of oil is not a very formidable production objective for new oil developments outside the OPEC countries, especially when one bears in mind that developments already under way in themselves create the potential for producing a high proportion of the net additional 6 million b/d of capacity needed by the end of the decade. For the first half of the decade at least, such developments already hold out a promise of new capacity which will increase faster than the rate of increase in demand.

Unfortunately, however, we remain so bemused by two recent, but now outdated, aspects of oil industry development that it becomes difficult to get the newly developing characteristics of oil supply and demand into perspective. First, there is a marked tendency still to think of oil demand in terms of the industry's 7% p.a. growth rate in the 25 extraordinary years up to 1973. But to appreciate just how significant the change since then has been, it should be noted that non-communist world oil demand is currently only just over 5% above its 1973 level. Had the pre-1973 trend of increasing use continued, then demand by now would have been over 50 per cent higher than it is. In other words, the world is now using only two barrels of oil for every three that were expected to be used by 1979, according to the forecasts of the industry in the early years of the decade. Nevertheless, plans were then confidently made for supplying the oil which the industry itself thought the world was certain to need.

Second, there is a tendency also to consider the oil industry as being necessarily – even inherently – international. This again, however, is nothing more than the result of the experience of the quarter of a century up to 1973 when a highly specific combination of the interests of the major international oil companies and the ready availability of low-cost oil resources from the Middle East and a handful of countries elsewhere in the world undermined the possibility of a more dispersed pattern of oil production. However, the politico-economic conditions which produced this phenomenon of a highly centralized non-communist world spatial concentration of oil production have now disappeared. Thus, a rapidly evolving dispersed production pattern from the geographically widely distributed oil resource bases may now be expected. This potential development will be demonstrated below.

The basic element in the changed situation lies in the fact that the price of oil has now been pushed, for reasons which are well-known, well above the long-run supply price of the quantity of the commodity that seem likely to be demanded over the whole of the medium-term future. This economic motivation to increase the supply is, of course, largely being ignored in the OPEC group of nations and, more surprisingly, it has also been partly ignored, as shown previously, in some of the world's industrialized countries. We shall return to look again in more detail at the latters' policies a little later, but, meanwhile, one observes that many other nations are certainly being motivated to respond to the economic opportunities presented by the oil price and demand situation. They thus appear to be evolving policies to meet the challenge – as follows:

The Soviet Union has extensive potentially petroliferous regions. Its resource base limitations are thus long, rather than medium, term. However, its opportunities to increase oil production rapidly enough to expand exports to non-communist countries are restrained by oilfield development and transport problems. Nevertheless, given the importance of oil exports to the West at the high prices obtainable in the market place, there seems likely to be a not immodest increase in Soviet involvement in satisfying western oil demand – albeit at the expense of deliberately constrained domestic and Eastern European demand.[3]

China is generally agreed to have a large potential oil resource base – both onshore and, more especially, offshore. Its new politico-economic strategy appears to include the deliberate use of oil as an export commodity, while maintaining an internal dependence on coal. One consequence of this has been its willingness to seek technological and other assistance from western oil companies for the development of its hydrocarbon resources. Another consequence has been its willingness to include oil exports in its international trade agreements – especially with Japan which China sees as a 'natural' market for its primary goods' exports. In light of these considerations, a net addition by China to the west's oil supplies in the 1980s seems to be a realistic expectation.[4]

In the non-communist world itself pride of place for a potentially much enhanced oil supply over the next decade must go to Mexico. Its very large potential for the rapid development of conventional, low-cost oil (and gas) resources is now very generally recognized. Its near-future potential to produce need only be linked to the intense speculation on the country's ultimate resources (of $200 \times 10^9$ barrels) in the context of these being available for the post-2000 period to back up the depletion over the next 20 years of its $40 \times 10^9$ barrels of proven and probable reserves. The latter in themselves are sufficient to sustain a production level of some 5 million b/d at which level Mexico would be second only to Saudi Arabia as an oil producer with considerable export potential (excluding, that is, the US and the USSR). Though Mexico has indicated that its policy in respect of oil production will be more constrained than this theoretical upper-limit based on known reserves, it does, nevertheless, have powerful politico-economic motivations to achieve high production and significant export levels. These emerge, first, as a consequence of its situation – both geographical and political, *vis a vis* the United States and, second, as a function of the country's need for the oil industry to act as a motor for ensuring the country's economic development.

Mexico may currently be the only Third World country with potential oil resources which could possibly rival those of the petroliferously richest OPEC countries. Nevertheless, almost all other countries in the Third World have, in the light of their oil dependent economies and the high price of international oil, even stronger motivations to seek to profit from more modest occurrences of oil and/or gas. This process can be seen at work already in countries as diverse and as geographically widely located as India, Malaysia, Egypt and Brazil. And it is happening even within the context of their generally nationalistic and/or statist approach to oil developments: an approach in which the rate of oil exploration and exploitation is curtailed. However, the now very much enhanced economic attractiveness of oil production appears to be modifying the earlier powerful political opposition in many such countries to the exploitation of their oil resources by private foreign companies. As a result, many new possibilities for new areas of oil production – especially offshore areas – are being opened up. Furthermore, the better prospect for new oil developments in such countries is being still further enhanced by the serious attention which is at last being given by the World Bank and other international financing institutions – both public and private – to the possibilities of investment in oil exploration and production in the Third World.[5] Overall, in this large and increasingly energy-hungry group of countries, the opportunities for oil production that are now being generated seem certain to be fully able to meet the increasing demands for oil in the Third World. Indeed, given the high price of international oil and the increasing difficulties which most developing countries are facing in respect of paying for their oil import needs, the motivation for the achievement of self-sufficiency by as many of these countries as the wide resource base makes possible, is so strong that their collective dependence on international traded oil seems likely to fall away very quickly. Help by the industrial nations for oil industry developments in the Third World countries could stimulate this development.

## VI. A Re-evaluation of Oil Production Policies by Industrialized Countries

We have already drawn attention to the unwillingness of some OECD countries to pursue policies which encourage the maximum possible production of indigenous oil – and of gas. Such policies are pursued not because it is uneconomic to increase levels of production, but because of fears for the scarcity of oil by the late 1980s and the 1990s and because of the lack of recognition of the seriousness of the short-term oil supply/demand situation and of the favourable influence that decisions for higher production levels by such countries could have on the situation. In respect of the latter consideration there appears to be a kind of fatalism in the West, *viz.* a belief that there is nothing the West can do in respect of oil and gas production which can affect the basic domination of OPEC supplies in the 1980s.

However, in terms of both demand and supply components for the 10–15 year outlook we have shown such fatalism to be misplaced. Indeed, the only 'component which is currently missing from the prospect for a radically different outlook (compared with that which is so pessimistically expressed in most western industrialized world analysis), is the willingness and ability of many of the industrialized countries to pursue policies which maximize the rates of production and of discovery of indigenous oil and gas resources. There are several different elements in this situation including, for example, the Australian and Canadian attitudes to the exploitation of their energy resources and the speed at which major exploration efforts can be initiated in the so far little explored potentially petroliferous regions of Western Europe.

There are, however, two aspects which stand out above all the others in respect both of the timing of the results that can be expected and of the quantities of potential production involved. First, there is the policy towards oil in the United States and most notably the question of allowing oil prices to rise towards international levels. The lack of sufficient incentive hitherto for companies in the US to maximize their production and exploration activities has severely restrained the degree to which the country's oil needs are being met from indigenous resources. A changed policy in this respect has recently been introduced by Presidential order and it now looks as though it can be made to stick, no matter what Congress may decide to do. Hopefully, it will succeed – albeit slowly, as the higher prices to be allowed for most domestic oil are only to be phased in over a period of almost three years. Hopefully, the oil companies will be prepared to use the 50 per cent of the 'windfall' profits that are to be left to them from the higher prices, for intensifying their upstream oil activities. Doubts must, nevertheless, remain about this until the precise form of implementation of the new policy is determined and until government and companies have established their relationships in the new order. Given success in implementing the policy, the US will become much better placed to meet an increasing share of its own oil needs in the 1980s and so help to take some pressure off the international market.

Second, there is the North Sea which, though an active petroleum province with a large number of discoveries and with major and expensive developments is, nevertheless, under-performing in terms of its likely level of output in the 1980s compared with its potential contribution to additional world oil – and gas – supplies. This is clearly so in respect of Norway and the Netherlands though for somewhat different reasons and in greatly different contexts.[6] It is also implicit, if not explicit, in the UK's policy partly because of a national unwillingness to conceive of the possibility of a rising level of production through the 1980s and partly because government/company relationships have not developed in such a way to ensure the fullest possible exploitation either of individual fields or of all the potential occurrences of oil and gas in the North Sea and adjacent areas.[7]

Unfortunately, unlike the now changing situation in the United State, North Sea countries' policies still remain uncommitted to highest possible levels of output. On the contrary, the continuing lack of appreciation of the opportunities in the North Sea plus a firm belief in a longer term absolute scarcity of oil, against the prospects for which potential production must be held back, combine to produce an attitude which undermines the significance of the North Sea's reserves. It is also helping to generate conditions in the world oil market in the 1980s which will not assist, to say the least, in respect of the ability of the western economic system to continue to prosper – or even to survive.

## VII. Conclusion

In the period since 1973 we have been made painfully aware of the significance of oil and of the power of the oil exporting countries in determining progress in the western economic system. A continuation of the imposition of oil price increases that we have had over the last decade – as illustrated in Figure I-5.3, which shows the real price of oil now to be approaching an order of magnitude higher than it was in 1970 and some 2.5 times its price in 1950 – and, even worse, a further period of continuing uncertainty about the availability of oil, as a consequence of the inherent insecurity which surrounds supplies from OPEC countries, together appear to pose a serious threat to the economic and political system of the west for the near-term future. In this context, alternative supplies of oil – and of natural gas which is able to substitute oil relatively easily in many uses – must be judged in terms of their contribution to ensuring a higher degree of stability in the western system. Thus, the nations of the west which have the opportunity to increase supplies of the commodities from their own resources have a particularly grave responsibility. Their continuing failure to produce as much oil and gas as possible may bring a feeling of satisfaction that their reserves are being conserved so as to ensure some residual warmth and light in a twenty-

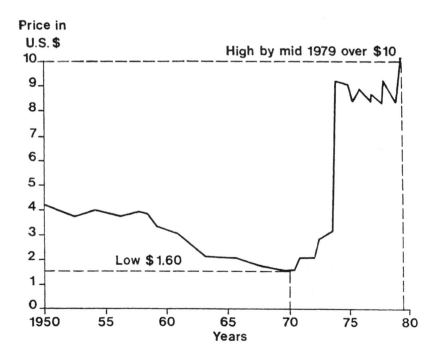

Figure I-5.3: The price of Saudi Arabian "marker crude" 1950–79 – in 1974 $ terms (this does not include the June 1979 price increase)

first century world which, it is feared, will otherwise be bereft of useable resources. Nevertheless, in the absence of any real evidence for such a scarcity of resources, within the context of a system which over two decades can adjust its energy use patterns to lesser energy intensive ways of doing things and of going places, any satisfaction from a failure to produce as much as could be produced – or as much as the market indicates ought to be produced – may be fleeting; given that a consequence of too little alternative oil (and gas) being made available to the world's consumers could mean that the whole system goes into a downward spiral of diminishing confidence and of increasing difficulties. It is in the context of such potential problems in the 1980s, emerging out of the political controls exercised by a small group of nations whose oil resources happened to have been discovered in the period between 1950 and 1970 and on which the world came unwisely to depend, that so-called oil and gas production conservation policies in so many nations of the industrialized world must be judged for what they really are; *viz.* policies that are contributing powerfully to the dangers of the western system's near-future demise.

## References

1      The issues involved in expanding the energy economies even of industrial nations based on nuclear electricity are explored in two other papers by the author, *viz.* "Europe and the cost of energy: nuclear power or oil and gas", *Energy Policy*, Vol.4, no.2, June 1976 and "The Electricity Sector and Energy Policy", in: C.Sweet (Ed.), *Energy Requirements and the Fast Breeder Programme*, Macmillan, London, 1979 (forthcoming).

2      These are resources which will become available from phenomena such as oil shales and tarsands, large occurrences of which are known to exist in many parts of the world.

3      See D.Park, *Oil and Gas in Comecon Countries*, Kogan Page, London, 1979 for a recent evaluation of Soviet export opportunities.

4      S.S. Harrison's, *China, Oil and Asia: Conflict Ahead?*, Columbia U.P., 1977, gives a clear picture of China's oil potential in general and its export potential in particular.

5      These, and related, issues are discussed in P.R. Odell and L. Vallenilla, *The Pressures of Oil*, Harper and Row, London, 1978.

6      This contrast between likely and potential production applies as much to natural gas as to oil. For an analysis of the situation in some detail see, P.R. Odell, "Constraints on the Development of Western Europe's Natural Gas Producing Potential in the 1980s", a paper presented to the

United Nations Economic Commission for Europe's symposium on *The Gas Situation in the E.C.E. Region*, Evian, 1978.

7    See P.R. Odell and K.E Rosing, *The Optimal Development of the North Sea's Oilfields*, Kogan Page, London, 1976.

# Chapter I – 6

# Third World Oil and Gas: Underexplored and Underdeveloped*

## Introduction

The prospect of a stable or even declining demand for oil over much of the rest of the century serves to emphasize the absence of any physical oil resource problem as a determinant of the evolving pattern of world oil supply. Proven oil reserves which may confidently be expected to continue to appreciate in size, as they have done over the whole history to date of the oil industry, are sufficient to cope with the demand for as far into the future as we need to plan for oil supplies. The real short to medium-term problem of the supply of oil can be defined somewhat differently: in the context, that is, of the acceptability and the accessibility of the reserves which have already been proven and which, as a result of a series of historical accidents arising out of recent imperial and colonial history, are unduly concentrated in a relatively small number of countries. In these mainly Middle Eastern countries, for reasons related essentially to the political relationships of the Great Powers in the 19th century and the early part of the 20th century, the oil industry was able to and, indeed, was strongly motivated to concentrate its attention. The Middle East, in other words, is not only unique – to date – in its geological characteristics as a habitat of oil, but it has also been well nigh unique as the happy hunting ground for the Anglo-American oil companies which, until very recently, were able to secure conditions for the exploitation of oil in the Middle East which were unachievable throughout most of the rest of the world. Elsewhere, factors of nationalism and of anti-colonialism succeeded in inhibiting the freedom of the companies to

* Reprinted from *Long Range Planning*, Vol.14, February 1981, pp.10–14

pursue their searches for low-cost oil resources – most notably in Latin America, but also in extensive parts of Africa and South East Asia.

It is thus essentially 'accidents' of political and economic history which determine the contemporary geographical pattern of the non-communist world's proven reserves of oil and which also produce the uncertain availability of those reserves, given their concentration in such politically unstable parts of the world. Indeed, the uncertainty of their availability is so high that the more than 500 billion ($10^9$) barrels of proven and probable reserves from the Middle East are unable to provide a guarantee of a continuing adequate flow of oil. This is so even though the volume of oil required from that area each year will now be a declining one. Over the rest of the century it need, in total, amount to no more than 200 billion barrels.

It is, of course, this uncertainty over the availability of oil from the Middle East's theoretically much-more-than-adequate known reserves which constitutes the basis of the short-term oil supply problem. It need not, however, constitute a post mid-1980s problem because alternative oil reserves, capable of sustaining the balance of the supplies needed over and above those which the Middle East countries are prepared to supply, can be developed and could be produced – given appropriate national and international policies towards the oil industry.

## Alternative Supplies

A complete survey of all the possibilities for geographically diversifying reserves and production potentials is clearly impossible in this paper. There are, however, several general points which need to be made.

a) The areas with the proven reserves which dominate the contemporary picture are a very small part, indeed, of the total potential regions for oil and gas occurrence. This can be clearly seen in Figure I-6.1 in which most of the world's potentially petroliferous regions (including those on the continental slopes as well as on land and on the continental shelves, but excluding Antarctica and the deep oceans) are defined.

b) In Figure I-6.2, however, we can see just how little oil exploration and development effort (as measured by the number of wells drilled in the different world regions) there has been in most parts of the world, suggesting that the industry has not progressed much beyond its infancy in terms of a world-wide search for opportunities for oil and gas occurrence.

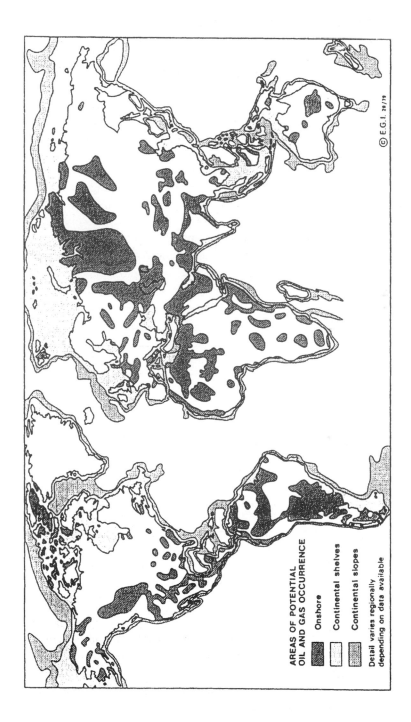

© E.G.I. 28/79

AREAS OF POTENTIAL
OIL AND GAS OCCURRENCE

■ Onshore

□ Continental shelves

▨ Continental slopes

Detail varies regionally
depending on data available

Figure I-6.1: The World's potentially petroliferous regions

c) Even in the developed world (on the left-hand side of the diagram in Figure I-6.2), most regions have enjoyed little petroleum development compared with the United States where, in spite of the long and intensive history of the industry, it is still generally agreed that there are considerable quantities of new oil to be found. It is this belief which, indeed, forms the basis of President Carter's crude oil price equalization policy. Without the frequently expressed convictions of US oil companies that higher prices for domestic oil would stimulate the search for and the discovery of new reserves, the price of US oil would surely have remained regulated. Given these expectations for new indigenous oil reserves of significance for the United States, with its demand for some 18 billion b/d of the commodity, it seems inappropriate to exclude the likelihood of additional large reserves of oil (relative to local demands) in other parts of the developed world. What is required in these regions to

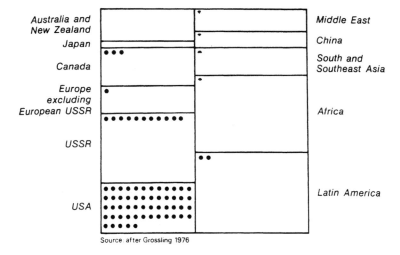

Source: after Grossling 1976

Figure I-6.2: The regional distribution of the world's potentially petroliferous areas, shown in proportion to the world total

*Within each region the number of exploration and development wells which have been drilled is shown – each full circle represents 50,000 wells drilled by the end of 1980. Segments of circles are included only for regions in which the total number of drilled wells to 1980 is less than 50,000. Relative to the US, all other parts of the world, but especially the regions of the Third World – are very little drilled for their oil.*

*Sources: B. Grossling,* Window on Oil: Survey of World Petroleum Sources, *Financial Times, London, 1976, p.83, for end-1975 situation. Grossling's data have been updated to end-1980 from the monthly US oil journal, World Oil, which each year in its August number publishes a table, 'Forecast of… World Drilling'.*

achieve that potential is the establishment of an acceptable *modus vivendi* between the governments concerned and the oil companies capable of undertaking the developments.

d) In particular, however, one notes (from the right-hand side of the diagram in Figure I-6.2) that the Third World of Latin America, Africa and South East Asia contains almost 50 per cent of the world's total potentially petroliferous regions (excluding Antarctica and the deep oceans). To date, however, petroleum development efforts in these regions have been minimal. They have amounted – and continue to amount – to less than 5 per cent of the world-wide effort. This absence of attention to the potential oil and gas resources of most of the Third World appears to constitute the most important gap in the development to date of the world oil industry – and the most important challenge for the industry in the future. I shall return to discuss the issues involved a little later in the paper.

e) The view taken of oil resources so far, however, relates only to conventional oil and thus excludes all of the world's so-called unconventional oil resources none of which have yet been tapped to any more than a minor – indeed, an experimental – degree. The oil shales and tar sands of North America, the shales of Brazil, Madagascar and India and the oil of the Orinoco oil belt of Venezuela constitute just part of this enormous alternative oil resource base. The quantities of these alternative resources of oil in place are an order of magnitude greater than the world's present proven reserves. However, the high costs and the great difficulties which are involved in developing them are well known so that it seems prudent largely to discount them until there is good evidence to show that the technological and environmental problems associated with their exploitation on a large scale are capable of being overcome. Let us not forget, however, that they do provide a hydrocarbon reserves' base of immense dimensions for 21st century utilization. Indeed, together with the even larger unconventional resources of natural gas – such as geo-pressured gas, gas in tight formations and gas hydrates in the Arctic and under the oceans, they could ensure the continuation of our economic systems based on the use of oil and gas through the 21st century, if, in the meantime, no acceptable forms of infinite energy sources (including coal) prove to be exploitable on a large scale – or only exploitable at a supply price well above that of the non-conventional oil and gas resources.

## The Potential of Developing Countries

To return, however, to the nearer term future – to consider the oil potential for the period between the mid-1980s and the end of the first quarter of the 21st century. In this temporal context, the world's most significant energy resources are the oil and the natural gas potential of the many unexplored, petroliferous regions of the Third World. Unfortunately, the potential opportunity for oil developments from these regions is not geographically coincident with the areas of operation of the international oil industry. On the one hand, the member companies of the industry have, over the past few decades, become decreasingly acceptable in most countries in the Third World and, on the other, the companies themselves have largely written off these parts of the world as regions in which they are prepared to risk their investments. The implication of this serious geographical mismatch are two-fold:

a)  By virtue of their virtual exclusion from such large parts of the world, the international oil companies have written down, or even written off, most of the developing countries as potentially major oil producers. They have, indeed, even failed to update their evaluations of the potential of these regions, consequent upon changing ideas on the patterns of oil and gas occurrence and in the light of rapidly developing geological knowledge. This is clearly demonstrated in Table I-6.1 in which the companies' outdated ideas of the Third World as a habitat for oil are compared with very recent figures produced by other entities with a high continuing interest in the hydrocarbon resources of the developing countries. Note, as an example, how ridiculous the oil industry's views on the potential resources of Latin America look in the light of recent events in Mexico. That single country's proven, probable and possible reserves in themselves are now approaching the oil industry's ultimate resources' figure for the whole of the continent, whilst Mexico's ultimate resources possibly exceed the industry's total figure for Latin America by a factor of two to three.

 An understatement of the Third World's potential is, of course, justifiable from the companies' point of view. To them, the effective resource base is the one which they expect to be able to exploit. If their relations with most countries of the world are either so bad or else non-existent, such that they cannot conceive of circumstances in which they could work in them, then the companies must discount the resources concerned. However, the international oil companies are institutions which are changeable, and/or substitutable and/or capable of being complemented. The fact that we currently have institutions which, by their present nature, cannot find and develop most of the world's oil resources does not mean that the latter can never be exploited. In other words, the fact that the oil companies, for their own completely justifiable reasons, are severely underestimating the oil resources of large parts of the world is no reason for the resources to be declared non-existent; and so entirely exclude the possibility of an expanded and diversified oil supply from such countries for use later in the century. By then, revised or entirely new institutional arrangements for tackling the oil and gas opportunities offered by the currently underexplored countries of the Third World could have been created.

b) Apart from this information and interpretation defect in our present understanding of the oil resource situation in the larger part of the potentially petroliferous world, there is a second and, arguably, an even more important aspect to the developmental prospects for a geographically more diversified oil reserves' position. This is related

### TABLE I-6.1
### Estimates of the ultimate oil resources of the third world (in bbls x 10$^9$)

|  | Oil Industry Views<br>(a) | Grossling<br>(USGS)<br>(b) | Min. of Geology<br>(USSR)<br>(c) |
|---|---|---|---|
| Latin America | 150–230 | 490–1225 | 686 |
| Africa | 120–170 | 470–1200 | 730 |
| South/SE Asia | 55–80 | 130–325 | 409 |
| Totals | 325–480 | 1090–2750 | 1825 |

(a)     Based on figures in R. Nehring, Giant Oil fields and World Oil Resources, Rand, Santa Monica, June 1978. Table on p.88 adjusted to give comparable geographical coverable with (b) and (c) above. Nehring also states (p.88) that his figures for these regions are 'roughly similar' to those published elsewhere by oil industry observers.

(b)     B.F. Grossling (United States Geological Survey), "In Search of a Probabilistic Model of Petroleum Resources Assessment", Methods and Models for Assessing Energy Resources (M. Grenon, Ed.), Pergamon Press, Oxford, 1979, pp.143–72.

(c)     V.I. Visotsky et al, Ministry of Geology, Moscow, "Petroleum Potential of the Sedimentary Basins in the Developing Countries", in R.F. Meyer (Ed.), op cit., pp.305–16.

to the availability of the management and technological expertise and of the financial resources which are required for major oil developments. Ninety per cent or thereabouts of such financial and human resources exist within, or are exclusively available to, the international oil companies and a small number of other entities – especially in the United States. As a consequence, the possibilities for the development of the oil and gas resources of the Third World are severely limited. Local state oil companies, or even state oil entities from other parts of the world, are not an adequate substitute, especially in managerial and technological terms, for the private sector oil companies, basically because of the contrast in size and in length of experience between the two sorts of entities.

## Implications of this Situation

This institutionalized dichotomy between those world regions with prospects for large-scale oil developments, on the one hand, and, on the other, the non-availability in these regions of the resources required to develop them, constitutes what appears to be the main barrier to the rapid geographical diversification of world oil reserves and the development of an enhanced supply potential. Moreover, the continuation of the present situation means, inevitably, that scarce capital and managerial resources for oil and gas exploitation will be used in relatively low-rates-of-return ventures in 'safe' countries, rather than in finding and producing the lowest cost oil and gas in parts of the Third World. A highly relevant and important example of the way in which this geographical misallocation of the factors of production works can be seen in the contrast between the ready availability of resources for the near-future development of high cost oil and gas resources of the United States and, on the other hand, the lack of adequate resources – both managerial and financial – for developing the low cost oil of some of the massive new fields in southern Mexico.

In the former, expertise and money will be absorbed in oil from shales and oil from coal projects or in working over, for the nth. time, regions that have already been thoroughly exploited for conventional oil over a period of many decades, simply in the hope of finding marginal extra quantities of producible oil. In the latter, on the other hand, development costs for the large-scale production of additional oil appear to be of the order of $2 per barrel and the number of potential opportunities in that country appears to be limitless in the context of the limited investments available to develop them.

Indeed, the development of southern Mexico as a possible second Saudi Arabia or even a second Middle East, if adjacent Caribbean potentials are also taken into account, underlines the sort of opportunities for the use of oil capital and management which still exist in the world. And, of course, it also illustrates the factors which inhibit their use; factors, that is, which relate essentially to the history of the international oil companies in a country still concerned for its effective national independence. It is highly unlikely that Mexico will be the only major prospect in the continuing undeveloped world of oil which could provide an element of desirable and necessary diversification of reserves' development and of production potential in respect of the world's relatively limited needs for oil over the next 50 years. During this period the well-proven use of hydro-carbons as the continued basis for economic development could thus be achieved.

## The Geographical Diversification of Reserves' Development and Supply Potential

One must first emphasize the great potential benefits which would flow from an approach to the future supply of oil and gas based on the geographical diffusion of the oil industry's activities. All potential new producers have a motivation, for sound economic development reasons, to produce any oil which is found so that each new producing country will add to the geographical diversity of supply. This will help not only with the world-wide strategic problems associated with dependence on a limited number of OPEC producers, but it will also help to reduce the expected continued upward pressure on prices and also serve to take the political pressures off OPEC member countries in respect of their existing too heavy a burden of responsibility in sustaining the international economic order.

Recognition of the validity of this approach to the future pattern of oil developments has already emerged. President Carter referred to it in his 'Energy Address to the Nation' in 1979 when he commented: 'There are several potentially abundant... sources of foreign oil and gas that have not yet been fully explored or developed... The Administration is developing a broader international strategy for increasing... exploration in these areas.' The World Bank has also launched a modest programme for investments in oil production in Third World countries. This involves some $450m over 5 years in a programme to accelerate petroleum production in the developing countries. Though small in relation to the opportunities and the needs, it does, nevertheless, represent a major breakthrough in that the World Bank for many years refused to invest in oil related activities at all. It could also be the means whereby private risk capital is attracted to the ventures sponsored by the World Bank. Some of the major oil companies – including Exxon, Gulf and Shell – have already made it known that they see good opportunities for productive investment in oil developments in such countries and that they would welcome suggestions as to how they might become involved so that the political problems and the high risks associated with such projects can be minimized. Finally, some of the Third World countries themselves now appear to be more willing to reconsider their previous hostility to, and rejection of, all involvement by the international oil companies in their oil sectors, providing they can be assured of the effective de-politicization of the companies in such activities. The new relationship of the companies to the members of OPEC – particularly in respect of the managerial and technical agreements and the companies' willingness to work for a flat per barrel fee in the context of such agreements – offers one possible approach to this difficulty.

In investment terms what appears to be needed for energy-related development in the Third World is an availability of up to $8000m per year, most of which for at least a decade would go into the oil and gas sectors. This finance, which is needed to pay for the necessary managerial and technical expertise, as well as for the hardware and the services involved, would enable the undeveloped and the under-developed oil and gas provinces of the Third World to be combed for their resources and would allow the development of large-scale productive capacity where the reserves discovered, as well as other factors, justified this.

One possible strategy involves an intergovernmental agency, embracing both OECD and OPEC's members and the utilization of the expertise of the international oil companies to ensure the technical and the commercial success of the schemes. The world banking industry, with its close connections with the international oil industry, might be able to evolve alternative lending mechanisms whereby the aims and objectives of the venture could be achieved. Or more radical alternatives might be considered more appropriate. Various options need to be examined so that a viable and acceptable solution to the opportunity offered to the future of the international economic system by these developable, but so far unattainable, oil and gas resources of the developing world can be found. The oil reserves and the production potential from the developing countries are needed to complement those which already exist as a result of the historical activities of the oil industry and those which are now being found in limited parts of the world, such as the North Sea, where the existing international oil companies are able to operate effectively. At stake, in meeting the challenge of a geographically more dispersed pattern of oil and gas developments, is the establishment of more favourable conditions for economic progress and for enhanced and more widespread prosperity in the system.

# Chapter I – 7

# The Future of Oil: a Complex Question*

## A. Introduction

Energy policies throughout the western world have come to be based on an assumption which has been elevated to the status of a self-evident truth, *viz*. that the world's need for oil will necessarily exceed the available supply. Over the next few years there is some probability that this could be the case as a result of the control that the member countries of OPEC exercise over the supply by their domination of the international oil market'[1] and the chances of interruptions to supply because of political and other factors. Such politico-economic considerations can, however, be set aside in respect of the medium- to longer-term relationships between supply and demand. For the period beyond the 1980s current predictions of a scarcity of oil are based on an entirely different set of factors.

First, there is a general acceptance of the idea that the total of ultimately producible reserves of oil in the world's crust are too small to enable production to expand much beyond levels already achieved; second, there is a belief that the annual quantity of oil which can be proved to exist and to be recoverable with existing technology and at going costs and prices will act as a constraint on the annual rate of production that can be achieved. The two considerations are not, of course, mutually exclusive. Indeed, they are presented as the complementary aspects of the oil scarcity

* Reprinted from *The Future of Oil; World Oil Resources and Use*, Kogan Page, London, 2nd Edition 1983, pp.23–52 (NB. The joint author of this book, Dr. K.E. Rosing, has given his permission for this reprint. This permission is gratefully acknowledged)

syndrome which, implicitly if not explicitly, is also presented as a function of a third major consideration – namely, high rates of increase in the rate of use of oil as the norm for the industry.

Given that resource availabilities, exploration and production successes and demand considerations all play a role in determining the future of oil, it is difficult to see how the necessarily complex future of the commodity can be viewed as one which is self-evident and about which there is little scope for argument, and even less room for divergent opinions. Indeed, the complexity which necessarily arises from the interrelationships of the variables involved suggests that any 'self-evident solution' ought to be treated with a healthy degree of scepticism. The need for such a sceptical reaction is strengthened when one notes the multi-faceted nature of each of the three basic variables. Each has physical, technical and economic (as well as political) aspects to its evaluation, and there is plenty of scope for widely differing opinions on the importance which should be attached to the various facets and to the ways in which they relate to each other.

## B. The Resource Base Consideration

The complexity of the oil resource base question can be demonstrated from the history of the industry in the period between 1950 and the early 1970s. This is true even though the period was one of relatively straightforward and continuing worldwide expansion of the industry, mainly under the aegis of the international oil companies, whose organization and motivations ensured the effective diffusion of knowledge and of advances in technology in respect of the exploration for, and the production and use of, oil in almost every country of the world.

Until the early 1970s the international oil companies expressed little or no concern for the question of the ultimate size of the world's oil resources. They simply considered the resource base to be so large as to make estimates of its size irrelevant to questions about the future development of the industry. Nevertheless, a number of estimates of the size of the resource base were made over those years. These are shown in Figure I.7.1. In this diagram we see, first, the generally upward changing views of the oil industry on the question. Estimates increased from about $500 \times 10^9$ barrels in the 1940s to over $3000 \times 10^9$ barrels by the early 1970s. Second, we may note how the changing size of the estimates can be closely correlated with the dates at which the different estimates were made. A correlation between the size and the date of a resource base estimate may, at first sight, appear to be of little interest and of even less significance. Time, however, in this instance is simply a proxy variable. It represents the impact of increasing knowledge about the occurrence of oil and of improving technology for its discovery and production. These, indeed, are the bases on which man's ability to understand and to exploit the world's hydrocarbon resources have gradually been developed and expanded. There is, as yet, nothing to

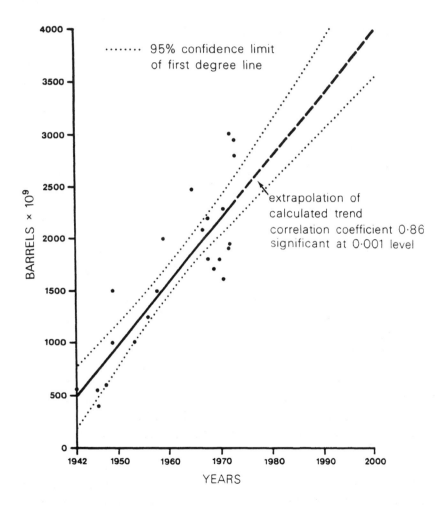

Figure I-7.1: Estimates of the World's ultimate oil resources, 1942–73

*See P.R. Odell and K.E. Rosing, '*Estimating World Oil Discoveries up to 1999 – a Question of Method*', Petroleum Times, Vol.79, No 2001, 7 February 1975, pp.26–9 for the source of the estimates and for further discussion of the issues. (NB. In this article, which as a matter of editorial policy, was not proof-read by the authors, the graphs in Figures 2 and 3 were transposed.)*

indicate that man has perfected either his knowledge of the world of oil or his technology for winning it and it is, thus, basically unscientific to assume that the most recent forecasts of the volume of the world's ultimately recoverable oil represent the last word on the subject. Quite the contrary can be argued when one observes that both knowledge and technology concerning the occurrence and development of oil resources continue to progress in many parts of the world. Examples of these processes at work can be seen in the contemporary reappraisal of more extensive and deeper geological habitats of oil and in the appraisal, for the first time, following new technological developments, of oil deposits lying beneath the extensive continental slopes of the world's land masses in water depths of 200 to 2000 metres. The very extensive and highly diffused geographical distribution of areas of oil potential is illustrated in Figure I-7.2.

There thus remain general geological and technical factors which will necessarily keep the estimates of the world's ultimately recoverable oil moving up. In addition, there are continuing opportunities for specific reappraisals of the volumes of producible oil. These arise in respect of recently discovered, and as yet little known major oil provinces. Of these, the North Sea and south-east Mexico are two important contemporary examples which will be followed by others in coming decades. The next 20 years thus appear to offer a high probability of the continuing evolution of ideas on the size of the world's ultimate oil resource base. There is, moreover, one important additional factor which will serve to encourage the process. This is the, as yet, seriously under-evaluated impact of recent major increases in the price of oil on the calculation of the percentage of oil in place in any field that is economic to produce.

Figure I-7.3 shows the formidable and continuing decline in the real price of oil (in constant 1974 $ terms) over the period 1950–70. By 1970 it was over 60 per cent cheaper than it had been 20 years previously. Yet, in spite of this powerful economic disincentive to seek new oil, or even to think seriously about future, less accessible and inherently more difficult habitats and their exploitation, there was an increase in the estimates of the size of the world's oil resource base from under 1000 x $10^9$ barrels in 1950 to at least 2000 x $10^9$ barrels by 1970. Figure I-7.3 also shows the formidable oil price increases since 1970. The real price of oil has, over the last decade, inceased by more than an order of magnitude in response to the revolution in the world of oil power.[2] By the end of 1981 the price of oil had increased to levels which open up the possibility of profitable investment in a wide range of enhanced recovery projects in

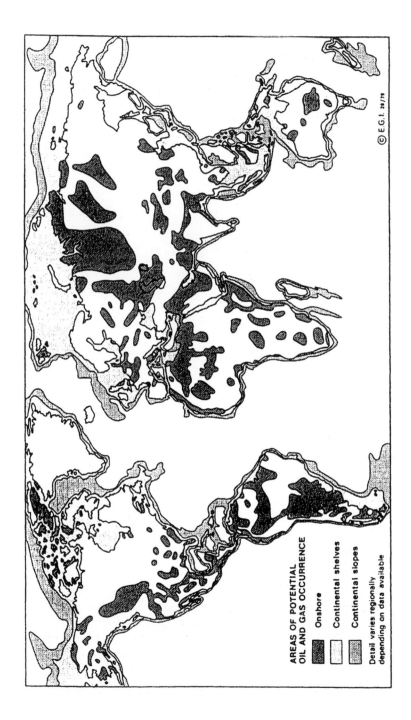

Figure I-7.2: The World's potentially petroliferous regions

fields which are already under development and from which the average rate of recovery is only about 30 per cent.[3] There is a similar economic motivation with higher oil prices for the exploitation of the world's less accessible resources. This is certainly reflected in the way in which most of the world's hitherto oil-short nations are anxiously seeking to test for the potential availability of indigenous oil resources in order to open up the possibility of eliminating the high cost of their oil imports and the penalties this inflicts on their prospects for economic developments.

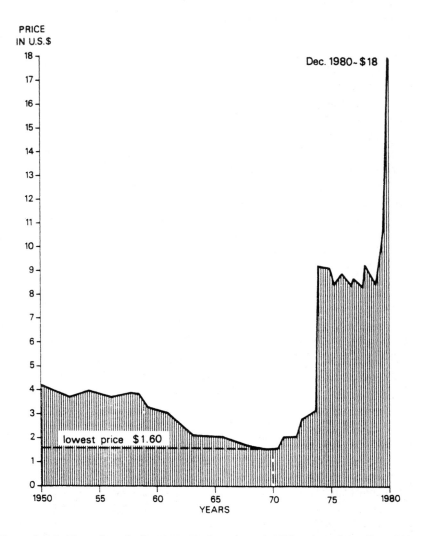

Figure I-7.3: The price of oil, 1950–80, based on the fob price of Saudi Arabian light crude oil expressed in 1974 $ terms

In the recently much changed economic circumstances a significant upward re-evaluation of the world's conventional oil resource base is thus mainly a question of the time which is required to organize and undertake the necessary increased and more extensive exploration efforts, and to make the multitude of reappraisals of the percentage of the oil in place which can be recovered from individual fields, given the much higher revenues which are now available to sustain higher levels of investment. A straight-line extrapolation, through to the year 2000, of the regression shown in Figure I-7.1 thus seems, under the radically changed economic conditions, more likely to represent a conservative rather than an optimistic statement of the future development of estimates of the world's ultimate oil resource base.

Estimates of the ultimate quantities of conventional oil do not, however, represent a comprehensive view of the total world oil potential. It is also necessary to add to these the potential for oil resources from unconventional habitats. Such unconventional oil is geographically very extensive. It includes the oil in the tar sands of the province of Alberta in Canada and in the USSR, the heavy oil belt of the Orinoco region of Venezuela and in Canada, and the oil shales of the United States, Brazil and Zaire. High production costs of such oil, coupled with low oil prices, have hitherto inhibited the inclusion of much unconventional oil in the total world oil resource figures. Now, developing production technologies and the very much higher market value of the oil that can be produced serve to convert large quantities of unconventional oil into an effective resource. The volume of this addition to the world's ultimate oil resource base is a minimum of $2000 \times 10^9$ barrels.[2] The maximum volume of unconventional oil remains, however, one of unknown dimensions because to date there has been no formal search for unconventional oil and no systematic evaluation of its occurrence on a world-wide basis. A figure of $5000 \times 10^9$ barrels, however, would not be out of line with the high estimates for such oil in those parts of the world where resources have already been proven.[3]

The rapid and continuing increase in the size of the estimates of the world's ultimate oil resource base over the past 40 years emphasizes the completely inappropriate nature of an approach to the future of oil which employs just one or two values for this important variable. The even more recent traumatic political and economic changes in the structure of the international oil industry justify even greater hesitancy in working with a single figure for the world's ultimately recoverable oil resources. Yet, in spite of such overwhelming reasons for extreme care in dealing with the value of this variable, there has been a marked tendency in many recent forecasts on the future of oil to work with but a single future oil availability component.

This is normally a figure of some 2000 x $10^9$ barrels of oil, inclusive of the oil which has already been used (about 400 x $10^9$ barrels).[4] The use of such an extraordinarily low figure, compared with the data on the oil resource base which we have outlined above and which we shall detail later in the chapter, appears to emerge from very recent international-oil company inspired views on the future of conventional oil.[5] Most of the international oil companies now appear to work – in public at least – with this 2000 x $10^9$ barrels oil resource base as a rule-of-thumb approach to the question, but this is a very recent specification by the companies. Moreover, as many spokesmen for the companies have themselves made clear, their use of this figure relates to very special considerations. In essence, it reflects their expectations of their ability and/or their willingness to explore for and to exploit oil resources in the future politico-economic conditions which they anticipate for various parts of the potentially petroliferous world. Their expectations in this respect are very pessimistic; justifiably so in view of the deterioration over the last 10 years in the opportunities and conditions for their pursuing oil exploration and production activities in many parts of the world.[6]

The companies have, for example, largely discounted the Latin American potential for hydrocarbons. This appears to be the result of a more than 50-year history of general hostility to the companies in most countries of the continent. Over this period the companies have generally been viewed as the prime agents of US economic imperialism and have been nationalized or otherwise severely constrained in their activities in general, and particularly in respect of their exploration and production work. As a result of these politico-economic factors the international oil companies' evaluation of Latin America's ultimate oil resource base has become unenthusiastic and outdated. Thus, in their world estimates of ultimate resources, Latin American resources are included at a level of only 150–230 x $10^9$ barrels (see Table I-7.1). This treatment of Latin America's potential must be compared with the alternative views of others such as those of Dr. B. Grossling, formerly of the United States Geological Survey (USGS), who gives a range for Latin America's ultimate oil resources of 490 to 1225x$10^9$ barrels. The estimates for Latin America made by the Soviet Ministry of Geology are for oil resources of 686x$10^9$ barrels.[7] Recent oil discoveries, which are of major significance even by world standards, in the southern part of Mexico, from which country the international oil companies were excluded when the industry was nationalized in 1938, indicate that Mexico alone may well have ultimate reserves in excess of the oil companies' current expectations for the whole of Latin America.[8]

Similarly, for the rest of the Third World the alternative estimates of its ultimate oil resources are several times higher than those currently specified by the oil companies for the regions concerned. These contrasts are detailed in Table I-7.1. The differences emerge from the same factors as in the case of Latin America, but they seem to be accentuated by the even greater paucity of oil industry efforts to find oil in these regions as shown in Figure I-7.4.

The figure for the world's ultimately recoverable resources of oil of $2000 \times 10^9$ barrels, though used almost exclusively in contemporary western world presentation on the future of oil, is certainly not, as we have seen, a generally accepted one. It must, indeed, be viewed as the figure which represents the most conservative view of the availability of

### TABLE I-7.1
### Estimates of the ultimate oil resources of the Third World (in bbls x $10^9$)

|  | Oil Industry Views<br>(a) | Grossling<br>(USGS)<br>(b) | Min. of Geology<br>(USSR)<br>(c) |
|---|---|---|---|
| Latin America | 150–230 | 490–1225 | 686 |
| Africa | 120–170 | 470–1200 | 730 |
| South/SE Asia | 55–80 | 130–325 | 409 |
| Totals | 325–480 | 1090–2750 | 1825 |

(a)    Based on figures in R. Nehring, Giant Oil fields and World Oil Resources, Rand, Santa Monica, June 1978. Table on p.88 adjusted to give comparable geographical coverable with (b) and (c) above. Nehring also states (p.88) that his figures for these regions are 'roughly similar' to those published elsewhere by oil industry observers.

(b)    B.F. Grossling (United States Geological Survey), "In Search of a Probabilistic Model of Petroleum Resources Assessment", Methods and Models for Assessing Energy Resources (M. Grenon, Ed.), Pergamon Press, Oxford, 1979, pp.143–72.

(c)    V.I. Visotsky et al, Ministry of Geology, Moscow, "Petroleum Potential of the Sedimentary Basins in the Developing Countries", in R.F. Meyer (Ed.), op cit., pp.305–16.

oil and so gives the lowest figure in a wide range of estimates of the world's ultimate oil resources. In an article published in 1977, Academician Styrikovich of the USSR and vice chairman of the Executive Council of the World Energy Conference showed that ultimate world oil resources could be as high as $11,000x10^9$ barrels.[9] This figure appears to give the upper limit to the current range of estimates on the size of the oil resource base, though it should be noted that Styrikovich himself, later in his article, refers to his estimate as a 'cautious' one.[10] Between these extreme values there have been many other estimates. In Figure I-7.1, for example, there are a number of estimates, made in the early 1970s, for a resource base of about $3000x10^9$ barrels.[11] More recently, the results of a delphi-type survey of oil resource expectations undertaken by the Institut Français du Pétrole in 1978 shown that there are four estimates of $4000x10^9$ barrels or more.[12] Elsewhere, an International Institute for Applied Systems Analysis (IIASA) 'workshop' on future oil and gas supply, attended by experts from many parts of the world, concluded that resources (of conventional oil) to be reckoned with are no longer around $3000x10^9$ barrels, as previously estimated, but rather 4000, 5000 or even $6000x10^9$ barrels, depending on the price which consumers can afford to pay.[13] In addition to the resources of so-called conventional oil, however, as indicated earlier in the chapter, there are those resources of oil which are potentially available from other habitats.

At the present time, production limitations, arising from technical, environmental and economic considerations, constitute effective short to medium-term constraints on the development of non-conventional oil resources. However, in looking at the issue of the long-term outlook for the supply of oil, it is quite appropriate to assume that these constraints will be diminished or even overcome. This is likely to happen over a period of the next 20 or 30 years, on the assumption that there will be a demand for oil in increasing quantities during this time. This means, of course, that potential ultimately recoverable availability of non-conventional oil can be treated as a valid input to a study of long-term future oil supply/demand relationships.

High estimates of the potential for non-conventional oil remain speculative, as insufficient geological and technological work has been done to date to substantiate the recoverability of such oil on a very large scale. There is, however, as already indicated, firm knowledge of the availability of 2000 to $5000 x 10^9$ barrels of recoverable unconventional oil. Successes which have already been achieved in recovering non-conventional oil in Canada,[14] Venezuela,[15] the USSR[16] and some 40 other countries[17] indicate a high probability that a volume of resources lying within this range will become economic to produce over the next half century.

In the light of the range of contemporary estimates of the ultimate availability of conventional oil, coupled with a reasonable expectation for the rapid evolution of an effective resource base of additional oil from unconventional habitats, there would seem to be adequate justification to simulate the long-term future of oil in a model in which the value of the resource base component rises to a level as high as $11,000 \times 10^9$ barrels. This upper limit to the range of potential oil resource bases emerges from totalling the highest contemporary western world

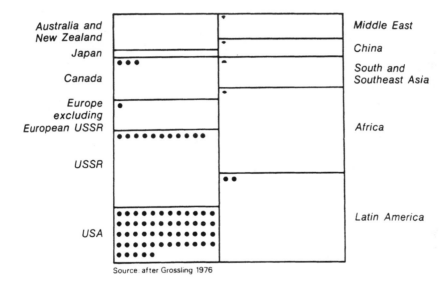

Source: after Grossling 1976

Figure I-7.4: The regional distribution of the world's potentially petroliferous areas, shown in proportion to the world total

*Within each region the number of exploration and development wells which have been drilled is shown – each full circle represents 50,000 wells drilled by the end of 1980. Segments of circles are included only for regions in which the total number of drilled wells to 1980 is less than 50,000. Relative to the US, all other parts of the world, but especially the regions of the Third World – are very little drilled for their oil.*

Sources: B. Grossling, Window on Oil: Survey of World Petroleum Sources, Financial Times, London, 1976, p.83, for end-1975 situation. Grossling's data have been updated to end-1980 from the monthly US oil journal, World Oil, which each year in its August number publishes a table, 'Forecast of... World Drilling'.

estimates for conventional oil (6000 x $10^9$ barrels) and unconventional oil (5000 x $10^9$ barrels). It is thus less than what must currently be viewed as the very low probability for an even larger resource base, as indicated, for example, in Soviet studies of the oil resource question. The elimination of these highest resource base estimates for this study of the future of oil, has its parallel at the other end of the range of resource bases. The few estimates[18] of total oil resources of less than 2000 x $10^9$ barrels have also been ignored in our simulation procedures.

## C. The Annual Rate of Additions to Proven Reserves.

The second major component in the basic complexity surrounding the future of oil is that of the annual rate of additions to proven reserves. This, too, is a component in considering the future of oil about which the post-1950 history of the oil industry has much to tell us. The inherent nature of the complicated and lengthy process of oil exploration, reservoir appraisal and the development of field production, means that the size of any field remains highly uncertain for long after its initial discovery. Sometimes the uncertainty lasts for 20 years or more. The first part of the process is that of the initial discovery of a field, but the discovery well (or wells) reveals only very limited information about the field. Thus the declaration of its reserves has to be small.[19] If for one reason or another, that field is not then appraised and/or developed immediately, its proven reserves necessarily remain declared at the minimum initial level.

With the second part of the process, *viz.* the appraisal of a field by additional wells designed to try to define the amount of oil in place and the percentage of oil in place which is likely to be recoverable, the reserves of a field automatically increase[20] and so provide the basis for the phenomenon of reserves appreciation. Thereafter, the continued development of a field over a long period of time (that is, the process of production) usually serves further to increase the estimates of the size of the recoverable reserves. This occurs as knowledge increases and as technology improves. Paradoxically, the very process of producing a field usually increases, rather than decreases, the volume of the remaining reserves in the field. This sometimes applies for only a few years but, quite frequently, the phenomenon persists for many years during the production phase so that the rate of production from the field can be increased beyond the initially specified level, in response to the greater opportunities offered by the larger-than-originally-expected reserves of the field.

This phenomenon can occur without any concurrent changes in the economic environment – in terms, that is, of changed production costs and/or oil price levels. However, an improved production technology or a higher than expected degree of recovery of oil in place, will lead to lower unit production costs so that more oil can be produced economically from a field. Similarly, an increase in the oil price will make it worthwhile to put more investment into a field's production system, thereby enhancing the level of recovery. Changes in the economic environment can also enable a reappraisal to be made of an earlier 'no-go' decision in respect of a discovered oil field. If development of the field becomes economic as a result of higher prices, then the earlier limited declared reserves (if any) will automatically be enhanced.[21] Such economic aspects of oil reserves appreciation were, of course, of limited importance in the 20-year period of declining real oil prices, from 1950 to 1970, as shown in Figure I-7.3. Over that period the possibilities for the appreciation of reserves were restricted to those arising from cost reductions through technological improvements. In today's conditions of much higher real oil prices (see Figure I-7.3) the reappraisal of earlier sub-economic finds has clearly become much more interesting. Given time, one can now anticipate an appreciation of the world's proven oil reserves based on the economic, as well as the technical, components in the process.

The development and application of new technology is already contributing powerfully to the appreciation of reserves through the upward modification of the percentage of the oil in place in a field which can be produced. An excellent example of this is the Schoonebeek field in the Netherlands. This field was discovered in 1943 and until recently, in the light of the undeveloped nature of the technology of heavy oil production, was expected to yield no more than 5 to 18 per cent of the 1 to 2 x $10^9$ barrels of oil variously estimated to be in place. In 1960, however, an experimental steam-drive system for enhanced recovery was installed in one part of the field. In 1980 the effects were carefully evaluated and the apparent success led to a second pilot project on another part of the field. The recovery factor was increased to up to 33 per cent and, given this favourable result, the technology has now been applied to a much larger area of the field. This 2.5 square kilometre area has had an average recovery rate, with primary and secondary methods, of 17 per cent, but Shell, the operating company, now expects to achieve an overall 39 per cent rate of recovery with the application of the new technology, to give a more than 125 per cent improvement in the recovery of the oil in place.[22] The extension of this to the whole Schoonebeek field would make it a large oil field instead of the relatively minor one which it has hitherto been.

The production of heavy oil is by no means new. Indeed, by the end of 1981 the cumulative production of such oil was about $50x10^9$ barrels – equal to 11 per cent of total cumulative world oil production since 1940.[23] What is new is the steadily advancing technology which enables recovery rates from these fields to be improved – as in the Schoonebeek example described above. A much more important example of this phenomenon at work is in the Belridge field in California. This was discovered in 1911 and by 1979 it had 2813 wells each producing an average of only 14 barrels per day with steam injection. Shell Oil acquired the field in 1980 for $3.65x10^9$ and began to apply a much improved and reorganized steam injection technology costing another $1.1x10^9$. Production from the field has already been doubled and it is expected to increase by another 50 per cent. Eventually a recovery rate of over 60 per cent is anticipated and the field is now expected to yield a total of more than 1000 million barrels – well above Shell's original expectations.[24]

There are also examples of developing technology which convert hitherto totally unproducible oil into effective resources. One example is in South Texas where Conoco has introduced a steam injection pilot project in a reservoir of oil with so high a viscosity that it hitherto totally inhibited production. In the area of the pilot project 50 per cent of the oil in place has been recovered within a period of 2.5 years and the project is now being extended to other parts of this very large field with 2 to $3x10^9$ barrels of oil in place.[25]

The motivation to develop the new technology and the ability to apply it so as to initiate or enhance the production of previously inaccessible oil is, of course, economic in character – arising, basically, from the relationship between the increased expenditure required to produce the additional oil and the higher revenues from the additional oil output. The results achieved on the Bellevue field in Louisiana have been reviewed in this context as a result of the financial involvement of the US Department of Energy, the rules of which require the disclosure of the economics of the project. Between 1976 and 1981 total costs incurred worked out at almost $11 per barrel, but the average pre-tax revenue was over $21 per barrel. After payment of taxes, the per barrel profit was almost $6. In 1981 both costs (at $17) and revenues (over $36) per barrel were higher with the differential between the two widening quite markedly. The apparently favourable economics of the project have, however, been undermined because of the imposition of a federal windfall profits tax, the impact of which has been to reduce net profits to under $2 per barrel. This has made the field no more than marginally interesting viewed from an economic standpoint showing clearly how the volumes of recoverable resources can be affected by government policies[26] – a consideration which, as we have shown elsewhere,[27] applies to high-cost offshore oil production as well.

The application of new and improved oil recovery technology also depends, however, on the availability of the necessary technical and managerial expertise, whereby the successful implementation of such schemes is made possible. These latter considerations will necessarily make the process at the world scale a slow one, given the concentration of most of the world's current and potential oil producing capacity in countries where such expertise is in short supply. It is from these same countries that the international oil companies – with their wealth of expertise – have largely been excluded through nationalization measures. Nevertheless, the technological developments will eventually be widely applied and ultimately they will have a formidable impact on the overall world-wide oil reserves situation. Conventionally, estimates of world oil reserves and resources are closely related to the historical ability of the industry to recover only a relatively small part of the total in place. This currently averages no more than some 30 per cent. At present levels of oil demand each 1 per cent increase in the recovery rate enhances the world's reserves by the equivalent of one and a half years' use.[28] Thus, an increase world-wide to an average of 40 per cent recovery would give more than another 10 years' supply of oil – even with an annual rate of increase in use as high as 5 per cent. Even higher recovery rates can, however, be expected from fields yet to be discovered (because they can be developed using the latest and most effective primary, secondary and tertiary recovery systems from the start), so that the minimum $1000 \times 10^9$ barrels of oil from new fields included in the estimates of world oil resources, based on an assumption of a continuation of the historic average rate of recovery, could well be increased by between 33 and 67 per cent (assuming future recovery rates of 40 to 50 per cent instead of the historic 30 per cent).

The appreciation of oil reserves in the medium-term future thus seems likely to be high – based on the impact of a combination of economic and technological factors. Even historically, when the economic factors did not apply, appreciation has been significant in increasing reserves above the levels which were originally thought to exist in the fields which had been discovered. The impact of the appreciation of reserves on the evolution of the size of the world's proven reserves from 1940 to 1970[29] is shown in Figure I-7.5. The lower dashed line in the diagram shows the size of the proven reserves as they were declared to be at the end of each year, while the solid line shows British Petroleum's 1978 estimate of proven reserves for each year when the contemporary reserves for each discovered field are allocated back to the year in which each field was discovered. The difference between the values for each year is a measure of the degree of reserves appreciation in each year over the period up to 1978. It shows just how formidable this process has been in terms of increasing the historically recorded reserves figures. For example, the less than $100x10^9$ barrels of declared reserves in 1950 have increased to over $350x10^9$ barrels while the $300x10^9$ barrels declared to exist in 1960 are already known to have been understated by over 65 per cent.

Appreciation has had a formidable impact in retrospective change on the reserves/production ratio for each year since 1950. This is shown in Table I-7.2 in which the R/P ratios for the period 1950–70, as calculated by BP in 1978,[30] are compared with the ratios which were

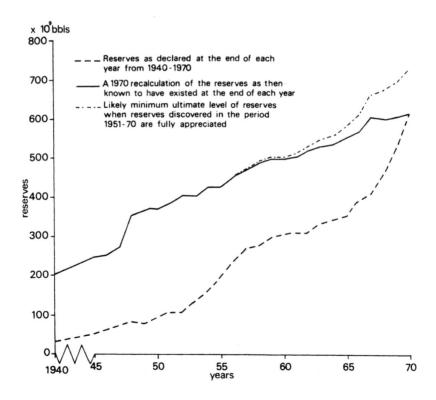

Figure I-7.5: Contemporary and retrospective estimates of world proven oil reserves

## TABLE I-7.2
### World proven oil reserves/production ratios, 1950–70 and 1981

| Year | 1978 Calculation with appreciated Reserves (years) | Contemporary Calculation from Reserves as declared (years) | Effective Ratio allowing for increasing oil use | |
|------|------|------|------|------|
| | | | Effective Ratio (years) | Rate of Increase in oil use* |
| 1950 | 95 | 24.7 | 19.2 | 5.41% |
| 1951 | 91 | 23.8 | 17.6 | 6.50% |
| 1952 | 90 | 25.9 | 18.9 | 6.74% |
| 1953 | 84 | 27.8 | 20.3 | 6.76% |
| 1954 | 84 | 31.3 | 22.7 | 6.97% |
| 1955 | 75 | 33.5 | 23.4 | 7.77% |
| 1956 | 75 | 37.6 | 25.6 | 8.28% |
| 1957 | 73 | 40.4 | 27.7 | 8.13% |
| 1958 | 75 | 41.6 | 30.2 | 6.95% |
| 1959 | 71 | 41.0 | 28.7 | 7.71% |
| 1960 | 66† | 38.9 | 27.6 | 7.38% |
| 1961 | 64† | 38.0 | 28.1 | 6.55% |
| 1962 | 63† | 35.3 | 25.5 | 7.08% |
| 1963 | 58† | 34.7 | 25.0 | 7.08% |
| 1964 | 55† | 33.0 | 23.3 | 7.51% |
| 1965 | 53† | 32.0 | 23.2 | 6.94% |
| 1966 | 51† | 32.0 | 23.3 | 6.88% |
| 1967 | 53† | 31.7 | 23.0 | 6.98% |
| 1968 | 46† | 30.6 | 20.7 | 8.37% |
| 1969 | 46† | 34.0 | 23.8 | 9.75% |
| 1970 | 45† | 32.4 | 22.4 | 7.96% |
| 1981 | Will not be known until 2001 | 32.8 | 31.6 | 0.04%** |

Notes: * Average annual rate of increase in oil use over previous 10 years.

   † These R/P ratios are likely to be understated as only a conservative view has been taken of the post-1970 appreciation of these reserves.

   ** Average annual rate of increase in oil use 1973–1981.

declared contemporaneously and which, of course, formed the data set on the basis of which the future of the oil industry was viewed from year to year. The third column of the Table then converts each contemporary nominal R/P ratio into an effective ratio by taking into account the average per annum rate of increase in oil use expected over the following decade (as represented by the average annual increase in oil use over the previous 10 years). By comparing this historic set of effective R/P ratios with the present reserves/production situation it is clear that the future of oil, reflected in the proven world reserves situation, has never been as good as it is now. The present nominal R/P ratio stands at a level of almost 33 years – somewhat lower than that of the nominal ratios as presented in the 1960s though still better than in any year between 1950 and 1955 and since 1964. However, present expectations are, at best, for a slow increase in oil use. This means that the effective R/P ratio now needs to be only a little smaller than the nominal one. The present effective ratio stands at 31.6 years and thus exceeds the effective ratio of known reserves to future demand for all previous years in the last 30 years' history of the industry.

The reality of the oil reserves position at any specific time is thus much more complex than any simplistic presentation would appear to indicate. It is, however, very important to understand this complexity since it contains an element which is crucial for an effective interpretation for the future of oil. This is the matter of the annual quantity of oil which needs to he added to reserves in order to maintain the lowest acceptable reserves to production ratio. It is generally accepted in the oil industry that effective forward planning of the industry's infrastructure necessitates proven reserves which are at least equal to the total amount of oil expected to be used in the following 10 years. If this were defined through the nominal R/P ratio then the required ratio would be a function of the expected rate of increase in the use of oil because the ratio, of course, rises increasingly steeply with higher rates of demand. For the average rate of increase in oil use between 1950 and 1970 of about 7.4 per cent per annum the required R/P ratio needed to be almost 15 years. Reference to Table I-7.2 shows that this was achieved throughout the 20-year period during which there was this rapid rate of increase in the use of oil. Since 1973, however, the rate of increase in the demand for oil has fallen dramatically to a rate which, in the intervening eight years, has only just been positive. At this low rate of increasing use, 10 years' availability of oil is achieved through an R/P ratio of only just over 10.0. This is a requirement which is very much lower than the current unadjusted reserves to production ratio of over 32 years.[31]

It is thus clear that the immediate future availability of sufficient oil to match the current level of world total annual use of some $21.8 \times 10^9$ barrels does not depend upon the continuing discovery each year of large volumes of new oil. Indeed, with a continuing low annual average rate of growth in oil use it would be after the year 2010 before the unadjusted R/P ratio was low enough to require new oil discoveries. This assumes, completely unrealistically, that there will be no appreciation whatsoever of currently known reserves in the meantime. With a rate of appreciation of reserves no higher than in the past, it would be well into the second quarter of the 21st century before any new oil discoveries had to be made.

This favourable contemporary oil reserves situation – when the outlook for availability is related to expected levels of demand – gives the oil industry a long breathing space in which to organize itself for the large-scale exploration and discovery of oil in new geographical locations (notably in the Third World and offshore) and/or in deeper formations, and/or from new habitats, such as the tar sands and the oil shales. This is an important element in respect of the non-communist world outlook for oil, given the current transition period through which the oil industry of the western world is going in terms of its ownership and organization.

As a consequence of the disturbing factors of this transition period, the western world's oil industry as a whole is now in a much less strong position to find new reserves than it was during the period between 1950 and 1974. At that time the large international oil companies dominated oil exploration and development efforts outside North America and were able to put their considerable resources of know-how and capital accumulation to work without having to worry too much about international and national political considerations. The relative decline in the importance of these historically significant institutions in the search for, and the development of, oil resources, as well as the lack of ability and/or motivation on the part of many of the newly formed oil entities to continue with the oil finding and developing processes, seems likely to have been the main factor which restricted growth in non-communist world oil reserves in the period from 1974 to 1980.

This is, however, certain to be only a temporary phenomenon as most of the member countries of OPEC will eventually have to search for and to prove new reserves of oil.[33] Moreover, many of the previously oil-poor countries which have attempted to find oil reserves in the post-1973 oil crisis period are already achieving some success in their efforts.[34] Nevertheless, it still seems prudent to assume that, for the next decade or so, additional new oil plus appreciation of oil will not, on average, be sufficient fully to replace the amount of oil which is used year by year. There could thus be a consequential modest annual decline in the unadjusted R/P ratio during the 1980s.[35]

Indeed, in the immediate future the annual additional amounts of oil which have to be discovered, or which must become available from appreciation, need be no more than modest. This is because all, or most of, the oil required to meet the now slowly increasing demand could come from the reserves that have already been discovered and appraised. That is, from oil that is already 'on the shelf' as far as the world oil industry is concerned. Nevertheless, if one assumes, first, that the R/P ratio is steadily and continuously run down to the minimum necessary level, and second, that the use of oil continues to grow, then increasing rather than decreasing quantities of oil will eventually have to be proved each year. It will be at some date after that, in the continuing development of the industry, that the industry's ability to achieve a high enough annual discovery rate will become a critical variable in determining whether or not oil production can continue to expand.

There has recently been much speculation as to what the maximum annual finding rate for new oil might be. There have been suggestions that this rate may already have been achieved in the 1950s and 1960s when the discovery of a series of super-giant fields[36] produced 'adjusted' additions to reserves of more than $100 \times 10^9$ barrels in a single year. These adjusted figures, however, involve the retroactive allocation of additions to reserves to a particular year, in the light of subsequent knowledge about the discoveries – and particularly the super-giant ones.[37] They were not the declared reserves for the fields concerned as recorded at the time of their discovery from the data then available. In the years between 1950 and 1979 such annual declarations of net additional reserves ranged from a low of $13 \times 10^9$ barrels to a high of $74 \times 10^9$ barrels. There was, moreover, a considerable degree of variation from year to year, as a result of the specific incidence of particularly large discoveries and/or a company's or government's timing in declaring them. In order to 'spread' the impact of these random events a five-year running mean of annual discovery rates has been calculated as shown in Figure I-7.6. This shows an upward trend in the rate of additions to reserves in the period 1960–70 with an eventual highest figure of $55 \times 10^9$ barrels. More recently the figure has fallen back to around $25 \times 10^9$ barrels but, as explained above, this appears to be due to changes in the ownership and the organizational structure of the industry since 1973, together with the fact that the severe retrenchment in the rate of increase in the demand for oil since than has undermined the earlier confidence of the industry in what it had hitherto come to accept as a process of inevitable expansion. Thus, large annual additions to reserves have become less important to the industry.

It is difficult, therefore, to judge what the maximum annual rate of additions to reserves will eventually turn out to be. Given the vast areas with petroliferous potential which have so far remained untouched or nearly untouched by exploration,[38] and given the now increasing motivation on the part of most countries to stimulate the process (as a result of the high price of imported oil), there seems to be no inherent reason why future annual additions to reserves higher than in any year in the past cannot be achieved. It is also possible that the importance of the issue will, in any case, soon be severely diminished in view of the forthcoming incorporation of reserves of non-conventional oil from tar sands and oil shales in the resource base. These non-conventional oil

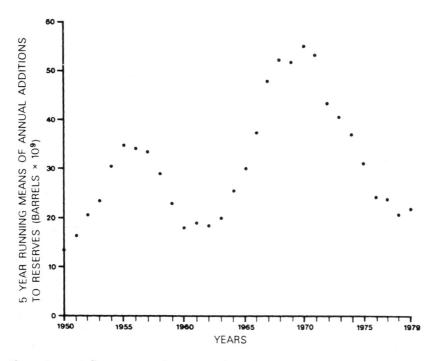

Figure I-7.6: A five-year running mean of oil discoveries, 1950–79
*(Based on contemporary declarations of proved reserves, ie with no allowance made for the later appreciation of reserves.)*

resources are known to exist in vast quantities from the geological information which is available on their occurrence. The process of gradually 'proving' the reserves with exploration, appraisal and development wells, as in the case of conventional oil the reserves of which can be proved only by the drill, is not required.[39] This possibility suggests that the future of oil could be modelled without a maximum annual finding rate constraint. However, in order to give a somewhat more pessimistic alternative view of the future of oil, annual additions-to-reserves constraints, related to what the industry has already achieved and in the light of the expected size of the resource base, must be introduced into the analysis. This must be done in case non-conventional oils only become available on a large scale at a later than currently expected period in the development of the industry, so that there is an interim period of pressure on the ability of the industry to find sufficient conventional oil to sustain increasing use.

It is apparent that the maximum annual rate of additions to reserves must bear some relationship to the size of the resource base. If more than half of the world's ultimate reserves of oil have already been found (as would be the case given the discovery to date of a total of some $1100 \times 10^9$ barrels and a total resource base of only $2000 \times 10^9$ barrels), then higher annual finding rates than those already achieved are most unlikely. This is because more and more effort will have to go into locating the remaining oil in increasingly difficult locations and habitats, so that the productivity of exploration investment is bound to fall significantly and more or less continuously. On the other hand, if only just over 20 per cent of the world's ultimate conventional oil has been found to date (as in the case of a resource base of $5000 \times 10^9$ barrels) then, unless finding rates increase markedly above those achieved to date, the oil in the ground is going to take well over another 100 years to locate and to prove. This is clearly very unlikely unless one assumes that technology will not develop, or that its availability will remain geographically constrained. Alternatively, one would have to make the assumption that the rate of growth in the use of oil will be so slow that the motivation to find more oil more quickly is never realized.

In other words, the maximum annual additions to reserves variable is also a variable to which it is necessary to attach a wide range of values in any comprehensive look at the future of oil. This has been done in this study with the values used ranging from 40 billion barrels, given a predicated resource base of 2000 billion barrels, up to 110 billion with a resource base of 11,000 billion barrels.

## D. The Use of Oil

The third main component which determines the future of oil is the evolution of the rate at which the commodity is used. In contrast with the other two main variables, however, this aspect of the outlook for oil cannot easily be related to the industry's experience in the period from the late 1940s to 1973. During that period the use of oil moved ahead at an exponential growth rate of over 7 per cent per annum under the combined stimulus of rapid economic expansion in most parts of the world, a declining real price for oil (see Figure I-7.3) which encouraged increasingly energy-intensive ways of doing things and of going places, and the substitution of coal and other sources of energy by oil in many parts of the world. Table I-7.3 illustrates the results of these factors in increasing the use of oil in a selection of industrialized countries,[40] whilst in Table I-7.4, the 1950–75 changes in energy/oil use in Latin America (as an example of changes in the Third World) are set out.[41]

Since 1973 both the general economic and the more specific energy situations have changed quite dramatically so that the evolution of the use of oil over this recent period bears no relationship to the earlier experience of the industry. World-wide, between 1973 and 1980, the annual average rate of growth in the use of oil was only 1.25 per cent and if the communist countries are excluded (their economies have been less seriously affected by the traumas of the 1970s), then the rate of increase in oil use over the period falls to a figure as low as 0.5 per cent per annum.

It would thus be quite wrong to think in terms of the evolution of the future use of oil as being likely to resemble the evolution of the oil use curve in the period up to 1973. There are several reasons for this; first, there is the order of magnitude increase in the real price of oil over the last decade – as shown in Figure I-7.3; second, there are the continuing economic difficulties in the whole of the western world; and third, one notes the impact of energy conservation policies in general, and of oil substitution policies in particular, in all the non-communist world's major energy consuming countries.

Gross uncertainties in the western world's system – both economic and political – plus technological changes in energy-using systems, are certain to make the future rate of increase in the use of oil much lower than in the period up to 1973. Moreover, within the context of this much lower rate of increase in the use of oil there seems likely to be accentuated temporal and spatial variations in the growth rates. These variations will arise from contrasting economic fortunes in different countries and regions, and from the unevenness with which energy/oil

## TABLE I-7.3
## Energy use in selected industrialized countries, 1952–72

| Year | Energy used | US | United Kingdom | West Germany | France | Italy | The Netherlands |
|------|-------------|-----|-----|-----|-----|-----|-----|
| 1952 | Total mtce* | 1176 | 232 | 145 | 89 | 25 | 22 |
| | Tons/capita | 7.5 | 4.6 | 2.9 | 2.1 | 0.6 | 2.1 |
| | % Coal | 34 | 90 | 95 | 79 | 43 | 78 |
| | % Oil | 38 | 10 | 4 | 18 | 35 | 22 |
| | % Gas | 27 | – | – | – | 8 | – |
| | % Other† | 1 | – | 1 | 3 | 15 | – |
| 1957 | Total mtce* | 1334 | 247 | 186 | 111 | 44 | 28 |
| | Tons/capita | 7.8 | 4.8 | 3.5 | 2.5 | 0.9 | 2.5 |
| | % Coal | 29 | 85 | 88 | 72 | 28 | 63 |
| | % Oil | 40 | 15 | 11 | 25 | 48 | 36 |
| | % Gas | 30 | – | – | 1 | 15 | 1 |
| | % Other† | 1 | – | 1 | 3 | 9 | – |
| 1962 | Total mtce* | 1546 | 265 | 221 | 122 | 71 | 35 |
| | Tons/capita | 8.3 | 4.9 | 3.9 | 2.6 | 1.4 | 2.6 |
| | % Coal | 23 | 72 | 71 | 58 | 18 | 47 |
| | % Oil | 41 | 28 | 27 | 33 | 62 | 51 |
| | % Gas | 35 | – | 1 | 5 | 13 | 2 |
| | % Other† | 1 | – | 1 | 4 | 8 | – |
| 1967 | Total mtce* | 1958 | 276 | 251 | 154 | 112 | 47 |
| | Tons/capita | 9.8 | 5.0 | 4.2 | 3.1 | 2.1 | 3.7 |
| | % Coal | 22 | 59 | 51 | 40 | 12 | 25 |
| | % Oil | 41 | 39 | 46 | 50 | 71 | 53 |
| | % Gas | 36 | 1 | 3 | 6 | 11 | 22 |
| | % Other† | 1 | 1 | 1 | 4 | 6 | – |
| 1972 | Total mtce* | 2426 | 302 | 333 | 215 | 152 | 76 |
| | Tons/capita | 11.6 | 5.4 | 5.4 | 4.2 | 2.8 | 5.7 |
| | % Coal | 20 | 40 | 35 | 21 | 7 | 6 |
| | % Oil | 43 | 46 | 52 | 66 | 75 | 36 |
| | % Gas | 36 | 12 | 11 | 10 | 14 | 59 |
| | % Other† | 1 | 2 | 2 | 4 | 4 | – |

*Source:* Based on UN Energy Statistics, Series J.

*Notes:* ★ millions tons coal equivalent.

† Mainly hydro-electricity converted to coal equivalent on the basis of the heat value of the output. 1972 figures also include some nuclear power converted on the same basis.

The most important energy source in each country for each year is underlined.

conservation policies are pursued and implemented by governments, either through the use of the pricing mechanism or by more directly interventionist and/or regulatory measures.[42] In these uncertain circumstance it is clearly necessary to employ a wide range of oil use growth rates in a study of the future of the industry.

The fact that a wide range of values has not been used in recent studies on the future of oil[43] reflects first, an apparent belief that a 10-year doubling rate in its use (as was achieved in 1950–60 and 1960–70 and was expected to continue thereafter) is an inevitable 'fact' of modern economic life; and second, an unrealistic view of the ability of the western economic system speedily to recover its economic equilibrium and its earlier propensity for strong and continuing growth. This, in part, reflects the difficulties of politicians and policy makers in coping with the implications of an enforced end the revolution of rising expectations by the populations of the industrialized countries. Thus, their forecasts of economic growth have to be higher than those justified by the realities of the changed international economic system. And as policy makers still accept, either explicitly or implicitly, the idea of a close link between economic growth and energy use, their forecasts of the latter are necessarily also higher than the changing conditions justify.[44] The consequence of the failure to recognize the very much more modest outlook for the future use of oil, compared with the world's experience in recent decades results, of course, in a serious overstatement of both

### TABLE I-7.4
### Energy use in Latin America, 1950–1975

|                          | 1950    |     | 1975    |      |
|--------------------------|---------|-----|---------|------|
|                          | mtoe*   | %   | mtoe*   | %    |
| Total energy use         | 68.0    | 100 | 231.1   | 100  |
| of which                 |         |     |         |      |
| Vegetable fuels          | 29.2    | 42.9| 41.4    | 17.9 |
| Oil                      | 28.8    | 42.4| 134.4   | 58.2 |
| Natural gas              | 2.9     | 4.3 | 33.7    | 14.6 |
| Coal                     | 5.5     | 8.1 | 11.0    | 4.8  |
| Hydro-electricity†       | 1.6     | 2.4 | 10.6    | 4.6  |

Notes:  *    mtoe = millions tons of oil equivalent
        †    Calculated on basis of heat value of electricity produced.
             1kWh = 3412 btu = 860 kcals

the rate of depletion of the oil resource base and the speed with which new reserves have to be found and proven. Demand considerations thus form an essential and uncertain element in the basic complexity of the future of oil so that an appropriately broad range of demand growth rates (from 0.65 to 5% per annum) have to be incorporated into the simulation of the future of oil in order to ensure that the component is given due weight in the overall evaluation of the possible futures for the world oil industry.

## References

1      See P.R. Odell, *Oil and World Power*, Penguin Books, Harmondsworth, 6th Edition, 1981, for a description of the international oil system in general and, in chapter 9, an analysis of the evolution of OPEC control in particular.

2      P.R.Odell, *Oil and World Power*, op. cit., Chapter 9.

3      Meyer and Fulton, for example, have estimated recoverable reserves of heavy oil and oil from tar sands at a minimum of $1000 \times 10^9$ barrels by taking into account only half of the currently known deposits while Donnell has estimated resources of oil from oil shales at $2800 \times 10^9$ barrels. See R.F. Meyer and P.A. Fulton, "Toward an Estimate of World Heavy Crude Oil and Tar Sands Resources", *Conference pre-print paper no VIII. 7, 2nd International Conference on Heavy Crude and Tar Sands*, Caracas, February 1982 and J.R. Donnell, "Global Oil Shale Resources and Costs", in R.F. Meyer (Ed.), *The Future Supply of Nature Made Petroleum and Gas*, Pergamon Press, Oxford, 1977, pp.843–56.

4      Examples of approaches to the future of oil incorporating an oil component of this size are given in Chapter 3.

5      Note, for example, the number of oil company estimates of ultimate oil availability at about this level in the delphi-type survey on the subject undertaken by the Institut Français du Pétrole for the 1978 World Energy Conference and the presentation by J.D. Moody (of the Mobil Oil Corporation) and M.T. Halbouty (at the 1979 World Petroleum Congress) of an oil reserves base of this size, See *Petroleum Economist*, Vol.XLVI, No.12, December 1979, pp.501–2.

6      See, for example, the comments by R.A. Sickler (Royal Dutch/Shell Exploration and Production Division) in *Methods and Models for Assessing Energy Resources* (M. Grenon, Ed.), Pergamon Press, Oxford, 1979, pp.137–40. In response to questions and discussion on the background to the presentation in his paper of the idea of a $2000 \times 10^9$ barrels ultimate oil resource figure, Sickler eventually agreed that he was not presenting an estimate of ultimately recoverable resources. His estimates, he agreed, 'have been constrained logistically in light of the climate for exploration and development available now and a scenario (for the increase in reserves) up to about the year 2000'.

7       See the notes to Table I-7.1 for the sources for these estimates.

8       See B. Netschert, "Mexico's Oil and Gas Potential", in *Latin America and Caribbean Oil Report*, Petroleum Economist, London, 1979; and B.F. Grossling, "Possible Dimensions of Mexican Petroleum", in J.R. Ladman et al. (Eds), US-Mexican Energy Relationships, Lexington Books, Lexington, Mass, 1981.

9       M.A.Styrikovich, "The Long Range Energy Perspective", *Natural Resources Forum*, Vol.1, No.3, April 1977, pp.252–3.

10      *Ibid.* p.254. He refers to an alternative Soviet view which puts recoverable resources of conventional oil at up to $15,000 \times 10^9$ barrels.

11      These are the estimates of the National Petroleum Council in Washington, 1972; H.R. Linden and J.D. Parent of the US Gas Association in 1973; and L.G.Weekes in a report prepared for the United Nations in 1973.

12      Institut Français du Pétrole, *op. cit.*, p.24.

13      R. Seidl, "Oil: The Picture is Changing", *Options*, International Institute for Applied Systems Analysis, Laxemburg, Winter 1977, p.4.

14      N.Strom, "Projection of Alberta Bitumen Synthetic and Extra Heavy Oil Developments", Conference pre-print paper III. 2, *2nd International Conference on Heavy Crude and Tar Sands, op.cit.*

15      See A. Volkenborn (Maraven, S.A. of Venezuela), "Venezuela's Heavy Oil Development Prospects and Plans", UNITAR Conference on *Long Term Energy Resources*, Montreal, November–December 1979 for a discussion of this process in respect of oil from the Orinoco oil belt.

16      R.F. Meyer and P.A. Fulton, "Towards an Estimate of World Heavy Crude Oil and Tar Sands Resources", Conference pre-print paper No. VIII. 7, *2nd International Conference on Heavy Crude and Tar Sands, op.cit.*

17      *ibid.*

18      See the results of the delphi-type survey of oil resource base estimates in *World Energy Resources 1985–2020, op.cit.*, p.24.

19      Indeed, in some countries/provinces, the reserves declaration has, by legislation, to be limited to the quantity of oil which can be produced by the discovery well. In all fields, other than the very smallest, this implies a declaration of reserves which is always less than the ultimate size of the field.

20      Except, of course, in the very limited number of cases when the initial expectations about the discovery from the exploration well – and these expectations underlie the appraisal and development decisions – turn out to be incorrect, either in terms of the quantity of oil in the reservoir or in terms of its producibility.

21      These and other issues relating to the economics of different levels of oil production from a field were discussed at length in P.R. Odell and

K.E. Rosing, *Optimal Development of the North Sea's Oil Fields*, Kogan Page, London, 1976.

22    P.J.P.M. Troost, "The Schoonebeek Oil Field. The RW-2E Steam Injection Project' Conference pre-print paper no.XVI. 5, *2nd International Conference on Heavy Crude and Tar Sands, op.cit.*

23    R.F. Meyer and P.A. Fulton, *op.cit.*

24    R. Dafter, *Winning More Oil*, Financial Times Business Information Ltd, London, 1981, and *Financial Times*, 12 February 1982.

25    W.L. Martin et al, "Conoco's South Texas Tar Sands Project", Conference pre-print no.XIII. 1, *2nd International Conference on Heavy Crude and Tar Sands, op.cit.*

26    C. Joseph and W. Pusch, "Engineering and Economics of the Bellvue In-situ Combustion Project", Conference pre-print no.XVI. 3, *ibid.*

27    P.R. Odell and K.E. Rosing, *The Optimal Development of North Sea Oil Fields, op.cit.*

28    On the assumption that the figure of $1050 \times 10^9$ barrels of used plus presently proven oil reserves emerge from an expected average recovery of about 30 per cent of the oil in place. A one per cent improvement in recovery rates would give an additional $35 \times 10^9$ barrels of oil. World oil use in 1981 was about $22 \times 10^9$ barrels.

29    There is, of course, absolutely no point in showing the situation post-1970 – over the last 12 years, that is. This is because fields discovered since then are still either undeveloped or in the very early stages of production (as, for example, the earliest North Sea fields) so that no realistic picture can yet have emerged of the size of the field. Even fields discovered between 1960 and 1970 and since developed are still likely to be appreciating to some degree, while many non-economic finds made in that period will now be worth further appraisal for possible development. The future minimum likely appreciation of the reserves discovered between 1960 and 1970 is shown in Figure I-7.5 by the topmost (dotted) line on the diagram.

30    See British Petroleum's publication, *Oil Crisis... Again?*, London, 1979.

31    This may be defined as 'unadjusted', given the certainty of appreciation of the reserves discovered to date – following the arguments set out above. Even if the historic trend of the process of appreciation is no more than repeated, this would make the current adjusted ratio more than 40 years. However, as also argued above, changed economic circumstances seem likely to make appreciation an even more important phenomenon in the future than it was in the period of decreasing real oil prices between 1950 and 1970.

32    This arises because they are mainly the state oil companies of the member countries of OPEC. Given the fact that most of these countries have rich existing known reserves, the state oil companies

have little incentive to prove additional new oil fields or even to reappraise the volume of oil in oil fields which are already known.

33    Some OPEC countries have already taken the necessary action to achieve new discoveries and reappraise old ones. These include Venezuela, Ecuador, Algeria, Nigeria and Indonesia where such developments are already important for the countries' medium to longer-term economic interests.

34    Large new reserves have recently been discovered in many of these countries including Mexico, Brazil, Chile, Egypt, Sudan, Malaysia and India. These successes, plus newly discovered oil in some OPEC countries and a number of industrialized nations (including the US), have taken annual additions to reserves back above annual use in both 1980 and 1981. Indeed, in 1981 twice as much oil was found as was used.

35    This is the way in which we have simulated the future of oil for the (variable) period post-1980 in which the nominal R/P ratio remains above the lowest level which is required to meet the next 10 years' oil consumption.

36    Internationally, super-giant fields are defined as those having at least $5 \times 10^9$ barrels of recoverable oil. In North America the same adjective is used to describe fields with more than $2 \times 10^9$ barrels of recoverable oil.

37    This is, for example, the approach used by British Petroleum for its publication, *Oil Crisis... Again?, op. cit.*

38    These characteristics of the world-wide occurrence of and search for petroleum resources are illustrated in Figures I-7.2 and I-7.4. See P.R. Odell, *World Energy, Needs and Resources*, Institute of Bankers, London, 1979, for a more detailed discussion of these considerations.

39    Oil reserves in such non-conventional forms have more in common with coal in respect of their reserves status than they have with conventional oil. The reserves are so large relative to the annual rate of production that the concept of an R/P ratio becomes ludicrous. Reserves development is now essentially a function of economic conditions and, of course, of the availability of the production technology; first, in terms of its evolution and second, in terms of its specific availability to a particular country or region.

40    This table is based on one from P.R. Odell's, *The Western Europe Energy Economy*, Stenfert Kroese, Leiden, 1976, in which the factors affecting developments in oil use in western Europe are discussed at length.

41    This is from P.R. Odell, *Latin America's Energy Prospects*, EGI Working Papers Series A No79–16, Rotterdam, 1979, in which the evolution of the energy/oil market in Latin America is also discussed. A shortened version of this paper (without the tables and other illustrations) has been published in the *Bank of London and South America Review*, May 1980.

42     See P.R. Odell, *The Western European Energy Economy*, and *The World's Energy Needs and Resources, op.cit.*, for more detailed discussion of these and related issues.

43     See, for example, OECD, *World Energy Outlook*, Paris, 1977; MIT Workshop on Alternative Energy Strategies, *Energy Global Prospect 1985–2000*, New York, 1978; British Petroleum, *Oil Crisis... Again?*, London, 1979; and Shell, *The Outlook for Oil 1980–2020*, London, 1979.

44     Such views also ignore the increasingly strong evidence that further economic growth in industrialized economies does not necessitate much, or even any, increase in energy use for the next 20–40 years. This is because the conditions of the 1950s and the 1960s encouraged such a wasteful use of energy that the elimination of waste could, even in a growth economy, inhibit the need for additional energy use. See, for example, R. Stobaugh and D. Yergin, *Energy Future*, Random House, New York, 1979 and G. Leach et al, *Towards a Low Energy Strategy for the UK*, Scientific Publications, London, 1978 for analysis of this economy/energy relationship in the US and the UK respectively.

# Chapter I – 8

# World Oil Resources: East-West Differ on Estimates*

Only a few years ago the idea of oil as an inherently very scarce commodity was widely presented and accepted. From this assumption it was concluded that oil prices would necessarily go on increasing even from the dizzy heights they reached in 1981/82. This conclusion formed a fundamental basis for energy policy and investment decisions throughout most of the Western industrialised world. The recent decline in the rate of increase in global demand has undermined the previously predicated immediacy of an oil supply problem, but even now spokesmen from institutions as diverse as the International Energy Agency and British Petroleum are still "warning the world" of early-1990s difficulties for global oil supply/demand relationships and, even more emphatically, of the certainty of the Middle East's renewed dominance in the required supply of oil for the West.

The essential elements used to support this view are; first, the purported consensus amongst petroleum geologists that the world has no better than an even chance of ultimately supplying 246 billion ($10^9$) tons (= about 1,800 billion ($10^9$) barrels) of economically recoverable conventional oil, of which almost 30% has already been used; and second, that the Middle East's original recoverable reserves not only account for 40% of the global total, but also, after taking into account the contrast in the amount of oil produced to date in the Middle East and in

* Reprinted from *Petroleum Economist*, Vol.LII, No.9, September 1985, pp.329–331. (NB. The joint author of this paper, Dr. K.E Rosing, has given his permission for this reprint. This permission is gratefully acknowledged.)

the rest of the world, the Middle East now has 54% of the oil resources that remain. Such a view of the limitation on world oil resources and their dominance by the Middle East was presented in depth by C.D. Masters of the United States Geological Survey at the 1983 London meeting of the World Petroleum Congress.[1] Since than the arguments and conclusions have been reiterated on many occasions and any challenge to this now-established conventional view on the availability of conventional oil is treated with disdain – to put it mildly.

In our book *The Future of Oil* (Kogan Page, London, 2nd Edition, 1983) we did attempt to challenge the validity of the inherent oil-scarcity/inevitability-of-mideast-domination-of-supply hypothesis, in part by drawing attention to a great deal of earlier and some contemporary Western world evidence of much higher global oil reserves. We also pointed out that there were a number of statements by Soviet scientists on likely global oil reserves which seemed to be markedly at variance with the conclusions of studies such as those by Masters and his USGS colleagues. In 1983 there were, however, difficulties in interpreting exactly what the USSR's geologists and their scientific colleagues were saying on global conventional oil resources as there appeared to be some confusion, in the English language presentation at least, between conventional and non-conventional oil. It was also difficult to make direct comparisons of Soviet and Western analyses because of incompatibilities over regional and habitat variables and also in respect of definitional problems. Nevertheless, one could conclude that the differences of opinion between Western (essentially US) and Soviet geologists working on oil resources issues were significant: and this in itself was remarkable because in most spheres of scientific endeavour such major East/West differences did not exist.

The depth of the difference between US and Soviet petroleum geologists has now become much clearer as a result of a major publication by the Soviet Scientific Research Institute on the Geology of Foreign Countries. This is a detailed survey, based on work extending over many years, of oil and natural gas resources in all countries of the non-socialist world. It was originally published in Russian[2] at just about the same time as Masters and his colleagues at the USGS presented the results of their years' long survey on oil resources to the World Petroleum Congress in London. The parallelism in the work and the publications appears to have gone unnoticed in the West where there seems to be no published commentary on the basic incompatibilities of the two studies. However, an English language presentation of the main elements in the Soviet study has recently been made available,[3] and it provides the basis for an effective comparison with the USGS' conclusions.

It is not possible, within the tight confines of this article, to describe and evaluate the content of the Soviet work in detail, but it should be made clear that in a number of significant ways the presentation is more comprehensive and more analytical than that of the USGS. There is a complete regional breakdown of global prospects (with the exception of the USSR itself and other socialist countries as these are areas which fall outside the remit of the Institute responsible for the study). In addition, the overall figures by region for ultimately recoverable conventional oil are broken down by methods of recovery (primary, secondary and tertiary); by field sizes (in six categories); by pay-zone depth (also in six categories); by physiographic conditions of resources' location (three categories); and, in the case of offshore resources, by water depth (in five categories). Finally, and most interesting of all – as it is an aspect which has not been tackled in any comprehensive Western work on a global basis – remaining resources of oil (broken down into proven reserves, additions to discovered reserves, undiscovered basic resources and additional resources) are detailed by region in terms of categories of production costs. There are four categories ranging from a low one, with oil costing less than $10 per ton to produce ($1.40 per barrel), to a high category, with oil at more than $80 per ton ($11 per barrel). Intermediate categories are $10–45 and $45–80 per ton.

The prospects opened up by the availability of this Soviet data for purposes of comparative analysis with the conventional Western view of conventional oil resources are legion. Here we wish to concentrate simply on a comparison of the resources-numbers at both the global and the regional levels. Specifically, one can compare the USGS's presentation of a range of probabilities for quantities of recoverable resources with the Soviet cost-related quantities of resource availabilities with the two lowest cost levels of up to $45 per ton taken together. This comparison is made on the assumption that the USGS implicitly – if not explicitly – relates the probability of ultimate recovery to variable cost-of-production considerations.[4] The results of this comparison are shown in Table I-8.1, where the data given in, or calculated from, the US and the USSR publications are followed by the calculated ratios between the two sets of data.

The Table shows that the Soviet estimates of the size of the non-communist world's oil resources and prospects are almost always larger than those of the so-called consensus Western view. The opposite is true in only two of the twenty-one comparisons that are made, *viz.* for the size of the lowest-cost oil resources (at less than $45 per ton) in North America and Western Europe. Otherwise, the Soviet data gives higher estimates – by factors (termed ratios in the Table) of up to almost 2.5. Both for illustrative purposes and, more importantly, to bring out the implications of the differences, we will concentrate our attention on a comparison of the USGS's "modal" view (the one, that is, that we referred to in the first two paragraphs of this article as providing the essential background to current conventional Western views on the future of oil), with the Soviet view of oil resources available at a production cost of no more than $80 per ton.

This is a cost of oil production equal to approximately $11 per barrel: a level to which very few analysts expect the oil price to decline so that, by definition, the availability of this oil will not be constrained by price/cost considerations. The USGS's figure for oil in this category in the non-communist world is 192.3 x $10^9$ tons. By contrast, the Soviet scientists indicate over 300 x $10^9$ tons as falling in this set of resources to give a ratio of 1.56 between the two figures. Viewed regionally, the Soviet figures are larger than those of the USGS by a minimum of 19% (in the case of North America) and up to a maximum of 141% (for Asia and Oceania). The difference for Western Europe – at 50% – lies in the centre of this range. The Soviet view of the resources of the Middle East compared with that of the USGS is very much in line with the contrast in the two sets of data at the global level – showing recoverable reserves (at a cost not exceeding $80 per ton) of 140 x $10^9$ tons compared with the 99 x $10^9$ tons indicated by the USGS. The Soviet scientists are even more bullish about the Middle East than their US counterparts, suggesting that Western world oil demand in future would be able to rely on the bountiful oil resources of that region – should this be considered desirable by the West.

## Resources outside the Middle East

However, the relatively even more bullish Soviet estimates of essentially low cost oil elsewhere in the Western world demonstrates that not even medium-term – let alone short-term – dependence on the Middle East is inevitable. Conventional Western views (as indicated above in the second paragraph) predicate such an inevitability basically because, as seen from the USGS data, there is thought to be a remaining ultimate availability of only about 60 x $10^9$ tons of non-communist world oil outside the Middle East so that the current annual rate of production of this oil, at about 1.2 x $10^9$ tons,

### TABLE I-8.1
## Comparison of estimates by the US Geological Survey and the USSR Research Institute of Geology of Foreign Countries of original recoverable resources of crude oil in the non-communist world
### (in $10^9$ tons)

| Region | USGS (probability estimates) | | | USSR (estimates by production cost) | | | Ratios of USSR/USGS Estimates | | |
|---|---|---|---|---|---|---|---|---|---|
| | (A) 95%* | (B) Mode* | (C) 5%* | (D) under $45/ ton† | (E) under $80/ ton† | (F) over $80/ ton† | (D/A) | (E/B) | (F/C) |
| North America | 36.3 | 39.8 | 47.4 | 34.1 | 47.5 | 99.9 | 0.94 | 1.19 | 2.05 |
| Latin America | 23.2 | 28.4 | 50.8 | 33.9 | 48.3 | 86.4 | 1.46 | 1.70 | 1.59 |
| Western Europe | 6.1 | 6.7 | 10.1 | 4.5 | 10.1 | 14.2 | 0.73 | 1.50 | 1.41 |
| Africa | 16.0 | 18.6 | 27.0 | 23.3 | 35.2 | 58.5 | 1.46 | 1.89 | 2.17 |
| Middle East | 91.4 | 99.0 | 129.2 | 96.3 | 140.1 | 172.7 | 1.05 | 1.42 | 1.34 |
| Asia & Oceania | 7.4 | 8.1 | 16.9 | 14.2 | 19.5 | 41.0 | 1.92 | 2.41 | 2.42 |
| **Total** | 146.2 | 192.3 | 282.7 | 206.3 | 300.7 | 472.7 | 1.41 | 1.56 | 1.67 |

★    *The USGS probabilities apply only to undiscovered resources and not to discovered (additional) reserves. We have here summed the (fixed) data on reserves and the probabilistic data on undiscovered resources. As the probabilities apply to only a portion of these totals the actual probabilities must be higher than stated at the top of each column.*

†    *Production cost per metric ton in constant 1980 dollars.*

simply cannot be enhanced to any significant degree for other than the very short term. By contrast, the Soviet geologists' figures indicate an availability of at least twice the volume of recoverable oil in the non-communist world outside the Middle East – a minimum of 120 x $10^9$ tons of oil and up to over 260 x $10^9$ tons if oil resources available at over $80 per ton are also included. Clearly, this potentially very much greater availability of non-Middle East non-communist world oil is of critical importance in undermining the hypothesis of renewed short-term Western world dependence on the Middle East. If the Soviet scientists are right, then the current widely accepted Western world predictions of little increase in the production of oil from non-Middle East sources in the short-term and an inevitable decline from present levels of production in the medium-term – leading to an early 1990s resurgence of Middle East control of the international oil market and a consequential inevitable rebound in the price of oil – are of doubtful validity. This could, indeed, be the data that lies behind the recently reported most likely prediction by Soviet analysts of the future behaviour of the international oil market, *viz.* continuing attrition in the price of oil and no likelihood of a tight market emerging this side of 2010.[5]

Given the enormous differences in the estimates of oil resources between Western and Soviet scientists – and given the importance of this difference for the future of the international oil market – it would seem necessary that the conventional Western view of conventional oil resources is opened up to the closest possible scrutiny and re-evaluation. Too much is at stake, both geopolitically and in terms of the Western World's macro-economic prospects, not to subject the current forecasts of the Western world's geologists to an effective comparison with the strongly contrasting alternative views of their peer group in the USSR.

## References

1    By C.D. Masters and D.H. Root (US Geological Survey), and W.D. Dietzman (US Energy Information Administration); *Distribution and Quantitative Assessment of World Crude Oil Reserves and Resources*, World Petroleum Congress, London, 1983 (in an invited paper to the Panel Discussion (PD 11) on *World Reserves of Crude Oil*).

2    As issues 40 and 41 of the *Transactions of the All Union Scientific Research Institute of Geology of Foreign Countries*, published by the Nedra Publishing House, Leningrad, in 1983.

3    M.S. Modelesky, G.S. Gurevich and E.M. Khartukov, *World Cheap Crude Oil and Natural Gas Resources*. Announced and made available in an unpublished form by Academician Styrikovich at the June 1985 meeting of the International Institute for Applied Systems Analysis Energy Workshop in Laxenburg, Austria.

4    This is not made explicit in Masters *et al's* presentation to the World Petroleum Congress, but there seems to be no other variable (other than the production cost variable) which would enable them to develop their probability curves for volumes of recoverable oil from a pre-defined oil-in-place calculation.

5    This computer-based model of the world oil system also shows lower – but still significant – probabilities of an oil price collapse with no rebound for the rest of the century.

# Chapter I – 9

# Global and Regional Energy Supplies: Recent Fictions and Fallacies Revisited*

## Introduction

Over the past 20 years a number of powerful perceptions concerning the prospects for the supply of various sources of energy, at both the global and the regional level, have strongly influenced governmental and inter-governmental energy sector analyses and policies. Unhappily, the perceptions that have so ruled supply-side questions have been based in part on a set of demonstrable fictions and fallacies. These have led to high costs and severe difficulties for the Western world's economy, most notably in respect of the massive increase in the price of oil since 1970. Much of this increase can, indeed, only be defined as a self-inflicted wound on the fabric of Western society arising from policies which reflected the unsubstantiated belief in an inevitable scarcity of oil. More generally, there has been an unwillingness to accept the energy supply potentials which were indicated by economic rationality. There was, instead, a propensity to follow non-economic interpretations of the nature of energy resources and of the processes of their exploitation whereby the efforts made to achieve efficient ways of meeting the energy demand of our economies and societies were undermined. A review of what went wrong – and why – may avoid equally, or even more costly, mistakes in the future.

---

* Reprinted from the author's Afscheid College (valedictory lecture) at Erasmus University, Rotterdam (EURICES Paper 91/Final), April, 1991 as published in *Energy Policy*, Vol.20, No.4, April 1992, pp.285–296.

## The Oil Scarcity Fallacy

Towards the end of the 1960s, after a period of more than 20 years of high annual growth rates in oil use (such that annual use had doubled each decade), increasingly strident fears were expressed for the impending near-future exhaustion of the world's oil reserves. This was partly an element in a generalized concern for the scarcity of resources[1] and in part it was a specific concern for oil as, for example, in analysis which purported to show that supply would fail to meet expected demand by the early 1980s and which 'calculated' an inevitable peak to global oil supply as early as the mid-1990s.[2]

Both the general and specific concern helped to create a climate of opinion in which the 50% increase in oil prices between 1970 and 1973, and their subsequent quadrupling in 1973–74 (see Figure I-7.3) were accepted as the initial stages of an inevitable process of a scarcity-driven oil pricing phenomenon. The stage was thus set for a second oil price shock in 1979–81 as a result of which the price of a barrel of crude oil reached a level more than an order of magnitude higher than just a decade earlier. By this time, moreover, the prospects for oil prices continuing to rise to $60–80 or even $100 per barrel were seriously presented. This is shown in Figure I-9.1 in which the result of an early 1980s survey of future oil prices are indicated. In this then-conventional view of oil price prospects, the long-term price elasticity of the demand for oil was assumed to be small, while the price elasticity of the supply of oil was often explicitly or implicitly presented as non-existent: because, it was argued, the world was literally running out of oil so that, no matter what the price, supplies could not be increased!

Some of us argued differently at the time; EURICES' contrasting contribution to the 1982 survey of oil price forecasts is, for example, also shown in Figure I-9.1. This reflected our conclusion that the conventional view was the result of essentially irrational interpretations of eminently straightforward supply-demand relations.[3,4] Sky-high prices not only engendered more efficiency in oil use and its substitution by other sources of energy, but they also stimulated the creation of more reserves from the world's generous – though almost always understated – oil resource base. At this point one should note the essential qualitative difference between resources which simply exist (as part of the natural environment) and reserves which are created by investments made in the search for relevant knowledge, in the development of new technology and in the required exploration activities.[5]

The creation of reserves from resources is, in other words, a continuing function of investment. Thus, since 1970, as shown in

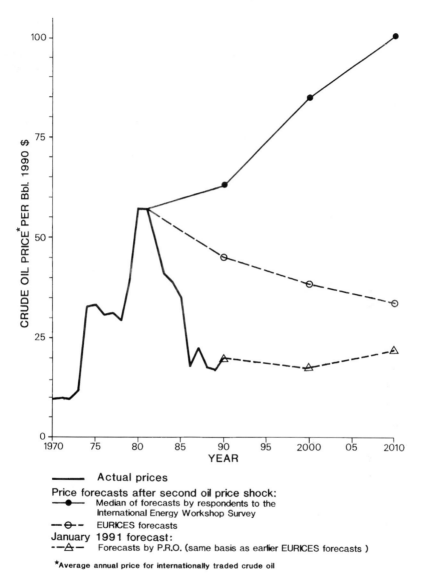

Figure I-9.1: Crude oil price forecasts 1990–2010

Table I-9.1, $860 \times 10^9$ barrels have been added to reserves. On the other hand, over the 20-year period only $424 \times 10^9$ barrels were used with the result that the ratio between reserves and annual production now stands at an all-time high of over 41 years. The process of reserves creation in relation to oil use is expressed graphically in Figure I-9.2. In this, the data shows that over the last 40 years the annual use of oil has only once exceeded the annual amount added to reserves by the global oil industry. These data effectively undermine the oft-repeated comment that more oil is being used each year than is being added to reserves (as recently, for example, in Heal and Chichilinsky[6]). On the contrary, the world is running into oil, not out of it. In light of this dramatic development in oil supply-demand relationships in the recent past, small wonder that the price of oil is now back (see Figure I-9.1) to a level below that of any year since 1973 – in spite of the recent Gulf War and the temporary – but still continuing – elimination of both Kuwaiti and Iraqi oil supplies from the world market.

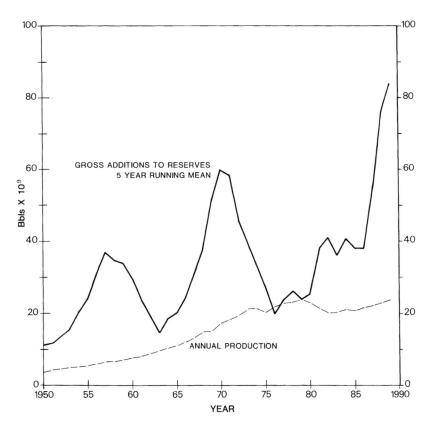

Figure I-9.2: Annual world production and the five-year running mean of gross additions to reserves, 1950–1989.

**TABLE I-9.1**

## Proven reserves, reserves-production ratio, oil production and net growth/decline in reserves over the 20-year period from 1970

(all data – except R/P ratios – in barrels x $10^9$)

|  | Proven reserves at beginning of year (R/P ratio – in brackets – in years) | Production of oil | Gross additions to reserves | Net growth (+) or decline (–) in reserves |
|---|---|---|---|---|
| 1970 | 553 | 17.4 | 62 | +45 |
| 1971 | 578 (33.2) | 18.3 | 40 | +22 |
| 1972 | 600 (32.8) | 19.3 | –4 | –23 |
| 1973 | 577 (29.9) | 21.2 | 35 | +14 |
| 1974 | 591 (27.9) | 21.2 | 32 | +11 |
| 1975 | 602 (28.4) | 20.2 | 31 | +11 |
| 1976 | 613 (30.3) | 21.9 | 4 | –18 |
| 1977 | 595 (27.2) | 22.6 | 16 | –7 |
| 1978 | 588 (26.0) | 22.9 | 45 | +22 |
| 1979 | 610 (26.6) | 23.7 | 22 | –2 |
| 1980 | 608 (25.7) | 22.8 | 34 | +11 |
| 1981 | 619 (27.1) | 21.3 | 67 | +46 |
| 1982 | 665 (31.2) | 20.1 | 30 | +10 |
| 1983 | 675 (33.6) | 20.0 | 21 | +1 |
| 1984 | 676 (33.8) | 21.1 | 44 | +23 |
| 1985 | 699 (33.1) | 20.5 | 30 | +9 |
| 1986 | 708 (34.5) | 21.4 | 67 | +45 |
| 1987 | 753 (35.2) | 21.9 | 113 | +91 |
| 1988 | 844 (38.5) | 22.8 | 99 | +76 |
| 1989 | 920 (40.3) | 23.5 | 72 | +48 |
| 1990 | 968 (41.2) |  |  |  |
| 1970–89 |  | 424.1 | 860 | +436 |

*Sources:*  *Reserves' developments based on data from the Annual Survey of World Oil Reserves in the* Oil and Gas Journal, *1970–90; from* World Oil, *1970–90 and from De Golyer and MacNaughton's* Annual Survey of the Oil Industry, *1975–83. Annual production data from the* Petroleum Economist, *1970–90.*

When supply/developments are viewed regionally, moreover, the state of the world oil market is not really surprising. There has, for example, been a tenfold increase in Western Europe's oil production since 1973 (from 20 to 200 million tons per year) and a 300% increase in output in the countries of the developing world outside the OPEC group of countries (from less than 160 million tones to almost 500 million tons per year). The fallacy of an absence of a long-run price elasticity of oil supply – as implied in conventional supply analyses in the late 1970s and early 1980s – could not be more clearly demonstrated. Indeed, even the 'low' oil price forecasts we made a decade ago (see Figure I-9.1) have turned out to be too high by a factor of over two. Our most recent forecasts, made in the context of no foreseeable element of potential oil scarcity, indicate somewhat more modest prices in 1995, 2000 and 2010 than the forecasts we made for the same dates a decade ago. The new forecasts indicate the long-run maintenance of real prices at levels which are close to those following the oil price collapse in 1986 for reasons which we have set out in detail elsewhere.[7] By 2010 prices seem likely to be no more than 25% of the $100 per barrel crude price which was forecast by those who accepted the oil scarcity fallacy of the 1970s and early 1980s.

## The European gas scarcity fiction

This syndrome emerged at a critical juncture in the evolution of Western Europe's energy economy, that is in the immediate aftermath of the 1973–74 oil supply and pricing crisis when, in theory at least, the objective of contemporary energy policy was to minimize dependence on imported oil. As an enhanced rate of exploitation of indigenous natural gas was self-evidently the lowest cost and quickest way to substitute the use of oil, the officially propagated fiction of gas scarcity was particularly unfortunate. And though the UK, Denmark and Norway each contributed to the fiction – from the constrained view they took of the potential for developing their own gas supplies, the fiction was most strongly expressed in the Dutch policy of restraining production at just the very moment in time when, from both the national economic and the wider European stand-points, a policy of expanding the Dutch gas supply should have been followed.

Though a number of other temporary factors were also involved in the constrained gas production policies, there was an underlying general cause – an over-simplistic interpretation of the volumes of gas reserves available for depletion. In essence, the ultimately available economically recoverable reserves from within the gas-rich province of north-west Europe were wrongly assumed to be no more than those reserves which had already been proven as a result of exploration successes to date and the experience-to-yesterday of producing those reserves. Thus, production decisions were related only to a declared minimum stock of proven reserves (in effect, no more than the equivalent of the limited shelf-stock carried by the street-corner grocer to meet anticipated short-term demands), rather than to a reasonable calculation of the reserves which would emerge from the continuing exploration, production and fields' re-evaluation processes. Had these processes not been certain to bring success, in terms of continuing reserves' development, then the oil industry as a whole would not have been willing – as, indeed, it was – to go on investing in them.

Thus, gas production in Western Europe was limited to volumes which were well below the potential to produce. In the case of the Netherlands, moreover, the economic irrationality was further compounded by the additional decision preferentially to produce high-cost rather than low-cost reserves. The latter it was argued, based on the false presumption of gas as a scarce commodity, would thus be 'saved' for use when gas was in short supply: or alternatively it was gas which could still be produced when real prices fell in the future to levels at which higher cost gas could not be profitably exploited. (In parenthesis it is worth noting that the two alternative explanations for 'explaining' the economic irrationality of the decision are themselves mutually inconsistent.)

Needless to say, as natural gas production continued, albeit at below economically optimum levels, new reserves have been more or less continuously found, while the reserves of oil fields have been regularly upgraded. For example, as shown in Figure I-9.3, the Groningen gas field had declared reserves of about $1600 \times 10^9 m^3$ when production started in 1965. Its proven reserves had increased to approximately $1750 \times 10^9 m^3$ by 1975 (even though it had by then already produced almost $500 \times 10^9 m^3$), but, in spite of this appreciation of the field's reserves, severe production constraints were imposed in the following year. Since 1975 about another $700 \times 10^9 m^3$ have been produced from the field, yet its proven reserves still remain about $1600 \times 10^9 m^3$; to give an original recoverable reserves figure for the field which is now in excess of $2700 \times 10^9 m^3$.

Meanwhile, of course, revenues from the gas production which was not permitted have been lost. They have, moreover, probably been lost for ever, when measured in real terms, given the fact that the limitations of production were at their most stringent when gas prices and hence the tax-take per m³ were at what is likely to have been an all-time high. Present real prices are little more than one-third of their highest levels a decade ago – and are unlikely to change much for the better! In addition, gas markets abroad have been lost to competitors; and other more expensive (and environmentally dirtier) energy alternatives have had to be used – both at home and abroad – to make up for the enforced, but entirely unnecessary, limitation on gas production.

As already indicated the Dutch lead in constraining gas supply was paralleled by similar actions in other north-west European gas-rich countries. In the UK the monopolistic state-owned gas corporation

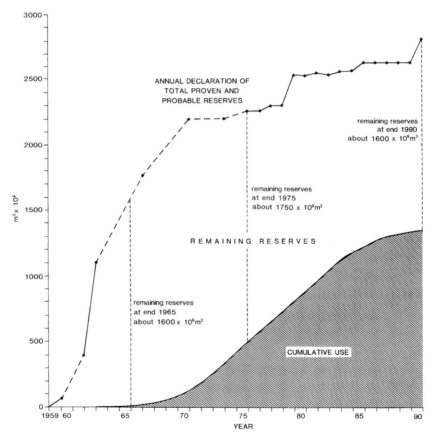

Figure I-9.3: The Groningen gasfield showing cumulative production (1965 to date) and the evolution of proven and probable reserves (from discovery in 1959 to date).

determined to serve only so-called high value markets, so eliminating gas use in power generation and severely limiting industrial use of gas (partly under government pressures designed to protect traditional markets for coal). In Denmark an inappropriate initial concession policy inhibited the production of discovered gas until 1984. Finally, in Norway the combination of a highly politicized field development decision programme – based on national and internal regional considerations, coupled with a state controlled export marketing strategy which, until the mid-1980s, assumed that the natural gas sector in Western Europe would be a permanent seller's market, served to limit the rate of build-up of Norway's gas supply, in spite of the discovery of massive reserves from the mid-1970s.

These indigenous gas supply constraints led to the near cessation of growth in the contribution of indigenous gas to post-1976 energy supply in Western Europe, in spite of energy market economic fundamentals which were highly favourable to its much expanded use. As shown in Table I-9.2, there had been rapid growth in gas use prior to 1976, but, thereafter, 80% of even the very modest further increases in Europe's gas use came from an increased volume of imports – from Algeria and the USSR. And this was, moreover, over a period (from 1976–86) when, as seen from Table I-9.3, West Europe's gross additions to proven gas reserves exceeded $5000 \times 10^9 \text{m}^3$. Over the decade less than one-third as much gas (about $1500 \times 10^9 \text{m}^3$) was used so that the reserves to production ratio

### TABLE I-9.2
## Natural gas production, imports and consumption in Western Europe

|      | Indigenous production $(m^3 \times 10^9)$ | Imports $(m^3 \times 10^9)$ | Total use $(m^3 \times 10^9)$ | % Dependence on imports | Number of gas using countries | Gas use as a percentage of energy use |
|------|------|------|------|------|------|------|
| 1961 | 16   | Negl | 16   | Negl | 5    | 1.8  |
| 1966 | 25   | Negl | 25   | Negl | 7    | 3.3  |
| 1971 | 104  | 2    | 106  | 1.9  | 8    | 9.7  |
| 1976 | 164  | 14   | 178  | 7.9  | 10   | 13.4 |
| 1981 | 177  | 20   | 197  | 10.2 | 11   | 14.7 |
| 1986 | 175  | 37   | 212  | 17.3 | 15   | 15.2 |
| 1989 | 183  | 49   | 232  | 21.1 | 15   | 15.0 |

Sources:    BP Statistical Review of the world oil industry, 1961–80; BP Statistical Review of World Energy, 1981–90: and BP Review of World Gas, 1990.

increased by 20 years – to close on 45 years of proven supply potential at the 1986 level of use. The strong contrast between reserves and use developments indicates just how far the policies related to the fiction of gas scarcity inhibited the expansion of a regional source of energy the use of which was clearly preferable not only from the standpoint of economics, but equally for security of supply and environmental reasons. Unhappily, in spite of some changes for the better over the past couple of years, Euro-gas supply developments remain constrained – compared with both the continuing expansion of reserves and the potential opportunities for additional gas use in the energy markets. The powerful forces which have restrained the supply of indigenous gas in Europe for the last decade and a half, to the detriment of the continent's economy, security and environment, have not yet by any means been broken.

## TABLE I-9.3
### Evolution of Western Europe's natural gas production and reserves 1956–96 (in m3x10$^9$)

| Year | Cumulative production to date | Cumulative production over previous decade | Year-end declaration of proven probable reserves | R/P ratio (years) | Total original recoverable reserves | Gross additions to reserves in decade | Net additions to reserves |
|------|------|------|------|------|------|------|------|
| 1956 | 50 | 35 | 500 | 40 | 550 | not known | not known |
| 1966 | 225 | 175 | 1900 | 76 | 2125 | 1575 | 1400 |
| 1976 | 1150 | 925 | 4350 | 27 | 5500 | 3375 | 2450 |
| 1986 | 2700 | 1550 | 7900 | 45 | 10600 | 5100 | 3550 |
| Forecast | | | | | | | |
| 1996 | 4450[a] | 1750[a] | 10250[b] | 55[a] | 14700 | 4100 | 2350 |

Notes:  a    *Assuming the continuation of present gas production and use policies*

   b    *On the basis of the most recent estimates of the reserves in fields already discovered in 1986. The estimates of the total reserves of these fields will almost certainly appreciate further by 1996. In addition, of course, many more fields have also been discovered since 1986 and there is also near-zero probability that no more fields will be found, given the continuation of an extensive and intensive exploration effort for natural gas in many Western European countries and their off-shore areas.*

Source:    *Author's research and estimates.*

## The fallacy of European coal supply expansion

It was the exploitation of Europe's coal reserves that supplied most of the energy for the continent's development for almost 200 years prior to the mid-1950s. Thereafter, however, the European coal industry went into trauma under the pressures of increasing costs (especially from rising real wages in what was still a labour intensive industry) and as a result of competition from imported oil. Between 1957 and 1972 annual coal production fell by almost 50% from just under 400 to a little over 200 million tons.

Higher oil prices after 1973, however, provoked much misguided enthusiasm for the initial stabilization and a later re-expansion of European coal production, based on the fallacy of 'the continent's 300 years of coal reserves'. Whilst it was certainly true that there were geological indications of quantities of technically-recoverable coal resources equal to 300 times the quantity of indigenous coal produced in 1973, these did not constitute 'proven reserves'. The definition of the latter involves an economic, as well as the geological, component: *viz.* that the coal is producible at present levels of costs and prices. In this much more limiting context, most of Western Europe's remaining coal reserves – including some of those technically recoverable from the then existing mines – had no economic significance as a potential energy supply.

Nevertheless, in all of the traditional coal producing countries much economically unjustified investment was committed to an effort to revive the coal industry. Thus, both individual countries – as well as the EEC and the IEA – wrote additional European coal production into their energy plans and forecasts. In 1974, for example, the EEC stated its 1985 objective for coal as 'the maintenance of production at 250 million tons per year involving support for the rationalisation of existing mines, the opening up of new production capacities, financial help for labour recruitment, investment in research and development and fiscal and regulatory measures to ensure... markets.[8] Between the publication of that report and 1979, Euro-coal production failed even to recover to 250 million tons per year in spite of continuing high oil prices. Nevertheless, the IEA in 1979 still perversely persisted in presenting European coal as a growth industry and anticipated a 25% increase in output by 1990.[9] Indeed, such views on the prospects for European coal were even enshrined in Western countries' heads of state declarations at their economic summit meetings in Tokyo and Vienna in 1979 and 1980. In other words, the unsustainable conclusions of the flawed analysis on the prospects for indigenous coal became incorporated into the strategies which were adopted at the highest political levels of the Western alliance.

In reality, even in the context of the widely-held contemporary expectation of continuing real increases in energy prices - and in spite of the financial commitments by governments – there was little chance that the European coal industry would be capable of re-expansion after a period of over 20 years of decline: in a set of economies in which real wages were still rising and in which coal mining was not an attractive employment opportunity, except at wages at an unpayable premium to those in the rest of the economic system. The predilection by policy-makers and others for the expansion of the indigenous production of coal gave completely inappropriate signals relative to the realities of the European energy supply prospects. Sustainable additional output subsequently proved to be largely impossible – so that most of the new investment has since had to be written-off. Meanwhile, much of the coal that was produced (and is, indeed, still being produced though in progressively smaller quantities) from pre-existing mines, often expensively refurbished, still had to be subsidized by taxpayers and/or by electricity consumers. The near-future near final demise of the Euro-coal industry (except for open cast production) will soon bring to an end the fallacious view of the region's coal supply prospects. Meanwhile, as with the too-high-price oil fallacy and the gas scarcity fiction, the costs to the European economy will have been high.

## The fiction of limitless, low-cost nuclear power

Pride of place in the economic costs' stake must, however, go to nuclear power. The industry in Western Europe got off to a slow and tentative development in the 1950s and the 1960s in spite of the euphoria which had attended the post-1945 'Atoms for Peace' propaganda and its accompanying international and Euro-level institutionalization (through the International Atomic Energy Agency and Euratom, respectively). Apart from the initial technological-cum-safety problems, the root cause of nuclear power's slow development was, of course, the falling real price of energy over the period to 1970 (under the influence of the downward trend in the prices of internationally traded oil – see Figure I-7.3). The oil price shocks of the 1970s were thus presented as opening up a massive opportunity for the rapid expansion of nuclear power – as a substitute for oil based electricity (which had become generally important over the pervious 20 years) and as a politically secure energy source.

Thus, in 1974 the EEC presented the nuclear opportunity as one of enormous significance. Its Energy Directorate's strategic document commented:

*"A special effort is required to promote the development of nuclear power... Compared with other energy sources, nuclear energy is* clearly *the best solution for... electricity generation... and for industrial uses... It is readily available, adaptable, easy to transport and store and it safeguards the environment... Of course nuclear power will have its difficulties, but none seem insurmountable... At least 50% of total energy requirements around the year 2000 could be covered by this source... It will be then be the privileged source of energy by reason of the use of breeder reactors and of fusion devices. It will satisfy a large proportion of our energy needs in electricity generation, process heat, gasification of coal, production of hydrogen and the propulsion of ships... Nuclear energy has now become economically competitive with all other sources of primary energy."*[10]

Clearly, within a few months of the first oil price shock and long before there had been an opportunity for a careful examination of the economic – and other – factors involved in such a fundamental change of direction for the European energy economy, the signals for nuclear power were quickly and unequivocally set to green for the supply of Western Europe's future energy demands. A simple test made at that time of the validity of the economic advantage assumption for nuclear power – in the context of mid-1970s alternative energies' cost levels – found the assumption to be unwarranted. Indeed, the study came to quite a contrary conclusion; *viz.* that the development of an energy economy in Western Europe based on nuclear power was likely to be between two and three times more expensive than a development based on the exploitation of indigenous oil and natural gas.[11]

Nevertheless, the official inter-governmental and governmental views on the prospects for nuclear power persisted; and they were, of course, strengthened by the second oil price shock (1979–81) and by the perception of oil scarcity and the rising real prices of fossil fuels (see Figure I-9.1). The EEC thus maintained its view that nuclear power expansion must be the topmost priority (even though progress in its development since 1974 had been slow compared with the expectations in that year). By this time, moreover, the International Energy Agency had also joined the pro-nuclear campaign. Its 1980 review, for example, forecast a more-than-doubling of nuclear capacity from 1986–90 and continued growth thereafter.[12] Relative to such expectations, nuclear power expansion has been limited, with only France, Belgium and Japan reaching or approaching the targets. Elsewhere, nuclear development has either been much delayed, has come to a halt or has not even be initiated.

Paradoxically, it has been the impact of non-economic factors – particularly safety fears and the lack of public acceptability – which have been most influential in constraining the expansion of the supply of nuclear energy: reflecting the greater public awareness of questions of reactor safety and doubts concerning the long-term storage and processing of nuclear waste. Doubts and questions of an economic character have much more readily and easily been kept under wraps, with essential information on costs and revenues either withheld or misrepresented so as to give the impression that nuclear power is economic.

On the revenue side, calculations have been based on the assumptions of high and increasing oil (and other fossil fuels) prices – as, for example, in the mid-1970s evaluation of its nuclear proposals by Electricité de France and at the public inquiry in the UK in the early 1980s on the government's plans for new nuclear developments. At the latter inquiry the case for nuclear power was predicated on the basis of a $60 per barrel oil price in 1990. The actual average price in 1990 was under $20 per barrel – in spite of the effect of the Gulf Crisis in driving prices up in the second half of the year! Protagonists of nuclear power have also viewed the average availability of, and the load demand on, nuclear stations optimistically, so further enhancing predicted revenues. Further revenue flows have, moreover, been discounted back to present values at rates well below the opportunity cost of capital, so once again exaggerating the benefits against which costs should be tested.

On the cost side, apart from a tendency to under-estimate both construction costs and running costs, there have been much too low calculations of the reprocessing and/or storage costs of spent fuel, of nuclear research costs (many of which have been incorporated into governmental, rather than the nuclear industry's, budgets) and of the costs of future decommissioning of the nuclear stations and other facilities (once their useful lives of about 25 years have expired).

These economic difficulties with nuclear power became evident in the privately-owned electricity generating companies in the USA more than a decade ago so that since then most of the planned and projected nuclear power stations have been abandoned – together with a number of near completed stations. Except in a small number of localities where fossil fuel costs are high, nuclear power in the USA ceased to be economically attractive once oil (and other energy) prices started to fall in 1982. In Europe this adverse outlook for nuclear power was effectively hidden; partly because of misinformation on costs and revenues and partly because the near-exclusively state-sector electricity industry in Europe faced less stringent economic tests than the mainly private-sector industry in the USA. The importance of this latter factor has been clearly confirmed in the aftermath of the recent decision to privatize the electricity industry in the UK. Data on costs and revenues which had to be revealed for the first time ever – in order to satisfy the rules of stock market launches – showed, first, that no single unit of electricity from any existing nuclear power station in the UK had even been produced at a profit; and second, that the true costs of developing the proposed new nuclear power stations were so high that they too could not reasonably be expected ever to be profitable. Consequentially, the whole of the nuclear electricity sector was excluded from the privatization of the industry and it has, instead, been kept in public ownership. In order, moreover, to make it possible for the existing nuclear power facilities to continue to operate and the construction of a new station to continue, the government introduced a requirement for the privatized regional electricity companies to buy a minimum (about 20%) amount of nuclear power and further required those companies to charge an 11% premium on the price of their electricity to help cover the nuclear sector's expected deficits. The fiction of so-called low-cost nuclear power in the UK was thus finally exposed. It seems highly likely that the past, present and future uneconomic reality of nuclear power in the UK would differ only in degree across the rest of Western Europe, except in the context of extraordinary local circumstances. Moreover, in the conditions as already described, of a Europe which is replete with relatively low-cost natural gas (which may now, following the removal of previous regulatory limitations on its use, be used in high-efficient combined cycle electricity generation stations with low-polluting characteristics), the probability of any new genuinely economic nuclear power plant anywhere in West Europe over the next two decades is minimal. We simply do not need to pursue any longer the fiction of low-cost nuclear electricity supply; until, that is, the industry can offer not only an inherently safe, but also an economically attractive, product.

## The emerging fiction of a potentially rapid switch to renewable energy sources

Pre-industrial revolution wind and water power developments and the regionally significant exploitation of hydroelectricity over the past 100 years provide evidence of the large-scale availability – under favourable circumstances – of renewable energy which is both economic and clean. In recent years the perception of fossil fuels resources' problems (non-existent though they really are), as well as fears for security of energy supplies from politically unreliable areas and concern for atmospheric pollution problems and even global warming and climate change from the increasing use of coal, oil and natural gas, have collectively led to promises of a global energy system based on a much enhanced production and use of various forms of renewable energy. Would that this were a lesser fiction than the others with which this paper has been concerned – but, unhappily, 'tis not so. The most that can be expected to 2020 is, as shown in Table I-9.4 and in Figures I-9.4 and 5, a steady, but modest, expansion of the contribution of renewable energy supplies to the slowly growing global demand for energy. Note, however, that the realities of nuclear power prospects – rather than the fiction to which we have been exposed over recent years – indicate a contribution by renewable energies which will become steadily more important than that of nuclear power. Given an enhanced commitment to R&D which remains to be done effectively to 'commercialize' the supply of renewable energies, in a Western economic system in which the real price of energy to producers will, in general, increase very little from present levels, renewable energy sources could, within a single generation, be poised to establish themselves in a position from which they can serve growing global energy supply demands for the second quarter of the 21st century and beyond.

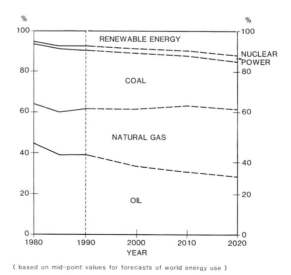

( based on mid-point values for forecasts of world energy use )

Figure I-9.4: The percentage contribution of energy sources to world energy supply, 1980–2020 (see Figure I-9.4 for derivation)

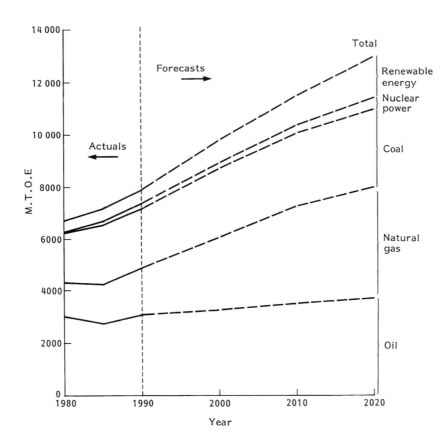

Figure I-9.5: World energy supply by sources 1980–2020: Actuals to 1990 and most recent forecasts to 2020.

Note:    Nuclear power and electricity component in renewable energy converted to mtoe at heat value equivalent.

Sources: Actuals from BP Statistical Review of World Energy: Forecasts by EURICES for Industrial Energy Workshop Annual Survey, December 1990.

## TABLE I-9.4
### Global economic/energy prospects[a] (1990–2020)

| | Actuals for: | | Forecasts for: | |
| --- | --- | --- | --- | --- |
| | 1990 | 2000 | 2010 | 2020 |
| Average price of internationally traded crude oils (in 1990 $) | 21.30 | 17.75 | 21.50 | 26.50 |
| World economic product (1990 = 100) | 100 | 130 | 175 | 235 |
| Average annual growth rate over previous decade | | 2.7 | 3.0 | 2.9 |
| Total primary energy use (mtoe) | 7900 | 9770 | 11,455 | 12,945 |
| Aagr[b] over previous decade | | 2.2 | 1.6 | 1.3 |
| of which: | | | | |
| Oil (mtoe) | 3100 | 3270 | 3250 | 3705 |
| Aagr (%) | | 0.5 | 0.8 | 0.7 |
| Natural gas (mtoe) | 1775 | 2760 | 3725 | 4270 |
| Aagr (%) | | 4.6 | 2.9 | 1.4 |
| Coal (mtoe) | 2275 | 2660 | 2810 | 2990 |
| Aagr (%) | | 1.6 | 0.6 | 0.6 |
| Nuclear power[c] (mtoe) | 180 | 225 | 280 | 400 |
| Aagr (%) | | 2.3 | 2.2 | 3.6 |
| Renewable energy[c] (mtoe) | 570 | 855 | 1110 | 1580 |
| Aagr (%) | | 4.1 | 2.7 | 3.6 |

Notes: a    The mid-point of a range of estimates based on a range of oil price forecasts
     b    Aagr = average annual growth rate
     c    Nuclear power electricity and primary electricity from renewables (eg hydro-electricity) converted to mtoe on the basis of the heat value equivalent of the electricity produced.

Source:    Derived from the most recent (December 1990) contribution by EURICES to the IIASA/Stanford University, International Energy Workshop Survey of Global Energy Prospects.

## References

1   D.H. Meadows *et al*, *The Limits to Growth*, Universe Books, New York, USA, 1972.

2   H.R. Warman, 'Future problems in petroleum exploration', *Petroleum Review*, Vol.25, No.3, March 1971, pp.96–101.

3   P.R. Odell, 'The future of oil: a rejoinder', *The Geographical Journal*, Vol.139, Part 3, October 1973, pp.436–454.

4   P.R. Odell and K.E. Rosing, *The Future of Oil: a Simulation Study of the Inter-relationships of Resources, Reserves and Use, 1980–2080*, Kogan Page Ltd, London, UK, 1980.

5   M.A. Adelman, *The World Petroleum Market*, Johns Hopkins University Press, Baltimore, USA, 1972.

6   G. Heal and G. Chichilinsky, *Oil and the International Economy*, Clarendon Press, Oxford, UK, 1991.

7   P.R. Odell, *Oil and World Power*, 8th edition, Penguin Books Ltd, Harmondsworth, UK, 1986.

8   Commission of the European Communities, *Towards a New Energy Policy Strategy for the E.C.*, COM (74) 550, Final, Brussels, May 1974.

9   OECD, *Energy Policies and Programmes of IEA Countries, 1979 Review*, Paris, 1980.

10  EEC, *op cit*, Ref 8. Emphasis by the author of this paper.

11  P.R. Odell, 'Europe and the cost of energy: nuclear power or oil and gas?' *Energy Policy*, Vol.4, No.2, June 1976, pp.109–118.

12  OECD, *Energy Policies and Programmes of I.E.A. Countries, 1980 Review*, Paris, 1981.

# Chapter I – 10

# The Geography of Reserves Development*

## Introduction

The under-exploration and limited exploitation effort in most parts of the world (as measured by the number of wells drilled) is graphically portrayed in Figure I-7.4 (see Grossling 1976). This shows the relationship, on a regional basis, between exploration/development activity, on the one hand, and the geographical extent of actual and potentially petroliferous areas, on the other. The United States' situation constitutes the basis for comparison and, in this context, it appears that no other part of the world has been more than "scratched" in the search for oil.

This is even true for the Middle East, where approximately two-thirds of the world's reserves are concentrated, so that almost any additional drilling there has a high probability of either finding new oil-fields or extended reserves from existing fields. Thus, since 1973, in spite of a less than systematic, continuing and intensive exploration and field development process in the region, remaining proven reserves of oil have increased from 350 to over 650 billion barrels; even though production over the 20 year period was more than 125 billion barrels. These data indicate an average annual gross additions to reserves of more than 21 billion barrels which, in round figures, is only a little short of the global average annual use of oil over the same period. Viewed in narrowly

* Reprinted from *The Energy Journal*, Special Issue on the Changing World Oil Market, Vol.15, 1994, pp.96–105.

economic terms, 85% or more of the Middle East's declared oil reserves are irrelevant; they simply have no net present value.

Elsewhere in the world (outside North America and the former Soviet Union), the areas of maturity in respect of oil exploitation are very limited: essentially to a few locations such as the Gulf coastal strip of Mexico, the periphery of Lake Maracaibo in Venezuela, parts of Indonesia and some small petroliferous zones in Europe. In each of these, the industry dates back to the 19th or the early 20th century. Otherwise, most upstream petroleum activities date only from the past 25 years or so, related principally to the motivation for exploration and exploitation created by the post-1970 rise in prices. In a few petroliferous areas, where access to appropriate technology, managerial know-how and investment has not been a problem, progress in upstream oil activities has been rapid in response to the highly profitable opportunities created by oil priced, in real terms, at four to seven times its 1970 value. Notable among such regions, mainly concentrated in the more industrialised countries, are the North Sea, offshore Mexico, Egypt, southern Brazil and parts of Australasia. In spite of the fact that most of these are major oil consuming areas – so giving a high motivation for discovered oil to be brought quickly into production – their remaining proven reserves have more than doubled in a couple of decades. In most countries the significant petroliferous regions that have been discovered have contained large – even giant – fields and, in response to an intensive search and exploitation effort have now become important elements in the world's upstream oil industry. Though well behind the Middle East (and other OPEC members), they have made an important contribution to the post-1973 expansion of the world's remaining proven reserves from 600 to 1000 billion barrels.

## The Industrialised Countries

For this group of countries it is self-evident that political, as well as the favourable geological, considerations have been influential in their petroleum exploitation history. In essence, they have all had political regimes which have been favourable to oil companies' aspirations: in terms of strong, stable and recognisably fair governments with respect to the industry's investments. They evolved petroleum exploration and exploitation regimes – in terms of concessions or contractual arrangements and of tax rules and rates – which generally generated enough confidence amongst the oil companies to persuade them to make large-scale investments with low (non-geological) risk factors. In most cases, moreover, they were industrialised countries so that there was, if need be, the opportunity for the immediate disposal of any indigenously produced oil into local markets. Table I-10.1 shows the reserves and production developments in the OECD countries (excluding the United States) over the period 1973 to 1993. It also shows the increasing importance of the indigenous (or regional) production to the total consumption of oil in the countries concerned. In the mid-1970s it contributed less than 14% of total oil requirements. Its share is now over 35%.

For such countries, the success of post-1973 upstream petroleum development is, with hindsight, hardly surprising. Nevertheless, the scale of the developments were not always foreseen. Notable amongst the highly pessimistic interpretations of the prospects by both companies and governments were those for the North Sea oil province.[1] Even after the first oil price shock raised net unit revenues to levels which justified much increased investments; and even though the latter, in turn, led to large additional discoveries and to technologically advanced production systems, there remained an unwillingness to accept the clear evidence of the high up-side potential which the region's petro-geology indicated – in terms of field size, recovery rates and high percentage success rates for continuing exploration.[2]

### TABLE I-10.1
### Oil reserves, production and use in the OECD countries
### (excluding the United States) 1973 to 1993 (in barrels x $10^9$)

| Period | Proven Reserves at Beginning of Period | Production in the 5 year Period | Use in the 5 year Period | Production as Percentage of Use | Gross Additions to Reserves in Period |
|---|---|---|---|---|---|
| 1973–77 | 24.9 | 5.43 | 38.89 | 13.96% | 16.73 |
| 1978–82 | 36.2 | 8.25 | 35.71 | 23.1% | 4.25 |
| 1983–87 | 32.2 | 11.31 | 33.11 | 34.2% | 11.41 |
| 1988–92 | 32.1 | 12.61 | 35.95 | 35.1% | 6.01 |
| 1993 | 25.5 | | | | |

Source: Sequential issues from 1973–1980 of BP's annual Statistical Review of the World Oil Industry and of its successor publication, Statistical Review of World Energy, 1981–1993.

As a result, the North Sea's oil prospects were generally presented as unlikely to contribute anything more than a modest small addition to Europe's oil supplies. At best, it was argued, there might perhaps be reserves enough to meet the region's additional requirements for oil for a limited period of time. Such views led to policy reactions which reflected fears for near-future reserves exhaustion. Thus, exploration and/or depletion constraints were introduced,[3] so threatening the expansion of production in the short term and the continuity of reserves' additions in the longer term.

In spite of these negative aspects of the history of the North Sea, the overwhelming attraction, in both economic and political terms, of North Sea oil to virtually all the world's largest private oil companies – combined with the positive impact of the state oil enterprises of Norway and the UK, as well as those of other European companies – eventually led to an exploration and development effort of such a magnitude that its results swamped the pessimism. Instead, a plethora of new finds of major significance, more cost effective technologies for their development, higher recovery rates than expected from most of the fields and the growing ability to exploit reservoir extensions and nearby fields, through the original investments in production platform complemented by sub-sea completion facilities and the falling real costs of servicing the fields and of transporting the oil to the onshore terminals, made the North Sea a petroliferous province of world significance.[4]

In the context of such a formidable range of positive geological and technological factors, neither the fall in oil prices from 1981, nor even their collapse in 1986, did more than create temporary short-term hiccups in the increasing levels of production. Indeed, in the medium term, the price falls motivated the companies to pursue major cost reductions[5] so as to maintain economic viability against the lower unit revenues. Moreover, because of the progressive nature of the tax regimes which Norway and Britain had imposed, the fall in revenues generated by declining prices, mainly impacted on the governments' tax-takes, rather than on the companies' profits.[6] Finally, in the more stringent economic climate created by lower prices, the governments took steps to modify their tax regimes, so as to make them less onerous for most companies, and thus maintain the latter's interest in continued investment.[7]

As a result, the North Sea oil province has come through to the early 1990s generally unscathed by an oil price which is now back, in real terms, to less than its pre-1979 level. Indeed, in late 1993, Norway and Britain, together producing more than five million b/d, became the world's fifth and sixth largest non-OPEC producers; and ahead of all member countries of OPEC, except Saudi Arabia, Iran and Venezuela. Moreover, The Netherlands and Britain are the world's fourth and fifth largest natural gas producers. Together with Norway and Denmark they supply over 200 Bcm per annum to the rapidly expanding West European gas markets.

Similarly, though quantitatively less important, with most of the other newly developed oil provinces that lie within the same category as the North Sea in geological and political terms, rapid expansion in the decade of rising prices has been followed by consolidation and more modest growth since the mid-1980s.

## Non-OPEC Developing Countries

Meanwhile, there is a larger set of countries in which post-1973 incentives to find and develop oil resources have had a significant impact on global production patterns and on reducing dependence on OPEC oil. This set consists of the developing countries outside OPEC. Prior to 1973 these countries, increasingly and, eventually, heavily orientated to oil use for energising their economic growth,[8] were largely dependent for their supplies on crude imported mainly from the OPEC countries; with the oil bought either in government to government deals or, more usually, through the supply systems of the international oil companies. Their own upstream oil industries' development had been thwarted by the low prices for internationally traded crude (which undermined the viability of investment in indigenous oil) and/or by their non-acceptance of the international oil companies as concessionaires for oil exploration and exploitation.[9]

The post-1973 oil price rises in the context of their heavy dependence on imported oil and their international indebtedness, coupled with adverse balance of payments situations, necessitated a change in attitudes and policies towards indigenous oil development potential. As Grossling (1976) has shown, their own considerable petroliferous potential had been under-evaluated – given the more rewarding alternative opportunities for the oil industry's investment in the OPEC countries – but for many of them there was a high probability that indigenous oil could be found and produced, given appropriate policies. These necessarily related to the ways in which sufficient expertise and investment could be attracted to get effective exploration programmes established.[10]

In a number of the countries, state oil companies succeeded in accelerating and expanding their exploration and development activities as governments allowed them to retain profits for investment and/or permitted oil to be priced realistically. In many others, however, governments improved the terms of concessions or contracts for oil exploration and production by private foreign oil companies. As a result reserves in existing oil provinces have been built up and other areas of oil potential successfully explored. Table I-10.2 indicates the achievements in respect of increased oil reserves, increased production and a reduction in dependence on imports over the 21-year period, 1973–1993.

Production, meanwhile, increased over threefold, from 3.5 to 10.9 million b/d, while oil consumption merely doubled. In 1983 the countries' net imports of oil were 2.8 million b/d, equal to 79.4% of indigenous production and to 44.2% of oil use. In 1993 imports were down to only 1.9 million b/d, equal to only 17.2% of indigenous production and 14.6% of oil use. The achievement had been relatively greater still in 1987 (when net imports were down to only 0.6 million b/d), but the deterioration in the relationship between production and use over the past six years reflects almost exclusively the recent much increased rate of growth in oil use in the rapidly expanding economies of the Western Pacific Rim and South East Asia. In this region the most important economies, *viz*. South Korea, Taiwan and Hong Kong have no oil production and, in any event, they have now achieved such strong economies – unlike most of the rest of the non-OPEC developing countries – so that increasing oil imports are not a burden. It is important to note that the deterioration does not reflect any significant reduction in the expansion of the indigenous oil production effort amongst the developing countries which have potential; their collective output has continued to grow at more than 4% per annum.

Similarly, as also shown in the Table, these countries have achieved a development of proven reserves equal to the degree required to sustain the increase in production noted above. Their proven reserves in 1973 totalled only some 35 billion barrels. Cumulative production since then has been more than 54 billion barrels, but, nevertheless, their proven reserves are now estimated[11] at some 95 billion barrels, indicating gross reserves additions over the 20-year period of some 115 billion barrels. The reserves to production ratio has thus been maintained at about 25 years, in spite of the three times greater production in 1993 compared with 1973.

With few exceptions, the developing countries outside OPEC still have oil industries which are in the earliest stages of development. The extent and intensity of exploration has, to date, been modest so that there is a very low probability that they are even approaching their peak

## TABLE I-10.2
## Oil reserves, production, use and import dependence in the Non-OPEC developing countries, 1973–1993

| Year | Reserves at Jan 1 (bill.bbls) | Production (mill. b/d) | Reserve/ Prod. Ratio (years) | Use (mill.b/d) | Net Oil Imports (mill.b/d) | Imports as % of | |
|------|------|------|------|------|------|------|------|
| | | | | | | i. Prod. % | ii. Use % |
| 1973 | 35.3 | 3.50 | 27.4 | 6.28 | 2.78 | 79.4 | 44.2 |
| 1974 | 35.0 | 3.44 | 27.9 | 6.46 | 3.02 | 87.8 | 46.7 |
| 1975 | 36.0 | 3.76 | 26.1 | 6.42 | 2.66 | 70.7 | 41.4 |
| 1976 | 35.2 | 3.94 | 26.9 | 6.72 | 2.78 | 70.6 | 41.4 |
| 1977 | 44.9 | 4.18 | 23.0 | 7.00 | 2.82 | 67.5 | 40.3 |
| 1978 | 44.5 | 4.56 | 27.3 | 7.30 | 2.84 | 63.7 | 38.9 |
| 1979 | 47.5 | 5.14 | 25.3 | 7.60 | 2.36 | 47.9 | 32.4 |
| 1980 | 57.1 | 5.70 | 27.4 | 7.98 | 2.28 | 40.0 | 28.5 |
| 1981 | 71.5 | 6.34 | 30.9 | 8.06 | 1.72 | 27.1 | 21.3 |
| 1982 | 83.0 | 7.00 | 32.3 | 8.48 | 1.48 | 21.1 | 17.5 |
| 1983 | 78.1 | 7.36 | 29.0 | 8.52 | 1.16 | 15.8 | 13.6 |
| 1984 | 82.0 | 7.88 | 28.4 | 8.82 | 0.94 | 11.9 | 10.6 |
| 1985 | 82.0 | 8.24 | 27.1 | 9.04 | 0.80 | 9.7 | 8.8 |
| 1986 | 89.0 | 8.25 | 29.5 | 8.94 | 0.69 | 8.2 | 7.6 |
| 1987 | 86.9 | 8.68 | 27.4 | 9.30 | 0.62 | 7.1 | 6.7 |
| 1988 | 96.4 | 9.02 | 29.2 | 9.96 | 0.94 | 10.4 | 9.4 |
| 1989 | 98.6 | 9.54 | 28.3 | 10.52 | 0.98 | 10.3 | 9.3 |
| 1990 | 90.5 | 9.98 | 24.8 | 10.98 | 1.00 | 10.0 | 9.1 |
| 1991 | 91.3 | 10.32 | 24.2 | 11.52 | 1.20 | 11.6 | 10.4 |
| 1992 | 92.2 | 10.54 | 24.9 | 12.26 | 1.72 | 16.3 | 14.0 |
| 1993 | 95.5 | 10.94 | 23.9 | 12.82 | 1.88 | 17.2 | 14.6 |

Sources: Oil and Gas Journal, *Annual Worldwide Reserves Report, 1972–1993* and British Petroleum, *Annual Review of World Oil (Energy), 1973–1994.*

production potential. They do, of course, have to maintain their motivation to continue to expand as oil producers, but this seems highly likely, even in conditions of relatively low oil prices, as their external economic relationships remain generally weak and especially exposed to balance of payments problems. Given these conditions the shadow-prices for their oil imports are well above the presently low international market price for oil.

## The (former) Centrally Planned Economies

Viewed in now somewhat outdated politico-economic terms, the remaining group of countries which have a significant part of the world's proven oil reserves portfolio is that of the formerly centrally planned economies, *viz. first*, the former Soviet Union and its dependencies in Eastern Europe and *second*, China. In 1973 this set of countries was estimated to account for about 37% of non-Middle East proven reserves (*viz.* over 100 billion barrels),[12] though note that this estimate probably tended to overstate the share because of differences between western and eastern definitions of reserves.[13] These countries' share of total non-Middle East production at that time was, however, much lower; at only about 15%. They were in 1973 thus relatively better off in respect of oil supply potential than other regions of high and increasing oil demand.

Since then, however, their position has quickly deteriorated. By 1980 their share of non-Middle East reserves was already down to 30%, while their share of production was up to 34%. Their proven reserves had, in absolute terms, fallen to a little over 86 billion barrels. By 1987, when oil production in the USSR and Eastern Europe peaked, their reserves had further declined to a little under 80 billion barrels while production had increased to almost 16 million b/d. By this time their share of non-Middle Eastern reserves was down to under 24%, while their share of production remained at about 34%. The reserves to production ratio of a set of countries which had elevated self-sufficiency in oil to the status of a principal planning objective, had thus fallen to only 13.6 years, compared with 29.9 years in 1973. Doubts already expressed some years earlier by outside analysts concerning the ability of the USSR's reserves – and, even more important, its annual rate of reserves discovery – to sustain the country's production policies[14] now became very evident. But 1987 turned out to be the peak year of production (after half a decade of increasingly desperate attempts to keep production above 12 million b/d). Since then, initially for reasons related directly to physical and economic parameters, but from 1989 also to political issues arising out of the difficulties and subsequent collapse of the Soviet regime, output has fallen year by year. By 1993 it was down to only 65% of its 1989 level, and had fallen to under 19% of the global total outside the Middle East. Meanwhile, declared proven reserves had stabilised at just over 80 billion barrels so that, together with the effect of the decline in production, the R/P ratio was now back up to about 20 years.

For the former Soviet Union, in particular, even at the significantly reduced rate of production over the last six years, the level of remaining proven reserves has been reduced by about two billion barrels. This appears to indicate that the oil industry has only survived the last six difficult years as well as it has by living, in part, off its accumulated capital (in the form of previously discovered reserves). In the short term, the deteriorating reserves position is not perhaps of great consequence in setting production limits as these are more immediately related to the continuing decline in oil use (which has fallen by over 35% since 1987) and to the inability of the industry to cope with the shortage of essential equipment, or even to pay the work force. The trend in reserves does, however, have important implications for the medium term ability of the oil industries of the post-Soviet republics to restore production to the higher levels of previous years, once the initial problems of the traumatic political changes have been overcome. Herein lies one of the less positive aspects of the prospects for global oil supplies beyond the year 2000.

Meanwhile, China is becoming relatively more important compared with the Soviet Union and its successor states. Its oil use, production and reserves have all continued to increase since 1973, but its long-standing export surplus has gradually been whittled away by an increase in the rate of consumption, consequent upon more rapid economic growth in recent years, above the increase in annual production. Nevertheless, China has succeeded in maintaining annual additions to reserves in excess of production and it still has a relatively high reserves to production ratio of more than 22 years. Geological opportunities and the political will to avoid becoming a significant oil importer suggest that Chinese policies toward the international oil industry will be adjusted, as necessary, to ensure the continuing expansion of production and of reserves discovery through the 1990s: though not quickly enough to avoid the country becoming a small net importer of oil over the next few years.

## References

1    See Arnold (1978), pp.55–72; Chapman (1976), pp.153–8; Davis (1981), Chapters 1 and 2; and MacKay and Mackay (1975), pp.57–65 for discussions on the North Sea's oil reserves.

2    See Odell and Rosing (1974 and 1976) and Odell (1979a).

3      The restraints imposed by governments on exploration and development of North Sea oil are evaluated in Dam (1976) and Andersen (1993).

4      See Odell (1979b)

5      Individual companies' efforts to achieve cost reductions in the North Sea have been synthesised into a collective effort by all the sectors of the UK offshore industry (through the UK Offshore Oil Operators' Association) in a project designated with the acronym, CRINE (Cost Reduction Initiative for the New Era). The CRINE report, published in 1993, details actions whereby capital costs can be reduced by at least 30%.

6      Shell UK's 1988 Annual report showed that while the company's profits increased in 1986, the year in which the average price of internationally traded oil fell by 45%, the company's tax payments to the UK government were reduced by almost 75% from over £1800 million to under £350 million.

7      This has been demonstrated by Kemp in his successive studies on the effects of British and Norwegian taxes on the attractiveness of oil companies' investments in North Sea developments. See Kemp (1984, 1985 and 1990).

8      By 1973 all but a handful of non-communist world developing countries were more than 50% dependent on oil and most were at least 75% dependent. See map on page 138 in Odell (1974).

9      For the politico-economic background to this see Tanzer (1969) and Odell (1974).

10      The United Nations National Resources and Energy Division initiated a programme of studies in the mid-1970s which aimed to stimulate appropriate action to overcome these restraints. See United Nations (1982A) and Khan (1987).

11      *Worldwide Reserves Report, Oil and Gas Journal*, December 26, 1993.

12      *Oil and Gas Journal*, 31 December, 1973.

13      These differences are reported and discussed by Madelevsky and Pominov (1979).

14      See, for example, Park (1979), Chapter 8.

## Bibliography

Andersen, Svein S. (1992). *The Struggle over North Sea Oil and Gas: Government Strategies in Denmark, Britain and Norway*. Oslo: Scandinavian University Press.

Arnold, Guy (1978). *Britain's Oil*. London: Hamish Hamilton.

British Petroleum (1973–80). *Annual Review of the World Oil Industry*. London: British Petroleum.

British Petroleum (1981–94). *Annual Statistical Review of World Energy*. London: British Petroleum.

Chapman, Keith (1976). *North Sea Oil and Gas: a Geographical Perspective*. Newton Abbott: David and Charles.

Dam, Kenneth W., (1976). *Oil Resources: Who gets What How?* Chicago: The University of Chicago Press.

Davis, Jerome D., (1981). *High Cost Oil and Gas Resources*. London: Croom Helm Ltd.

Grenon, Michael (Ed.), (1979). *Methods and Models for Assessing Energy Resources*. Oxford: Pergamon Press Ltd.

Grossling, B., (1976) *Window on Oil: A Survey of World Petroleum Sources*. London: Financial Times Publications Ltd.

Kemp, Alexander G., and David Rose (1984). "Investment in Oil Exploration and Production: the Comparative Influence of Taxation" in Pearce, David et al. (Eds). *Risk and the Political Economy of Resource Development*. London: The MacMillan Press Ltd.

Kemp, Alexander G., and David Rose (1985). "The Effects of Petroleum Taxation in the United Kingdom, Norway, Denmark and the Netherlands: a Comparative Study." *The Energy Journal* 6 (Special Tax Issue): 109–124.

Kemp, Alexander G., (1990) "Twenty Five Years on: an Assessment of UK North Sea Oil and Gas Policies." North Sea Study Occasional Paper No.30. Aberdeen: University of Aberdeen.

Khan, Kameel I.F., (Ed), 1987). *Petroleum Resources and Development: Economic, Legal and Policy Issues for Developing Countries*. London: Belhaven Press Ltd.

MacKay, D.I., and G. A. Mackay (1975). *The Political Economy of North Sea Oil*. London: Martin Robertson & Co. Ltd.

Madelevsky, M.Sh., and V.F. Pominov (1979). "Classifications used in other Countries" in Grenon, *op. cit*, pp.98–105.

Odell, Peter R., (1973). The Future of Oil: a Rejoinder." *The Geographical Journal* 139, Part 3, October: 436–454.

Odell, Peter R., (1974). *Oil and World Power: Background to the Oil Crisis*. 3rd Edition. Harmondsworth: Penguin Books Ltd.

Odell, Peter R., (1979a). *British Oil Policy: a Radical Alternative*. London: Kogan Page Ltd.

Odell, Peter R., (1979b). *The Future Supply of Indigenous Oil and Gas in Western Europe*. Paris: International Energy Agency.

Odell, Peter R. and Rosing, K.E., (1974). *The North Sea Oil Province: a Simulation Model of its Exploration and Exploitation*. London: Kogan Page Ltd.

Odell, Peter R. and Rosing, K.E., (1976). *Optimal Development of North Sea Oilfields*. London: Kogan Page Ltd.

*Oil and Gas Journal*, (1973–94). Annual Surveys of World Oil Reserves.

Park, Daniel, (1979). *Oil and Gas in Comecon Countries*. London: Kogan Page Ltd.

*Petroleum Economist* (1989–1992). Surveys of the Oil Industry in India, Vol. LVI, No.8, August 1989; Vol. LVI, No.12, December 1989; Vol. LVII, No.7, July 1990; Vol.58, No.7, July 1991; Vol.59, No.7, July 1992; Vol.59, No.8, August 1992.

Tanzer, Michael, (1969). *The Political Economy of International Oil and the Underdeveloped Countries*. Boston: Beacon Press.

UKOOA (1993). *The CRINE Report*, United Kingdom Offshore Oil Operators' Association, London.

United Nations (1982A). *Petroleum Exploration Strategies in Developing Countries*. London: Graham and Trotman Ltd.

World Oil (1973–92). Annual Surveys of World Oil Reserves.

# Chapter I – 11

# Oil and Gas Reserves: Retrospect and Prospect*

## Retrospect: the Conventional View

Between 1950 and 1970 oil use quadrupled – from a mere 4 billion barrels per year in 1950 to almost 16.5 billion barrels by 1970. This was mainly the consequence of the expanding need for oil to sustain post-World War II rehabilitation and the subsequent rapid growth of the world's industrialised economies, most notably in Western Europe and Japan: in a situation in which their economies' previous dependence on coal could not be sustained.

As the process of increasing oil use in the industrialised world accelerated in the late 1960s, it was accompanied by the onset of a rising demand for oil in the Soviet Union and its allied countries and in parts of the developing world. Forecasts of horrendously-large global oil requirements became common place: most often from the simple extrapolation of the 7½% per annum historic growth rate which became accepted as the conventional wisdom on the future of oil demand at the time. With this assumption, the "expected" cumulative use of oil over the last three decades of the 21st century was calculated as almost 1750 billion barrels, with use in the year 2000 indicated at about 130 billion barrels (Warman, 1972).

Such a prospective development not only carried enormous implications for the expansion of the production, transportation and

* Reprinted from *Energy Exploration and Exploitation*, Volume 16, Numbers 2/3, 1998, pp.117–124.

distribution infrastructure, but also equally worrying implications for the availability of global oil reserves and resources. The 400 billion barrels of oil which had been used from the beginning of the industry in the third quarter of the 19th century to 1970, plus the 1971 declaration of proven reserves (about 520 billion barrels), indicated a total resource of less than 1000 billion barrels. In order to sustain the anticipated expansion of the industry to 2000, additional reserves of some 4000 billion barrels of oil would, it was calculated, be required, if a minimum desirable reserves to production ratio of 15 years was to be maintained (Warman, 1972). Needless to say such a massive expansion of the industry's exploration and reserves' development effort could not be realistically envisaged, irrespective of the world's ultimate resource base of conventional oil; but with the latter then estimated generally at under 2000 billion barrels, the outlook appeared impossible.

Hence, the global oil industry appeared to be en route to a relatively near-future crisis: defined as a prospective inevitable scarcity of oil, as a result of which there would, at the very least, be significant increases in prices and, at worst, given oil starvation, a collapse of the world's economic and social systems. Though this "scarcity-view" of oil by the early 1970s was by no means the first oil reserves' crisis scenario (Williamson et al. 1963), it was the first which had emerged in the context of oil as a global industry and the world's single most important source of energy. These fears in themselves helped to produce a propensity for price increases. Thus, the oil price shocks, though due principally to oil politics (Odell, 1986), were accepted as inevitable. Consequential energy sector policy decisions in many countries involved the investment of many thousands of millions of dollars in what generally turned out to be a futile short-term search for alternatives to oil.

## Retrospect: The Reality

For a few of us at that time, this conventional view of a future for oil beset with such problems seemed not to be realistic. Our counter arguments sought to introduce an element of economics into both the demand and the supply components and insisted on the continuing positive role of increasing knowledge and advancing technology in both oil production and the technical efficiency of its use (Adelman, 1993: Odell, 1973).

As and when prices did increase, so that the demand for oil was curbed, it seemed that the 1971–2000 requirement for additions to reserves would be no more than 1000 to 1500 billion barrels rather than the 4000 billion hypothesised in the conventional wisdom. Moreover, of the additional oil required, much would, we argued, emerge from the continuing process of appreciation of the amounts of oil in already known fields. Thus, the rate of new discoveries of oil would need to be no better than the rate already achieved prior to 1970. Given the newly developing technologies of exploration and development then under way and the extensive areas of the world which remained unexplored – particularly offshore, where developments to 1970 were minimal – then, we concluded, the task of finding and exploiting the oil required to keep the industry expanding at the much more modest rate which could be anticipated at higher prices, did not appear to be very formidable.

The evolution of the oil industry from 1970–1996 has proved to be much closer to this "reality" prospect than to the conventional wisdom. Indeed, production from 1971 to 1996 totalled only 576 billion barrels (see Table I-11.1). Thus, use over the whole period 1971–2000 will be well below even the modest 1000 billion barrels we suggested, and barely one-third of the 1750 billion hypothesised by conventional wisdom. In spite of the very restrained growth in demand over the period (so limiting the exploration effort), more than 1150 billion barrels of oil have, nevertheless, already been added to reserves (see Table I-11.1). Thus, the proven reserves figure is now 1098 billion barrels, compared with only 521 billion in 1971. In essence, rather than a world which is "running out of oil", as was so widely forecast in the late 1960s/early 1970s, the world has been "running into it". This is a direct result of the enhancement of knowledge and know-how in the intervening period – as we forecast. Thus, even by now, in spite of the demands on the industry since 1970 – and the limitations on supply developments imposed by the politics of the industry over the last 25 years – very little non-conventional oil (such as ultra heavy oil and oil from sands and shales) has yet been developed and exploited. There has simply been no motivation to do so. All the world's very large resources of such oil remain to be used – if and when the occasion arises (Odell, 1997).

Meanwhile, proven reserves of natural gas have grown dramatically: from some 40,000 Bcm in 1971 to almost 150,000 Bcm at the end of 1996. This growth occurred, moreover, in spite of a cumulative use over the 25 year period of more than 55,000 Bcm. The current global reserves to production ratio is over 62 years, compared with only 30 years in 1971. The global gas reserves question has thus been one of more or less continuously decreasing concern. Indeed, demand for gas has been restrained by factors other than fears for the adequacy of reserves – most notably, in many parts of the world by the high investment costs involved in delivering the gas to markets.

There have, from time to time, been some regional concerns for supplies relative to markets. For example, in North America in the 1970s and in Europe in the late 70s/early 80s. These have, however, always proved to be unjustified. Reserves developments stimulated by new policies have quickly restored conditions for supply expansion. The irrational fears did, nevertheless, sometimes lead to a significant loss of revenues for countries (such as The Netherlands) which unnecessarily imposed quantitative restrictions on sales to avoid a perceived "scarcity" of gas (Odell, 1992).

## Prospects: Conventional Oil

It is surprising, in the light of the real-world developments in oil reserves since 1970 and the clear-to-see reality of the continuity of the processes of improved finding and production technologies, that some voices are once again raised to warn of an impending near-future oil sector disaster arising from a shortage of the commodity (Campbell, 1996 and 1997). Even more surprising is the similarity between the arguments which lie behind this recent warning, and those of the scarcity fears of 30 years ago; arguments, that is, that have both inappropriate demand and supply components. However, most surprising of all, are the arguments which are based on an analysis of reserves prospects. These ignore not only the dynamics of the processes whereby reserves evolve, but also the central role played by economics in equilibrating the markets. Such irrational warnings are best ignored.

Instead, in evaluating the prospects for oil over the next 30 to 50 years, one needs only to reiterate the realities, as described above, of the last 30 years, *viz.* of a world which still remains only incompletely explored for conventional oil. Around the globe, both onshore and, even more so, offshore, there are almost 100 frontier regions' basins whose hydrocarbons potential has to date been little, if at all, tested (Petroleum Economist, 1993). Some of these remain untested because of their location and others because of political conditions. Mostly, however, their potential resources remain unexplored because of the absence of any motivation for risky investment in them, given a world which is already replete with proven oil, relative to the current level of demand.

As already indicated above, over 1150 billion barrels of oil have been added to reserves during the past 26 years. Moreover, in only three of those years (all of which were in the 1970s) have additions to proven reserves been less than the volumes of oil used (see Table I-11.1). Over

**TABLE I-11.1**

## Proven reserves, reserves/production ratio, oil production and net growth/decline in reserves over the 26 year period from 1971

(in barrels x10$^9$)

|  | Proven reserves at the beginning of year (R/P ratio in brackets – in years) | Production of oil in year | Gross additions to reserves in year | Net growth (+) or decline (–) in reserves in year |
|---|---|---|---|---|
| 1971 | 521 (28.3) | 18.4 | 38 | + 20 |
| 1972 | 542 (27.9) | 19.4 | 54 | + 35 |
| 1973 | 577 (27.2) | 21.2 | 35 | + 14 |
| 1974 | 591 (27.9) | 21.2 | 32 | + 11 |
| 1975 | 602 (28.4) | 20.2 | 31 | + 11 |
| 1976 | 613 (30.3) | 21.9 | 4 | – 18 |
| 1977 | 595 (27.2) | 22.6 | 16 | – 7 |
| 1978 | 588 (26.0) | 22.9 | 45 | + 22 |
| 1979 | 610 (26.6) | 23.7 | 22 | – 2 |
| 1980 | 608 (25.7) | 22.8 | 34 | + 11 |
| 1981 | 619 (27.1) | 21.3 | 67 | + 46 |
| 1982 | 665 (31.2) | 20.1 | 30 | + 10 |
| 1983 | 675 (33.6) | 20.0 | 21 | + 1 |
| 1984 | 676 (33.8) | 21.1 | 44 | + 23 |
| 1985 | 699 (33.1) | 20.5 | 30 | + 9 |
| 1986 | 708 (34.5) | 21.4 | 67 | + 45 |
| 1987 | 753 (35.2) | 21.9 | 129 | + 107 |
| 1988 | 860 (39.3) | 22.8 | 83 | + 60 |
| 1989 | 920 (40.3) | 23.5 | 87 | + 63 |
| 1990 | 983 (41.8) | 23.8 | 26 | + 2 |
| 1991 | 985 (41.4) | 23.7 | 65 | + 41 |
| 1992 | 1026 (43.3) | 23.9 | 46 | + 22 |
| 1993 | 1048 (43.9) | 23.7 | 31 | + 7 |
| 1994 | 1055 (44.5) | 24.0 | 29 | + 5 |
| 1995 | 1060 (44.2) | 24.3 | 48 | + 24 |
| 1996 | 1084 (44.6) | 25.5 | 40 | + 14 |
| 1997 | 1098 (43.1) | | | |
| Totals 1971–97 | | 576 | + 1154 | + 578 |

*Sources:* *Reserves' developments based on data from the annual survey of world oil reserves in the* Oil and Gas Journal, *1970–96 ; from* World Oil, *1971–97 and from De Golyer and MacNaughton's* Annual Survey of the Oil Industry, *1975–83. Annual production data from the B.P.* Statistical Review of World Oil/Energy, *1971–1997*

the period, the upward revaluation of reserves from known fields has been more important than new discoveries: partly, because this is a lower cost way of enhancing reserves and, partly, because so little exploration work has been undertaken in the OPEC countries as a result of, first, the reduced demand for their oil and, second, the lack of investment funds available to the state oil companies. The main reason, however, is a function of new and improved reservoir evaluation methods, together with the application of major advances in production technology in many other parts of the world. These developments have increasingly added to recoverable reserves – most notably in North America, the North Sea and in other industrialised countries where the international oil companies have been centrally concerned with increasing their reserves of equity oil, following the nationalisation of their assets in most OPEC countries (Odell, 1994). The impact of reserves' re-evaluations overall is shown in Fig.I-11.1. A backcast in 1970 showed, for example, an 80% increase of reserves declared to have been discovered by 1955: while a 1996 backcast (derived from Campbell, 1996) shows a 105% increase in the reserves as declared in 1955. For 1970 the 1996 backcast shows a 55% higher reserves figure than that declared contemporaneously.

It is self-evident that not only will the technological developments which have produced these results be further intensified in areas where they have already been successful, but also that they will diffuse into the remaining producing areas where they have not yet been applied; most especially in the OPEC countries, in Russia and in other former Soviet republics. Their impact on reserves growth will remain significant for at least another 25 years, even if the price of oil (in real terms) does not rise above its present level. Figure I-11.1 also shows a forecast of a 2020 backcast of reserves declarations back to 1945. This indicates that today's declaration of proven reserves will be too low by over 350 billion barrels (equal to almost 14 years of supply at 1996 levels). Taken together, the more efficient and comprehensive exploitation of fields already in production, plus the new oil discoveries which will continue to be made as investment in exploration continues at a high level, will lead to a declaration of recoverable reserves of conventional oil of about 3000 billion barrels (Shell, 1995). The depletion of the remaining ±2000 billion barrels of these reserves will sustain the international industry's growth until after 2025 – as demonstrated in Fig.I-11.2 (Odell and Rosing, 1980).

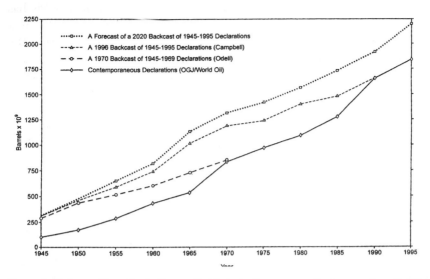

Figure I-11.1: The Evolution of Global Reserves of Conventional Oil, 1945–2020

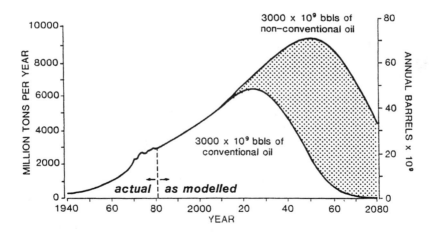

Figure I-11.2: A prospective depletion curve for the world's conventional and non-conventional oil to 2080.

*The production and use of the latter complements the former and extends the potential for growth of oil production (at an annual rate increase of just under 2% per annum) into the second half of the 21st Century.*

## Prospects: Non-Conventional Oil

Nevertheless, during this 30 year period there will also be some limited opportunities for the exploitation of alternative sources of oil. Least controversial in this respect are the reserves of the tar-sands of Athabasca in Western Canada and those of the heavy oil belt of the Orinoco region of Venezuela. Indeed, their exploitation has already started on a limited scale (Odell, 1997). Increasing knowledge and improving technology will produce significant real cost reductions in their production and in time will lead to their large-scale processing, assuming there is a requirement for them to help meet global – or regional – demands for oil, (notably in the Western hemisphere). Oil from oil shales (in the US, Brazil, Zaire, Madagascar, India and other countries) could, if still more oil is required in the markets, follow in due course (post-2025).

There has not to date been any comprehensive and systematic evaluation of the locations and distribution of non-conventional oil (an effort in the 1970s by the UN to do so was thwarted by the decline in the oil price after 1982 and by the ready availability of oil from conventional habitats), but estimates of volumes of its potentially availability range up to 40,000 billion barrels (Meyer, 1977). As shown in the Figure I-11.2, the depletion of only 7½% of this total potential of the world's non-conventional oil would, if necessary, further extend the expansion of the international oil industry until after 2050 (assuming a rate of increase in demand of about 2.0% per annum, as compared with the much lower average growth rate in use of only 1.3% per year over the past 25 years). The industry would, by 2050, under this scenario then be three times its present size.

Finally, there is one other potentially important element for the prospective evolution of reserves. Fears of scarcity and forecasts of reserves running out have all been related to the presently accepted hypothesis in the West for the origin of oil and gas, *viz.* from organic materials which accumulated on the sea-bottom, were rapidly buried by the deposition of fine grained material and then converted to hydrocarbons by subsequent processes involving heat and pressure. There is, however, an alternative hypothesis long since generally accepted in the former Soviet Union (with the world's largest oil and gas industry until the disruptions caused by the counter-revolution), *viz.* "that oil and natural gas have no intrinsic connection with biological matter originating near the surface of the earth, but are primordial materials which have been erupted from great depth." (Kenney, 1996). For this, it is argued "the most direct and most convincing proof... is the existence of commercial petroleum accumulations in crystalline and metamorphic basement rocks" (Porfir'yev, 1974).

Under this theory the world's hydrocarbon resources are, in essence, unlimited in relation to any conceivable evolution of demand. The habitats of oil and gas become much more widely distributed than the world's limited number of sedimentary basins which are perceived to be petroliferous under the 'Western' hypothesis. This alternative view of hydrocarbons occurrence is said to be "the guiding perspective in petroleum exploration throughout the FSU" (Kenney, 1996) and it is claimed that there are already hundreds of oil fields in production there which produce oil from the crystalline basement rock. There is thus, it is argued, no reason to worry about, and even less to plan for, any predicted demise of the petroleum industry, based on a fear for disappearing reserves. If additional oil and gas should be required (beyond that discovered in sedimentary basins at shallow depths), then that will compel "additional investment and development in the technological skills of deep drilling, of deep seismic measurement and interpretation, of the reservoir properties of crystalline rocks and of the associated completion and production practices which should be applied in such non-traditional reservoirs." (Kenney, 1996). Even the output of the prolific conventional reservoirs of the Gulf could, it is suggested, eventually be supplemented by even larger potential volumes from the exploitation of these additional sources (Mahfoud and Beck, 1995).

Given, on the one hand, the reserves/resources prospects and, on the other, the growing propensity for environmentally-orientated constraints on the future use of oil, then the overall long term prospects for the industry seem much more likely to be demand, rather than supply, constrained. As in recent decades, when demand limitations have kept large resources of producible coal unmined, so with oil by the mid-21st century: much of it will eventually be left *in situ*; unwanted, unmourned and hence unrecovered!

## Prospects: Natural Gas

Part, at least, of that unrecovered oil will still be there because of oil's increasing substitution by natural gas, a much more environmentally-friendly energy source. The more rapid recent growth in gas use compared with oil (25% and 12%, respectively, over the past 10 years) is not, however, likely to create any potential gas availability problems. Already proven reserves of gas (150,000 Bcm) are sufficient to cope with anticipated increases in demand for at least the period to 2030. The increased production activities themselves will, meanwhile, serve to enhance the exploration efforts and virtually guarantee reserves' volumes from contemporary areas of gas supply well beyond current levels; as has already happened, for example, in the doubling of the volumes of ultimately recoverable reserves of gas from the Groningen field over the past 30 years (Odell, 1991). Finally, the U.S.G.S.'s modal value for the world's ultimate resources of conventional gas (Masters, 1994) is currently more than double the proven reserves figure. This takes the prospects for production growth through to the 22nd century! And beyond that, if need be, there remain the prospects for the growth of the industry related to the gasification of coal, gas recovery from coal-measures and, ultimately, to the possibility of primeval (abiotic) methane (Kenney, 1996).

## References

Adelman, M.A. (1993), *The Economics of Petroleum Supply*, MIT Press, Cambridge.

Campbell, C.J. (1996), "Oil shock", *Energy World*, No.240, June pp.7–12.

Campbell, C.J. (1997), *The Coming Oil Crisis*, Multi Science Publishing Company, Brentwood.

Kenney, J.F. (1996), "Impending Shortages of Petroleum Re-evaluated, *Energy World*, No.240, June, pp.16–18.

Masters, C.D. et al (1994), "World Petroleum Assessment and Analysis," *Proceedings of the 14th World Petroleum Congress*, J. Wiley and Sons, Chichester, pp.529–541.

Mahfoud, R.F. and Beck, J.N. (1995), "Why the Middle East Fields may produce Oil forever." *Offshore*, April 1995, pp.58–62 and 2106.

Meyer, R.F. (Ed.) (1977), *The Future Supply of Nature Made Petroleum and Gas*, Pergamon Press, Oxford.

Odell, P.R. (1973), "The Future of Oil: a Rejoinder", *The Geographical Journal*, Vol.139, No.3, pp.436–454.

Odell, P.R. (1986), *Oil and World Power*, Penguin Books, Harmondsworth, 8th Edition.

Odell, P.R. (1992), "Global and Regional Energy Supplies; Recent Fictions and Fallacies Revisited, *Energy Policy*, Vol.20, No.4, pp.285–296.

Odell, P.R (1994), "World Oil Resources, Reserves and Production," *The Energy Journal*, Special Issue on the Changing World Petroleum Market, pp.89–114.

Odell, P.R. (1997), "Oil Shock: a Rejoinder", *Energy World*, No.247, March, pp.11–14.

Odell, P.R. and Rosing, K.E. (1980 and 1983, 2nd Edition), *The Future of Oil*, Kogan Page Ltd., London.

Porfir'yev, V.B. (1974), "Inorganic Origin of Petroleum," *The American Association of Petroleum Geologists Bulletin*, Vol.58, No.1, pp.3–33.

Petroleum Economist (1993), *Frontier and Underexplored Basins of the World*, Map Series no.24, London.

Shell International Petroleum Company (1995), "Energy in Profile," *Shell Briefing Service*, No.2., London.

Warman, H.R. (1972), "The Future of Oil," *Geographical Journal*, Vol.138, No.3, pp.287–297.

Williamson, H.F. et al (1963), *The American Petroleum Industry; 1899–1959: the Age of Energy*, Northwestern University Press, Evanston.

# Chapter I – 12

# The Global Energy Market in the Long-Term: the Continuing Dominance of Affordable Non-Renewable Resources*

## Introduction

Regularly recurring fears of impending scarcity of non-renewable energy have, as previously shown, all proved to be groundless (McCabe, 1998; Odell, 1973; Odell and Rosing, 1983). Nevertheless, the issue of the world's potential supplies of coal, oil and natural gas seems to remain one of concern, reflecting humankind's continuing overwhelming dependence on them for its energy needs. This concern is, however, misplaced in the light of recent important changes on both the supply and demand sides.

## The Demand for Energy

Since 1973 the rate of increase in the global use of energy has, as shown in Fig.I-12.1, reverted back to its long-term 1860–1945 trend of only about 2% per annum, compared with the ±5% per year growth rate which occurred from 1945–73. The probability of a future return to the latter 28-year long much higher annual rate of growth is now close to zero, given that it reflected an inherently temporary combination of a set of conditions which cannot re-occur (Odell, 1989). Thus, a contemporary realistic consideration of long-term non-renewable energy supply requirements has to be orientated to a modest growth rate

* Reprinted from Energy Exploration and Exploitation, Vol.18, No.2 and 3, 2000, pp.131–145: and in a corrected version in Vol.18. No.5.

in use, even without taking into account the impact on energy use of the recently much enhanced concern for $CO_2$ emissions and the now accelerating pace of growth in the use of renewable energies emanating from rapidly evolving technologies of direct and indirect solar energy production.

Thus, the long-term future supply prospects for fossil fuels must be put in the context of a maximum 2% per annum growth as the highest likely requirement. Moreover, the year 2000 base from which we must now evaluate 21st century hydrocarbons' demand is very much lower than the conventional and widely-accepted forecasts of 30 years ago indicated would be the situation. Oil production, for example, will this year be about 3.6 gigatons (Gt) compared with expectations in the early 1970s of over 19 Gt (Warman, 1972; Odell, 1973). Likewise, the cumulative use of oil from 1970–1999 has been less than 90 Gt, compared with the close to 240 Gt anticipated. Thus, of the 155 Gt of additions to proven oil reserves since 1970 over 140 Gt remain to be used in the 21st century. Likewise for coal and gas, albeit to a lesser degree. It is thus hardly surprising that views on the future availabilities of fossil fuels' resources and the supply potential associated with them can now be so very much more relaxed that they were in the 1970s (WOCOL, 1980).

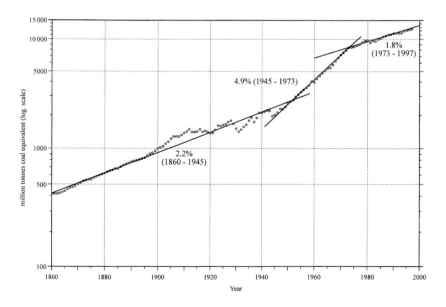

Figure I-12.1: Trends in the Evolution of World Energy Use (1860–1997)

## Global Energy Production

Figure I-12.2 shows a trend in the annual production of energy supplies by source through the 21st century – in response to energy demand growth that is sustained at ±2% per annum. Non-renewable energy sources can be seen to remain dominant until 2060. Alternatives to fossil fuels do not exceed 20% of energy supplied until then, while it is 2079 before they account for 50% of supply. Even cumulatively over the century, as shown in Figure I.12.3, renewable energy sources supply only 35% of total energy used and of this almost two-thirds is supplied in the last two decades. Unless and until the governments and peoples of the world not only accept the desirability of a much faster switch to renewable energy, but also take the necessary steps to implement the change, global energy use in the 21st century will remain heavily orientated to a combination of coal, oil and natural gas. As there are, to date, no serious signs of these requirements for change being met, global energy markets can be predicated at remain dominated by non-renewable resources for decades into the future.

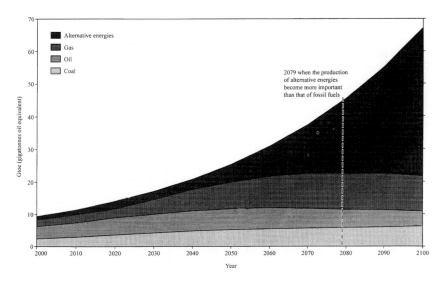

Figure I-12.2: Trends in the Evolution of Energy Supplies, by Source, in the 21st Century

## Coal's Relative Unimportance

Nevertheless, for both economic and environmental reasons, the future pattern of non-renewable energy supplies will show marked changes from the contemporary situation, while the present quite widely-accepted conventional wisdom of constrained supplies in the medium term will be seen to be based on misconceptions concerning potentially available resources (IEA, 1999). These usually indicate coal resources as an order-of-magnitude greater than those of oil or gas and thus assume, implicitly or explicitly, that coal must become, at least, more important than oil and gas and, most likely, the dominant component in the 21st century fossil fuel supplies (Grübler et al, 1999). Given coal's general lack of acceptability and, even more so, as shown in Table I-12.1, its highly geographically concentrated global patterns of both reserves (almost 53% in only 3 countries and over 90% in 10 countries) and production (54% in merely 2 countries and 91% in 10 countries), this prospect is very unlikely. The earlier expressed likelihood of coal as "the fuel of the 21st century", in the aftermath of the oil price increases of the 1970s, (WOCOL Report, 1980) has, over the past decade, been effectively undermined by a combination of local, regional and global environmental concerns over coal production and use.

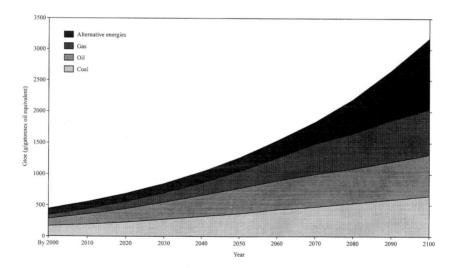

Figure I-12.3: Cumulative Supplies of Energy by Source in the 21st Century

Instead, as shown in Figure I-12.2, coal's share of global fossil fuel production fails to increase over the century. It makes a 21st century contribution of only just over 25% to cumulative non-renewable energy use (Figure I-12.3).

It is thus oil and natural gas which must, between them, continue to supply the bulk of the world's energy supply – at least until the mid-2060s. This implies an overall three-fold increase in their joint contribution to the annual supply of energy in the 21st century (see Figures I-12.2 and 3). The relative contribution of oil and gas to the total supply of hydrocarbons does, however, change very radically, as shown in Fig.I-12.4. As will be demonstrated later in the paper, this change is, in part, a function of possible long-term constraints on oil supplies and, in part, a reflection of the inherent advantages for natural gas in respect of both supply and use considerations. Natural gas supplies are thus indicated to continue to expand to 2090, when global production is predicated output 5.5 times its year 2000 level. On the other hand, as oil's

### TABLE I-12.1
### Global rank ordering of the ten leading countries' coal reserves and production, 1998

| Country | RESERVES Share of Global Reserves | Cumulative Reserves' Share | Country | PRODUCTION Share of Global Production | Cumulative Production Share |
|---|---|---|---|---|---|
| U.S.A. | 25.1 | 25.1 | China | 28.0 | 28.0 |
| Russia | 15.9 | 41.0 | U.S.A. | 26.4 | 54.4 |
| China | 11.6 | 52.6 | India | 6.6 | 61.0 |
| Australia | 9.2 | 61.8 | Australia | 6.6 | 67.6 |
| India | 7.6 | 69.4 | South Africa | 5.3 | 72.9 |
| Germany | 6.8 | 76.2 | Russia | 4.7 | 77.6 |
| South Africa | 5.6 | 81.8 | Poland | 3.4 | 81.0 |
| Kazakhstan | 3.5 | 85.3 | Germany | 2.7 | 83.7 |
| Ukraine | 3.5 | 88.8 | Canada | 1.8 | 85.5 |
| Poland | 1.4 | 90.2 | Ukraine | 1.8 | 87.3 |
| Next 20 Countries | 6.6 | 96.8 | Next 20 Countries | 10.2 | 97.5 |

*Source:   BP Amoco Statistical Review of World Energy, London, 1999*

output is anticipated to start slowly declining from the 2050s, its contribution to the total hydrocarbons' supply ultimately falls from its present contribution of 65% to 44% by 2050 and to under 29% by 2100 (see Table I-12.2 and Figure I.12.4). Over the century as a whole, its share of the cumulative use of 1256 Gtoe of hydrocarbons is, at 540 Gt, only 43% (see Table I-12.3).

## Natural Gas: the Fuel of the 21st Century

Natural gas will have overtaken coal as a global energy source by the 2020s, or even well before that if gas' substitution of oil use accelerates. Its rapidly expanding production in recent years reflects the more than doubling of proven global gas reserves since 1980 and the expansion of European and other markets for gas over the period since 1990. It thus enters the 21st century with an R/P ratio in excess of 60 years. The expansion of its production in the early decades will thus continue to be demand limited (through infrastructual restraints), rather than resources-related. Indeed, the reserves of already discovered fields could in themselves serve to keep global gas production growth at 3+% per annum until 2025, location and demand permitting. But the continuation of large additional discoveries is a certain prospect, given, first, the geographically broadening base of exploration activities and, second, the continuing opportunities of the more intensive exploitation of existing gas-rich provinces, including some hitherto thought to be well matured (eg in the Gulf of Mexico and the North Sea). The mid-point of a range of estimated additional reserves (from 197 to 303 Gtoe) indicates a volume which is about 30% greater than that of the gas used to date plus currently proven known reserves. These data, regionally defined, are set out in Table I-12.4. The indicated proven plus additional reserves are sufficient to support the conventional gas production curve set out in Figure I-12.5A.

Superimposed on the 200 years curve of conventional gas supply shown in Figure I-12.5B is the complementary production curve for the supply of non-conventional natural gas. This is assumed to begin in 2020, with the shape of the curve emerging from the depletion of some 650 Gtoe of the estimated ultimately recoverable reserves of 837 Gtoe, as shown in Table I-12.4. These emerge from the combined potential availability of gas from coal-bed methane, tight formation gas, gas from fractured shales and gas remaining in place after conventional production. It excludes, however, the possibilities of any recovery from gas hydrates, the energy value of which is tentatively estimated to be some 30 times greater than all other non-renewable resources taken

**TABLE I-12.2**
### The changing contributions of oil to the total supply of hydrocarbons, 2000–2050 and 2100

| Period | Total oil and gas supply (Gtoe) | Total oil supply (Gtoe) | Oil as a share of the total |
|---|---|---|---|
| 2000 | 5.84 | 3.79 | 64.9 |
| 2010 | 6.95 | 4.43 | 63.4 |
| 2020 | 8.24 | 5.17 | 62.7 |
| 2030 | 10.58 | 5.76 | 54.4 |
| 2040 | 12.85 | 6.28 | 48.9 |
| 2050 | 14.86 | 6.54 | 44.0 |
| ↓ | ↓ | ↓ | ↓ |
| 2100 | 15.45 | 4.45 | 28.8 |

**TABLE I-12.3**
### The cumulative contributions of oil and natural gas to the energy supply in the 21st century

| Period | Cumulative oil and gas (Gtoe) | Cumulative oil (Gtoe) | Oil's share of cumulative total (%) |
|---|---|---|---|
| Pre-2000 | 176 | 120 | 68.2 |
| 2000–2009 | 66 | 42 | 63.6 |
| 2000–2019 | 143 | 90 | 62.9 |
| 2000–2029 | 237 | 147 | 62.0 |
| 2000–2039 | 355 | 210 | 59.2 |
| 2000–2049 | 495 | 276 | 55.8 |
| 2000–2059 | 647 | 341 | 52.7 |
| 2000–2069 | 806 | 403 | 50.0 |
| 2000–2079 | 943 | 440 | 46.7 |
| 2000–2089 | 1108 | 493 | 44.5 |
| 2000–2099 | 1256 | 540 | 43.0 |

together (H.A. Rogner, 1996). Even without taking any of this potential from gas hydrates into account, non-conventional gas production starting in the 2020s seems likely to become the more important component in global total gas supply within 50 years and, thereafter, its production can continue to grow beyond the end of the 21st century. Natural gas overall in 2100 is predicated to supply over 50% of the year's non-renewable energy production (see Figure I-12.2) while, over the 21st century as a whole, it supplies about 41% of the cumulative total of non-renewable energy (see Figure I-12.3).

Gas will thus undoubtedly be the fuel of the 21st century (as coal was of the 19th and oil of the 20th), even within the framework of the limiting assumptions on gas resources' exploitation as specified above (*viz.* the exploitation of only 75% of ultimately recoverable non-conventional gas and no gas at all from gas hydrates). These limitations do, indeed, inhibit the ability of gas supply to grow sufficiently post-2050 so as to sustain a 2% per annum expansion in overall non-renewable energy use. This predicated annual "shortfall" in the supply of non-renewable energy rises from about 4.5 Gtoe in 2060 to almost 25 Gtoe in 2100. As can be seen in Figure I-12.2, it necessitates a rapidly rising supply of alternative energies during the last quarter of the 21st century.

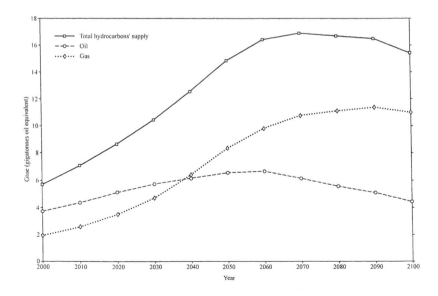

Figure I-12.4: Oil and Gas Supplies in the 21st Century

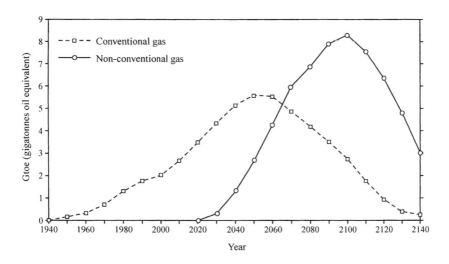

Figure I-12.5A: Conventional and Non-Conventional Gas Production, 1940–2140

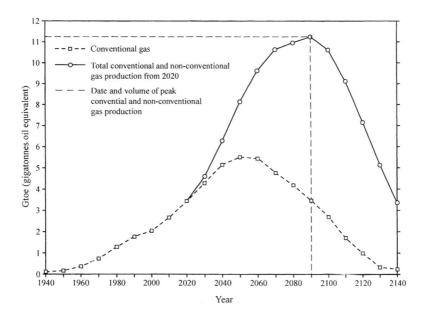

Figure I-12.5B: The Complementarity of Conventional and Non-Conventional Gas Production: giving a Higher and Later Peak to Global Gas Supplies

There then seems likely, however, to be a better than 50% probability for additional gas supplies from the initial exploitation of gas from hydrates. Sixty more years of continuing scientific advances and engineering capabilities would seem to provide time enough to enable a small part of that massive potential source of energy, in the form of the world's preferred fossil fuel for both commercial and environmental reasons, to be brought to the market. Such a development could take care of the "shortfall", assuming, of course, that the exploitation of gas hydrates is then able to compete in economic terms with the alternatives of renewable energies and is, moreover, not excluded by global warming considerations.

## TABLE I-12.4
### World gas reserves and resources by region (Gtoe)

| Region | Conventional Gas | | | Non-Conventional Resources | |
|---|---|---|---|---|---|
| | Cumulative Production to 1999 | Remaining Proven Reserves | Estimated Additional Reserves | Coal-bed Methane, Tight Formation Gas, Gas from Shales and Gas Remaining after Conventional Production | Gas Hydrates and Geopressured Gas |
| North America | 26.9 | 7.5 | 30–52 | 22 | 6,100 |
| Central & South America | 3.0 | 5.6 | 7–22 | 91 | 4,571 |
| Europe (excluding FSU) | 7.3 | 4.7 | 5–14 | 36 | 765 |
| Former Soviet Union | 16.1 | 51.0 | 95–110 | 159 | 4,208 |
| Middle East | 3.9 | 44.6 | 28–50 | 99 | 203 |
| Africa | 2.1 | 9.3 | 5–14 | 29 | 383 |
| Asia/Pacific (excluding FSU) | 3.6 | 9.2 | 25–41 | 203 | 2,528 |
| Total | 62.9 | 131.9 | 197–303 | 837 | 18,758 |

*Source:*  *H-H Rogner, An Assessment of World Hydrocarbon Resources, IIASA,1996.*
*BP Amoco: Statistical Review of World Energy, 1999. Author's estimate of 1999 production.*

## Oil's Relative Decline in Importance

The future for oil lies squeezed – to a greater or lesser extent, depending on environmental and economic considerations – between the prospective solidity of 2.3 to 6 Gtoe of annual global coal supply, on the one hand, and the dynamics of the 21st century gas industry, on the other. Oil's future thus seems likely to become demand-side limited, so that potential supply-side (= resources) limitations represent only a least-likely prospect. This is shown in Figure I-12.6A in which the production curves indicate most of the 200 years' period of actual and potential oil supply from 1940–2140. Conventional oil production is, of course, already well into its life cycle, but nevertheless, it still has, as shown on the graph, some ±30 years to go to reach its annual output zenith at ±4.36 Gt in the early 2030s (compared with 3.3 Gt in 2000), when about 240 Gt of such oil will have been used. Thereafter, the eventual depletion of the remainder of the presently estimated ultimately recoverable reserves (URR) of conventional oil of 410 Gtoe (= $3005 \times 10^9$ barrels) will take another century. This figure of conventional oil's URR is close to the year 2000 value of the calculated upward trend, based on 33 estimates over the past 50 years of the conventional oil resource base as presented to the 1997 World Petroleum Congress (see Krylov et al, 1997). It is also very close to the mid-point of Shell's 1995 presentation of an estimated 2675 to 3275 billion barrels range for ultimately recoverable conventional oil (Shell, 1995). These data are shown in Figure I-12.7.

By contrast, non-conventional oil production has barely started. It is now being developed more rapidly but, using a conservative assumption of only $3000 \times 10^9$ barrels of recoverable reserves (within a resource base many times larger), it will, under restraints imposed by costs, environmental and demand considerations, take 80–90 years to reach its potential peak production (see Figure I-12.6A), at a level which seems likely to be a little lower than that reached by conventional oil in the 2030s. As in the case of conventional and non-conventional gas, however, the two types of oil, though designated by the nature of their occurrence, are essentially complementary in respect of satisfying market demand. Customers are indifferent as to the sources of the crude oil from which their demands for products can be derived; their interests lie only in the utility to them of the oil products they need. Thus, Figure I-12.6B shows the production of both types of oil in an integrated way. From 2000–2030 non-conventional oil accounts for only about 12% of total supply, so merely modestly supplementing increasing availabilities of conventional oil.

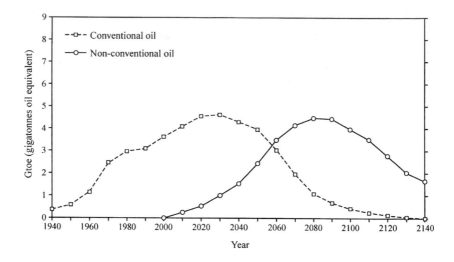

Figure I-12.6A: Production Curves for Conventional and Non Conventional Oil, 1940–2140

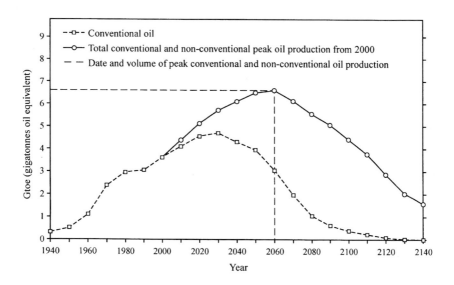

Figure I-12.6B: The Complementarity of Conventional and Non-Conventional Oil Production: giving a Higher and Later Peak to Global Oil Supplies

Thereafter, its relative importance to total supply rises sharply; and by 2060 it becomes the more important component in overall supply. Oil supply in 2100 is predicated to be over 90% non-conventional. The near-100 year period suggested for the full change from conventional to non-conventional oil can be interpreted as reflecting a slow, but continuing, process, based on the joint influences of economic considerations and technological developments.

The oil component in the 21st century's energy production potential is, however, by no means a small or short-lived one. The supply increase shown for the first half of the century is at a rate made possible by already known reserves, reserves appreciations and new discoveries of conventional oil, plus the steadily rising flows of non-conventional oil. Nevertheless, the peak of production in about 2060 may be seen as both later and lower than it would otherwise have been in the absence of competition from other energy sources. As a consequence of this increasing competitive tendency for oil demand in the first half of the 21st century, the decline rate of oil supply after 2060 is relatively slow. Thus, even in 2100, as shown in Figure I-12.6, an oil industry which is still larger than that of 2000 can be predicated. By then, however, in the context of potential resources' limitations and the increasingly

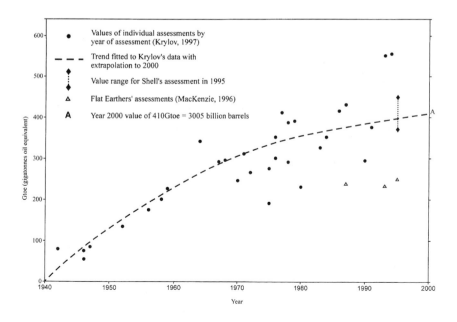

Figure I-12.7: Assessments of Total World Initial Oil Reserves over Time

intensifying competition from other energy sources, oil will contribute, as shown in Figure I-12.2, rather less than coal to global energy supply: while it will, of course, have become less important than natural gas during the 2030s (see Figure I-12.4). Oil's geo-political importance will undoubtedly continue into the early decades of the 21st century, but, thereafter, it really will become just another energy source of steadily decreasing importance in the world's energy demand: and, moreover, one which is available from a broader geographical diversity of locations than hitherto.

## Trends in Production Costs and Prices

In this paper implicit references have been made to trends in costs and prices, but these observations do need to be brought together into an integrated overview, as questions of production costs and of returns on investment constitute a central set of parameters in resource evaluations. No significant differences in prices can emerge in the context of competitive markets between the alternative resources – or between regions – except as temporary phenomena. When such divergences do emerge they inevitably lead either to compensating changes in supply schedules, so that the equilibrium is re-established, or to the need for subsidies or other protective measures to sustain expensive production. Contrasting transportation costs in getting production to markets (as, for example, between oil and gas) produce contrasting netbacks for different fuels in the same regions and between regions for the same fuel, but in a macro-study of the overall global situation over a long period of time, these contrasts seem unlikely to mount to much more than fine-tuning elements in the global energy system.

The starting point of this long-term study of the future availability of coal, oil and natural gas is the declaration of proven reserves for each fuel. Such declarations are, by definition, of reserves that are economic to produce at current levels of costs and prices. As the total quantity of the reserves declared remains much more than adequate to serve the world's slowly expanding energy markets, then competition will generally keep prices from escalating in real terms. In past decades, traumatic events, such as US oil import quotas in the 1960s and 1970s and supply limitations imposed by OPEC and others in the 1970s, 1980s and over the past year, have upset the equilibrium from time to time. Similar relatively short-term influences on oil prices will undoubtedly occur in the future, but, given their essentially ephemeral nature they are irrelevant to an attempt to predict overall long-term cost and price trends.

The potential supply schedule forecast for oil – for long the price leader in the fossil fuel market – suggests little or no pressure of demand on supply for at least the next 20 years. For this period there is thus no reason why oil prices in real terms should rise much and no reason why any significant volumes of reserves of oil, gas or coal which involve higher costs should be produced. The inter-quartile average annual price range of $16.85 to $19.50 per barrel (1999 dollars) for internationally traded crude oil over the past 15 years seems to indicate the most likely price-range for oil for the medium-term future. Technological developments have already brought production costs down in many areas of production (Econ Centre for Economic Analysis, 1997) and this process can be expected to continue.

Sometime in the 2010s, however, upward price pressure on the oil market seems likely as the attempt to maintain growth in the production of conventional oil leads to a requirement for higher investment costs. An increase of 10–20% in unit exploration or development costs will then be confirmed by the required new and heavy investments in the first substantial exploitation of non-conventional oil. Such cost increases will need to be passed through into the general level of prices. The equilibrating price of $17–20/bbl of oil in the meantime thus seems likely to be converted by 2020 to $19–24/bbl (in 1999 dollars). At this modestly higher real price level, the requirement for the highest cost oil producer to sell into the market and earn sufficient profits will be satisfied. Neither gas nor coal seem likely to 'need' the price increase, but the producers will accept it as a means of enhancing their rate of return on investment. In the case of coal, such price rises could enable the industry to absorb the relatively higher carbon taxes to which coal by then seems certain to have become subject. Equilibrium in the global energy market will thus be maintained in a situation in which alternative energies will, in general, still require a subsidy and so be unable to bring downward pressure to bear on prices (World Energy Council, 1997).

The next likely future non-renewable energies price crunch, when upward pressures generated by relative scarcity again exceed the downward pressures engendered by technology, will emerge as conventional gas output approaches its period of maximum production in the 2040s. By then gas will have overtaken oil as the world's single most important source of energy and will thus also have become the price leader in the global energy market. In order to sustain the growth of the industry into a period of the profitable exploitation of non-conventional gas, a further modest price increase (in real terms) will then be required to generate the higher level of investments in exploiting such gas. This could take the oil equivalent price to a level of $21–26/bbl (in 1999 dollars).

By that time, however, yet higher taxes on the use of coal for environmental reasons will be tending to price the fuel out of many geographically disparate markets. Technological improvements in coal production notwithstanding, coal will not have the ability generally to influence energy prices, although it may still exercise downward pressures on energy prices in a limited number of coal-rich and coal-dependent countries. Oil will also have become largely a price taker – rather than a price maker – and the industry will be anxious to benefit from the revenue flows generated from the upward movement of gas prices, in order to secure sufficient returns for financing investments in the continuing expansion of non-conventional oil production to 2080 (see Figure I-12.6A). Likewise, a rapidly expanding delivery of renewable energies to the market will be made possible from the upward price movements in the market. These will enable governments to reduce or even eliminate the subsidies hitherto required for sustaining the growth of the industry.

From 2040 to 2060, as the supply of fossil fuels becomes increasingly orientated towards natural gas, global energy supply seems likely to enter the age in which gas supply and users' technologies are honed towards 'perfection', so that costs do not rise in real terms. Gas prices also seem likely to be stabilised by the sharpening competition created by the lower real costs of delivering renewable energy supplies under conditions of increasing economies of scale.

From 2060 this paper has suggested that increases in the supply of gas may fail to reach the level required to sustain an annual growth of 2% in the supply of non-renewable fuels. Cost increases required to maintain gas production at the high global level then achieved may become a problem, with particular respect to the exploitation of gas hydrates. Whether or not such cost increases will become significant before 2100 is impossible to judge. Even if they do there is, however, no certainty that they could be passed on in higher prices as the improving economies and technology of energy supplied by renewable energies could by then be setting the general price level. Nor is it impossible that the very large resources of relatively low-cost coal remaining in 2100 may finally become a replacement fuel. This could result either from the failure of global warming to occur (so that $CO_2$ emissions are no longer considered a problem), or from the emergence of technologies which enables coal to be used in a way which is environmentally friendly (Williams, 1998).

Indeed, by the end of the century, in the context of continuing technological progress, a fall in the real costs of supplying renewable energy seems more likely than a situation of inherently rising costs. If so, then this will bring downward pressure to bear on the prices indicated above for non-renewable fuels. Such a development could conceivably accelerate the latters' already declining contribution to world energy supply, as they become unable to compete effectively in the market-place.

## Are Oil and Gas Fossil Fuels?

Finally, a word of caution on the essential fragility of a study on the very long-term future for the world's energy supply which accepts without question the validity of the original 18th century hypothesis that all oil and gas resources have been generated from biological matter in the chemical and thermodynamic environments of the earth's crust. There is an alternative theory – already 50 years old – which suggests an inorganic origin for additional oil and gas (Porfir'yev, 1974; Kenney, 1996; Gold, 1999). This alternative view is widely accepted in the countries of the former Soviet Union where, it is claimed, "large volumes of hydrocarbons are being produced from the pre-Cambrian crystalline basement" (Krayushkin et al, 1994). Recent applications of the inorganic theory have, however, also led to claims for the possibility of the Middle East fields being able to produce oil "forever" (Mahmoud and Beck, 1996) and to the suggestion of *repleting* oil and gas fields in the Gulf of Mexico (Gurney, 1997). More generally, it is argued, "all giant fields are most logically explained by inorganic theory because simple calculations of potential hydrocarbon contents in sediments shows that organic materials are too few to supply the volumes of petroleum involved." (Porfir'yev, 1974).

The significance of the alternative theory of the origin of additional oil and gas potential is self-evident for the issue of the longevity of hydrocarbons' production potential and production costs in the 21st century. Instead of having to consider a stock reserve already accumulated in a finite number of so-called oil and gas plays, the possibility emerges of evaluating hydrocarbons as essentially renewable resources in the context of whatever demand developments may emerge. If fields do replete because the oil and gas extracted from them is abyssal and abiotic (based on chemical reactions under specific thermodynamic conditions deep in the earth's mantle), then extraction costs should not rise as production from such fields continues for an indefinite period. Neither do estimates of reserves, reserves-to-production ratios and

annual rates of discovery and additions to reserves have any of the
importance correctly attributed to them in evaluating the future supply
prospects under the organic theory of oil and gas' derivation (Campbell,
1997). In essence, the "ball park" in which consideration of the issues
relating to the future of oil and gas has hitherto been made would no
longer remain relevant.

## References

Campbell, C.J., *The Coming Oil Crisis*, Multi-Science Publishing Company,
     Brentwood, 1997.

Gold, T., *The Deep Hot Biosphere*, Copernicus Press, New York, 1999.

Grübler, A. et al., 'Dynamics of Energy Technologies and Global Change,'
     *Energy Policy*, Vol.27, 1999, pp.247–280.

Gurney, J., 'Migration or Replenishment in the Gulf,' *Petroleum Review*, May
     1997, pp.200–203.

International Energy Agency (IEA), *World Energy Outlook*, London, 1999.

Kenney, J.F., 'Impending Shortage of Petroleum Re-evaluated,' *Energy World*,
     No.240, June 1996.

Krayushkin, V.A. et al, 'Recent Applications of the Modern Theory of
     Abiogenic Hydrocarbons Origin,' *Proceedings, VIIth International
     Symposium on the Continental Crust*, Sante Fe, 1994.

Krylov, N.A. et al, 'Exploration Concepts for the Next Century,' *Proceedings,
     15th World Petroleum Congress*, Beijing, 1997.

Mahmoud, R.F. and Beck, J.N., 'Why the Middle East Fields may produce Oil
     Forever,' *Offshore*, April 1995, pp.56–62.

McCabe, P.J., 'Energy Resources – Cornucopia or Empty Barrel,' *Bulletin of the
     American Association of Petroleum Geologists*, Vol.82, No.11, 1998,
     pp.2110–2134.

Odell, P.R., 'The Future of Oil; a Rejoinder,' *The Geographical Journal*,
     Vol.139/3, 1973, pp.436–454.

Odell, P.R. and Rosing, K.E., *The Future of Oil*, Kogan Page Ltd, London, 1982.

Odell, P.R., 'Draining the World of Energy,' R.J. Johnston and P.J. Taylor
     (Eds), *A World in Crisis*, Blackwell, London, 1989, pp.79–100

Odell, P.R., *Fossil Fuel Resources in the 21st Century*, Financial Times Energy,
     London, 1999.

Odell, P.R., 'Dynamics of Energy Technologies and Global Change: a
     Commentary,' *Energy Policy*, Vol.27, 1999, pp.737–742.

Porfir'yev, V.B., 'Inorganic Origin of Petroleum,' *AAPG Bulletin*, Vol.58, No.1,
     1974, p.3–33.

Rogner, H-H., 'An Assessment of World Hydrocarbon Resources,' *IIASA, Laxemburg, WP-96-56*, 1996.

Shell Briefing Service, *Energy in Profile*, No.2, London, 1995.

Warman, H.R., 'The Future of Oil,' *The Geographical Journal*, Vol.138/3, 1972, pp.287–297.

Williams, R.H., 'A Technological Strategy for making Fossil Fuels Environment and Climate Friendly,' *World Energy Council Review*, September 1998, pp.59–67.

WOCOL Study, *Future Coal Prospects*, Ballinger Publishing Co., Cambridge, Mass., 1980.

# Section II

# International Oil Markets

# Chapter II – 1

# The Global Oil Industry in the late 1960s*

## Introduction

A description of the world's oil industry demands the use of many superlatives. By any standards it is the world's leading industry in size; it is probably the only international industry that concerns every country of the world; and, as a result of the geographical separation of regions of major production and regions of high consumption, it is first in importance in its contribution to the world's tonnage of international trade and shipping. Because of these and other attributes, a day rarely passes without oil being in the news. Often the significance of such news items is not apparent in isolation or without some background knowledge of the way in which the industry is organized internationally and of its impact upon individual countries and groups of countries in which its operations and interests lie. This book aims at providing such a background by describing and analysing the oil industry's affairs and relationships around the world.

We shall be concerned almost exclusively with the period of some twenty-five years since the end of the Second World War, when growth in oil has exceeded growth in all other large-scale economic activities. Already by 1950 twice as much crude oil was produced as had been produced in 1945. Ten years later, in 1960, production had doubled again to 1,000 million tons. The rate of expansion continued, and it took only

* Reprinted from *Oil and World Power*, Penguin Books, Harmondsworth, 1st edition, 1970, pages 11–32.

five years more for the next additional 500 million tons annual output to be achieved. In 1968, only three years later, another leap of 500 million tons in annual output to a total of 2,000 million tons was achieved and today the prospect of an annual output of 3,000 million tons by 1974, and of 4,000 million tons by 1980, is considered to be as near certain as any forecast can be. Thus, the greatest possibilities for growth still lie ahead. In light of this, more petrol tankers on the roads and railways, new pipelines to major consuming centres, additional refineries on major estuaries and elsewhere, more and even larger tankers for moving oil around the world, and continued news items about oil exploration and development efforts in hitherto unexpected places can be expected for as far ahead as we can judge.

But generalizations about the world oil industry are somewhat misleading, particularly as it is not a fully interlocking system. There is, to start with, and as one would expect from the division between the communist group of countries and what the Americans call the 'free world' (a description often used in oil-company literature), a clear break between the industry of the communist nations and that of the rest of the world – though, as we shall see later, there are certain interrelationships which appear to be of increasing significance. On the other hand, within the communist world the increasingly deep cleavage between the Soviet Union and its European allies, on one side, and China and some of its friendly neighbours, on the other, is also reflected in a decreasing degree of contact between the oil industries of these countries. A few years ago, increases in Soviet oil output were planned on the assumption of an increasing import requirement by China. Today oil trade between the Soviet Union and China has greatly diminished and it is now probably a matter of only a short time before China's efforts at securing a rapid expansion of its own oil industry eliminate entirely its reliance on the Soviet Union.

In the rest of the world, in the post-war period, a split has emerged between the oil industry of North America (mainly the USA, but also including Canada and, to a lesser degree, Mexico) and that of the remaining non-communist world. Before the war, and up to 1945, in as far as wartime supply routes made this possible, the USA was the leading exporter of petroleum products to Europe and to other parts of the world, but it is no longer in that position. Instead, its oil industry can now be considered as a separate entity, with such differences in price levels and in organization from the rest of the world as to necessitate an increasingly autarchic policy on the part of the USA, whose oil industry would greatly diminish in size if it were subjected to competition from outside. In this respect at least, the USA and the USSR have something in common, for the latter also protects its domestic industry through the exclusion of imports.

## The International Majors

Outside North America and the communist world we have the territory of the so-called 'international oil industry'; but 'international' used in this context does not imply an industry owned and/or controlled by the world's many nations. Instead it refers simply to the fact that this is an industry which operates internationally, with a complex network of relationships connecting most countries of the world. Its ownership and direction lie mainly in the hands of a very small group of companies known, in oil-industry terminology, as the 'international majors'. These companies are responsible for something like 80 per cent of all oil production in the world outside the communist countries and North America. In the same areas they own or control over 70 per cent of the total refining capacity and they operate either directly, or indirectly through long-term charter, well over 50 per cent of the tonnage of internationally operating tankers. Just a few years ago these percentage figures were even higher.

The companies concerned are international only in the senses that their operations are world-wide, in their employment of nationals of many countries and in their having locally (that is nationally) registered subsidiaries in many countries of the world. But their ownership and ways and methods of working are limited to those of just three countries, with the USA as the dominant element. Thus, of the seven 'international majors' no fewer than five have their headquarters and the overwhelming majority of their shareholders in the USA, which also provides all the top management and a high proportion of the lower echelons as well. The largest is Standard Oil of New Jersey, which through most of the world trades under the Esso ('tiger in your tank') sign – except, ironically, in the USA, where the nationally trading subsidiary is, rather amusingly in the light of the group's strength and size, known as Humble Oil! Another Standard company (almost every state in the USA secured such a company when the coast-to-coast 'empire' of Rockefeller's Standard Oil was broken up by anti-monopoly legislation earlier in the century), Standard Oil of New York, is the parent company of many subsidiaries around the world trading as Mobiloil. The other Standard Company which has achieved a place amongst the international 'majors' is Standard Oil of California, which, after many years of operating abroad jointly with one or other of the other majors, is now 'going it alone' under the Chevron sign. Gulf Oil, with headquarters in Pittsburg, has for long been a major producer of crude oil in the Middle East, having secured a 50 per cent interest in the extremely prolific Kuwait oilfields in the late 1930s, and is now increasingly diversifying into refining and marketing operations throughout Europe and in parts of the Far East and South East Asia. Finally Texaco, which, as its name suggests, operates out of Texas, is an aggressive company with major overseas interests in the Caribbean and South America and with important shares in joint producing companies in the Middle East. It is also involved through a wide range of subsidiaries in refining and/or marketing oil products in some forty countries.

Though these companies have – and even deliberately foster – different corporate images, they do have one overriding attribute in common: their Americanism. Although many of their subsidiaries have a management consisting mainly of nationals of the country concerned, it is seldom necessary to dig very deeply to find the key US personnel whose job it is to ensure that American professionalism and expertise in oil reaches into the furthest parts of the company's empire and who, at the same time, will be responsible for ensuring that the policy of the subsidiary is in line with the authorized interpretation of centrally taken decisions. The communications infrastructure of these USA centred 'international' firms certainly rivals that of the foreign services of the majority of the world's nations.

This is also true of the two remaining 'international majors', the Royal Dutch/Shell group and British Petroleum. The former is an Anglo-Dutch enterprise with the shareholding interest of the Dutch parent making the Dutch element the more important in the ratio of 60:40 (reflected also in a 4:3 split of its managing directors in favour of the Dutch side), but with operational and commercial headquarters in London. It is second only to Standard Oil of New Jersey in the size and complexity of its operations, and there are only a handful of countries in the world in which there is no local Shell Company. The latter, B.P., is a wholly British-owned enterprise (except for a little foreign ownership of its shares) in which, incidentally, the state has a 49 per cent holding dating from 1913. However, this dominant state holding, much to the disbelief of foreigners, does not result in the British government as such exercising a dominant role in the company. In fact, successive governments have elected to exercise no role at all, beyond that of nominating two directors to the seven-man board. These directors are then quite free apparently to act as they themselves think best, without even a formal obligation to report back to the government or to seek its advice when participating in company policy making. In other words, the state is a sleeping partner and is content to let B.P. operate just as the other six private-enterprise international oil companies do. The state's interest lies only in its drawing not inconsiderable dividends from the company, which, like Gulf Oil, has had the advantage of access to massive supplies of low cost crude oil in the Middle East in the post-war period and has thus had the incentive to diversify into refining and marketing operations in an increasing number of countries. These include the U.S.A., where, after a series of small incursions to test the market, B.P. made up its mind in 1968 to go into refining and marketing in a very big way with take-over bids for a number of medium-sized oil companies. Favourable decisions by the U.S. Department of Justice (which is responsible for investigating take-over bids in the light of the country's anti-trust legislation) gave the green light for the implementation of this major policy development, which will ultimately have the effect of making B.P. about the ninth or tenth largest oil company in the United States.

These then are the seven 'majors', and, as the figures given earlier indicate, they make up the dominant element in the international oil industry. Before the war they collectively constituted a near-cartel – having agreed on market shares. Effects of the war and of U.S. anti-trust legislation eliminated the pre-war formal agreements between them, but until recently – because of their organization and control over supplies – they were able to have the industry work the way they wanted it, with mutual understanding between them over production and pricing policies ensuring high profits for them.

## The Independents

It should be noted, however, that their ability to dominate has been falling in recent years and is likely to continue to fall. This change is a result of the growth of new elements in the oil business. One is the rapid growth of important European enterprises such as Compagnie Française des Pétroles (C.F.P.) of France and Ente Nazionale Idrocarburi (E.N.I.), the Italian state company. Their interests, moreover, have not remained confined to Europe – though this is where they are strongest – but have spread other parts of the world, where they are participating not only in production ventures, but also in refining and marketing.

Even more important are the activities of U.S. oil companies which used to have interests only within the U.S.A. but which, since the early 1950s, have sought both oil production and oil markets abroad.

They did this at first to obtain lower-cost crude-oil supplies than were available in the U.S.A., with a view to shipping these back home and thus increasing the profitability of their domestic refining and marketing operations. But since the U.S. government imposed oil import quotas in 1959, limiting the amounts of oil the companies could send from their production overseas to the U.S.A., they have had to diversify into refining and selling oil in other parts of the world – notably Western Europe – in order to obtain some revenues from the crude-oil supplies that they have discovered and developed, often at a high capital cost. In a world of generally booming markets for oil they have not found such expansion too difficult, but they have, of course, expanded partly by taking business that might otherwise have gone to the major companies. In the future the growth of markets for oil in Western Europe and Japan will slow down and eventually even cease, and a more difficult period will begin for these new companies in the international business as they face the competition of the 'majors', which will then be unwilling to risk the loss of any further percentage of the total markets available.

At that stage the continued success on a world-wide scale of the new companies will depend on their willingness to get involved with refining and marketing in Asia, Africa and Latin America. So far there is little sign of this happening – conditions of doing business are perhaps too different from what they are used to in the U.S.A. or Western Europe. But, if and when they do overcome their hesitation in this respect, they could well take advantage of the increasingly strong reaction in the developing world against the control effectively exercised over the whole of their oil business by the seven major international companies and the subsidiary companies they have established in most countries of Latin America, Asia and Africa, largely to market crude oil and products from their larger-scale producing and refining operations in the major oil-producing areas.

## The Impact of Nationalism

The reaction arises from an enhanced degree of economic and political nationalism in a period in which colonialism in any guise or form has been highly suspect; and many countries have viewed the control over their own oil industries by one or more of the major international companies very much as a form of economic colonialism incompatible with the status of a sovereign nation. The result has been an increasing tendency for such nations to exercise their authority over the companies concerned through import quotas, price controls, insistence on the employment of nationals rather than expatriates, the imposition of unfavourable taxes and other regulations, etc. Some countries have even established and encouraged alternative systems for securing their supplies of oil from overseas and it is in such cases that new entities and companies can find opportunities for development. E.N.I. and French state-sponsored entities have already done so in Africa and Asia, and American 'independents' (that is, U.S. companies other than the five American majors) may well follow suit.

This nationalist reaction against the major companies arose first amongst the larger Latin American nations, which, after a century or so of political independence, also sought their economic independence from what they considered to be an element of American imperialism. Thus, countries such as Chile, Brazil and Argentina have gradually restrained the freedom of action of these companies to operate within their territories. The companies have, for example, been refused permission to expand their activities beyond those existing at a certain date, or they have been obliged to integrate their own operations into a framework established by state control and direction. Moreover, to supplement and intensify these measures the countries concerned often created state-owned oil entities which have then been given the overriding responsibility for ensuring the provision of the countries' needs for petroleum. Today, only Ecuador, Paraguay and recently independent Guyana of the thirteen nations of South America are without their state oil companies. This trend has quickly been taken up by many more recently independent countries in Asia and Africa, and today over twenty countries in these two continents have state-owned oil entities with responsibilities extending from exploration to final marketing. This trend towards state control and/or ownership appears to be likely to continue amongst the developing nations, where suspicions about the wisdom of permitting the international companies to look after the oil sector still appear to be increasing.

But, except for oil exploration and production, it is in the countries of Europe, together with Japan, that the principal post-war expansion of the international oil companies' activities has occurred. Even in these areas the question of allowing the companies freely to supply, refine and market the oil required has been increasingly raised, and countries such as Finland, France, Italy, Austria, Spain and Portugal have all decided to place oil partly, at least, in the public sector through the establishment of a state-owned oil enterprise. And where state companies have not been established, other European governments have, almost without exception, involved themselves in attempts to persuade the companies concerned to pursue policies, as regards refining and the pricing of products, etc., which accord with what the country considers nationally desirable. It is perhaps Japan which has taken this degree of persuasion through national supervision and control over oil further than any other industrialised country. There, through the powerful Ministry of Trade and Industry, national control is exercised at all significant points over the activities and decisions of the international companies selling oil to or in Japan.

Thus, the post-war period, particularly the last decade, has been one of severe limitations on the freedom of the international oil companies to take decisions solely in the light of their own corporate interests and to follow up opportunities for expanding their activities in refining and marketing oil. They have moved ahead less rapidly than would otherwise have been the case and have thus had their degree of dominance in these activities somewhat curtailed.

At the same time they have also faced limitations on their ability to do as they would commercially and corporately have liked to do in the main areas of oil production. Not only have all major producing countries insisted on a larger share of the profits from oil production, but they have also insisted in some cases that the companies' operations should either be made part of a joint venture with a state-owned entity, or be brought much more directly under the general overall control of a Ministry of Petroleum or similar official body. And often, other non-major companies have been invited into a country to participate in, or even to initiate an oil development programme, to the detriment of the major companies' share of total production.

## The Oil Producing Countries

Producing countries took these lines of action against the companies working concessions in their territories, in the light of what they saw their economic and political interests to be. In the early postwar period of rapidly increasing demand for oil, at price levels much in excess of long-run supply prices, the governments were usually able to secure their objectives because, no matter what they demanded, the business remained very profitable for the companies, which could increase prices sufficiently to offset the additional payments. But, since the late 1950s, this highly favourable situation for the producing countries has deteriorated under the impact of a world surplus of oil-producing capacity and consequent price weakness in many markets. Producing countries soon found that the international companies, with the flexibility given to them by their operations in a number of major producing areas, were able to concentrate their activities in areas where there was least interference from the governments concerned. This flexibility was a factor in the enthusiastic development by the companies of newly discovered oil resources in Libya and Nigeria, in the 1950s and early 1960s, and in Australia and Alaska in more recent years. The companies were aware that developments in new areas would enable them to withstand unwelcome pressures from the governments of the longer-established producing areas. The producing countries were made painfully aware of this danger in the late 1950s, when a potential surplus of oil supplies and the need to stimulate demand persuaded the companies to announce significant overall reductions in the prices they 'posted' for crude oils. Such reductions served to decrease the revenues which host governments received from the oil industry for, in most cases, such revenues were calculated on the basis of the posted prices.

The result of this shock was the formation in 1960 of the Organization of Petroleum Exporting Countries (O.P.E.C.), whereby the oil producers sought collectively to enhance their bargaining power by standing together in order to prevent the companies from playing one country off against another. O.P.E.C. has grown in stature quite quickly in the short period of its existence and has succeeded in preventing further reductions in posted price levels, so ensuring the maintenance of government revenues per barrel of oil produced. It has also successfully negotiated technical changes in the methods whereby company profits are calculated, and, as a result, government revenues per barrel have even been increased by a few cents. As we write, however, O.P.E.C. has not yet succeeded in introducing a mechanism for controlling and regulating the output of oil in member countries. In the absence of such controls there still exists a potential danger of further reductions in crude oil prices.

Paradoxically, it is the earlier encouragement that the producing countries gave to companies other than the 'majors' to participate in oil exploration and development that stands as the major hurdle against the achievement of this objective. If the countries concerned had only had the seven major international companies to negotiate with, then some form of regulation of oil output, acceptable to both sides, would most probably have been achieved. However, the 'majors' are now forced to battle for markets with the 'independents' and the competition is becoming so acute, even amongst themselves, that the prospect of the international pro-rationing of oil production – whereby production levels can be tied in to estimated demand levels at a given price (which is the sort of system that is worked within the major producing areas of the U.S.A.) – now seems to be more remote than it was at the time of O.P.E.C.'s formation!

## International Interactions

Thus, today, the international oil industry consists of a number of interacting elements. There are not only the major international companies, such as Shell and Esso, which formerly ran the industry much as they pleased, but also a range of additional companies also interested in producing oil and selling it on the markets of the world, together with the oil interests of over a hundred national governments throughout Europe, the Middle East and the developing world. The large companies continue to dominate both the stage and the play as the producers, refiners and sellers of most of the world's oil, and they are brought into virtually daily contact with the governments of those countries in which they both choose and are allowed to have interests. To some degree these companies also act as intermediaries between the conflicting oil interests of producing countries on the one hand and consuming countries on the other. Thus they operate not only as commercial enterprises attempting to maximize their profits, but also as diplomatic channels attempting to keep the oil flowing around the world. But their function in this latter respect may well be a short-lived one, for, if state interests and direct involvement in oil continue to expand at the rate at which they have expanded in the post-war period, then, in another decade or so, it seems more than likely that the oil companies will no longer be required as intermediaries. The economic and political negotiations needed to get oil from the points at which it is produced to the points at which it is to be consumed will then be undertaken by the governments of the countries directly concerned.

When and if this happens, the 'international oil industry' will be international in a very real sense, with all that this implies. On the one hand, there are the dangers that such a situation will pose as a result of the confrontation of conflicting national interests, but on the other, there are also the opportunities that such a situation will present for the effective internationalization of this most important, internationally traded commodity. This could conceivably be under an agency responsible for the production, refining, transporting and marketing of the oil within the framework of the most rational system that it is humanly possible to conceive: a system the basis for which already exists in the increasingly sophisticated logistical systems that are currently being developed by the separate international companies, not only for the sake of their own profitability, but also because their managers are responding to the intellectual challenge of such approaches to the organization of the world's largest industry.

## Bibliography

An earlier book by the author, *An Economic geography of Oil* (Bell, 1963), analyses the locational patterns of the world oil industry on a function-by-function basis.

E.T. Penrose's *The Large International Firm in Developing Countries* (Allen & Unwin, 1968) in itself provides a more than adequate introduction to the economics and organization of the oil industry, but this may be supplemented by *Oil, the Facts of Life* by P.H. Frankel (Weidenfeld & Nicolson, 1962) and 'The World Oil Outlook' by M.A. Adelman in *Natural Resources and Economic Development*, edited by M. Clawson (Johns Hopkins Press, 1964).

Two books by authors whose work as specialized journalists put them closely in touch with the working of the oil industry fill in some of the industry's political and historical backgrounds. These are *Oil Companies and Governments* (Faber & Faber, 1967) by J.E. Hartshorn, and *Oil: the Biggest Business* (Eyre & Spottiswoode, 1968) by C. Tugendhat.

In contrast, H. O'Connor's *The Empire of Oil* (Monthly Review Press, 1955) and *World Crisis in Oil* (Elek Books, 1963) provide a markedly different interpretation of the history and politics of the industry. The reader seeking objectivity will want to read both points of view.

# Chapter II – 2

# The Significance of Oil*

## The Middle East Oil Industry –
## Its Early Twentieth Century Beginnings

The Middle East without oil would be a very different region. In addition to changing the face of large parts of the area, oil has helped shape the policies and alignments of all the countries of the Middle East – not only with each other, but also with the world's great powers. These outside powers, in their turn, have found their oil interests and their oil ambitions in the Middle East spilling over into their more general relationships with each other, thus helping a process which has made the region a centre of international tension over long periods of time.

Early British oil interests in the area merely reflected late nineteenth-century British interests of a more general nature. These arose out of imperial connections with India and the Far East which demanded the establishment of coaling stations and of territorial enclaves to protect both them and routeways such as the Suez Canal. As a result of these imperial developments, Britain then took an earlier interest than any other outside power in the possibilities of oil from the Middle East. Even so, the region was well behind other parts of the world in the development of oil resources. By the time the Anglo-Persian company finally discovered oil in Persia in 1908, and started to work the discoveries under the stimulus of participation by the British

* Reprinted from the *Journal of Contemporary History*, Special Middle East Issue, Vol.3, No.3, July 1968, pp.93–110.

government, which sought British owned and controlled oil for its navy, the oil industry elsewhere in the world had almost half a century of history behind it. Britain's strategic and military domination of all areas of the Middle East outside Turkish control acted as a restraint on oil activities, other than those blessed officially by Britain, and thus from 1908 to 1919, apart from the steady development of resources in Persia, little in the oil sector took place to disturb the traditional economies of the Middle East.

After the first World War, the division of the former Turkish empire into British and French spheres of control led to an understanding between the two countries to divide the spoils as far as oil was concerned. Thus, although Persia remained wholly British in its oil development, neighbouring Iraq, where the possibilities of oil appeared good, lay on the border of British and French influence and hence necessitated the participation of both British and French capital in a consortium created to initiate oil exploitation. Those parts of the Middle East wholly within the French sphere turned out to offer little by way of oil possibilities – given the technology of the period – and hence French oil interests were limited to a share of Iraqi operations and, at a later period, to participation in the crude oil pipelines that were built from Iraq to the Mediterranean. On the other hand, Britain gradually extended its oil interests to other parts of the Persian Gulf where its earlier informal political authority was formally confirmed in 1919. In the inter-war period British firms discovered oil resources along this coast, particularly in Kuwait. The demand for oil at that time, however, was not great enough to give a strong enough incentive to develop these facilities. Persia and Iraq were quite capable of producing whatever oil could be sold out of the Middle East.

## Inter-War Restraints on Development

The world oil-pricing system, dominated by the United States, and the existence, informally before 1928 and formally after that, of an international oil cartel, effectively inhibited the exploitation of the potential producing areas of the Middle East. The USA, Mexico, and, to a lesser degree, Venezuela and the Dutch East Indies, maintained their dominant role in world oil markets so that expansion of known Middle East oil resources was restrained. But attempts to make Middle East oil more important in an international context also turned on the political and commercial efforts of Washington and of American companies to win access to the oil wealth of the region. Eventually, as a result of both political and commercial pressure, the US secured an entry to Middle

East producing areas, and although the British government managed to keep American interests clear of its most important producing and refining facilities in Persia, it was forced to allow them into Iraq and the Persian Gulf states. In Iraq, the consortium, in which British and French companies retained an interest, came to be dominated by a group of US oil companies. In Kuwait, the concessionary company became 50/50 British and American. Thus, by the outbreak of war in 1939 the Untied States had won for itself a position of virtual parity with British interests, and one potentially stronger than that of the longer-standing French oil influence in the region.

In this period up to 1930 one can view the structure of the Middle East oil industry as lying wholly within the framework of a classical colonial situation. While the countries there were not colonies like British or French possessions elsewhere in the world, nevertheless British and French political control over virtually all the territories concerned was practically complete. In addition, oil wealth provided the basis for economic colonialism, while the power of the United States, in general, and of its oil industry, in particular, brought in a third outside power as an economic force of significance. As shown already, however, one should not overstress the significance of oil in this period. Middle East oil remained relatively unimportant in world markets, and even by 1939 it was contributing only 5 per cent of total world oil production, while its exports were limited to countries within the immediate region and, via the Suez Canal, in western Europe. But even in western Europe, oil from the Middle East accounted for under 20 per cent of total imports, which remained dominated by those from the United States and Mexico. The real significance of pre-1939 developments in the Middle East, however, is that they established the framework for the post-1945 expansion of oil activities in the region. With the cessation of hostilities, British, French and United States oil companies rushed in to take advantage of the concession areas which they had tied up for themselves before 1939.

## Post-1945 Expansion

All companies concerned participated in the rush to expand Middle East oil production and exports. The post-war world was short of energy for various reasons. First, US oil was in demand at home; second, Mexican oil was required to service the increasing demands of Mexico itself under the process of industrialization; third, though Venezuelan production had expanded rapidly towards the end of the war, it was physically unable to meet the rising demands of both the Americas and western Europe; and, finally, in the Far East, in the formerly important oil-exporting areas of the Dutch East Indies and Burma, the industry had been destroyed in the war.

War-time oil development in the Middle East had been limited, but there had been sufficient work to demonstrate that the area was potentially prolific, and there was little hesitation in investing capital in Iran, Iraq, Kuwait and Saudi Arabia, because of the virtual certainty of discovering new oil resources and finding outlets in world markets. The rush to invest was not limited to the development of producing facilities, but was accompanied by expansion of the area's refineries and, more significantly, by arrangements for more effective transport facilities for crude oil to the markets of western Europe. Investment in the latter respect included expenditure on the Suez Canal to ensure its usefulness for larger tankers which were now being built as a result of war-time design developments, and also on the construction of pipelines from the producing areas of Iraq and the Gulf to Mediterranean ports in the Lebanon and Palestine – pipelines which, in the period before the advent of large tankers, reduced the cost of getting the oil to western Europe and hence improved the profitability of the companies concerned. Thus, in the Middle East as a whole, production increased from under 50 million tons in 1945 to almost 100 million tons by 1950 and 250 million tons ten years later. At the same time the declared reserves discovered by the drill rose from the estimated 3000 million tons in 1945 to nearly ten times this level by 1960. Moreover, it is well known that these declarations of reserves by the companies concerned represented only a percentage of the total volumes discovered, for some of the companies were, and, indeed still are, under no obligation to reveal the extent of the resources they have discovered. From providing only about 10 per cent of the world's total oil production in 1945, the Middle East's importance increased steadily throughout the 1940 and 1950s; by 1960, it was responsible for no less than 25 per cent of the world's total. This breathtaking expansion of the physical facilities to produce and transport Middle East oil fell almost entirely within the framework of the operations of the seven major international oil companies which, before 1939, had constituted an international cartel and which in the post-war period 'understood' each other so well that they could be relied upon not to take decisions which might operate against the collective interests of the group.

The Middle East thus achieved a significant status in the rapidly growing international oil industry. The economies of western European countries and of many countries in other parts of the world became dependent upon Middle East oil exports. The investments by the companies grew by hundreds of millions of dollars every year, but this increasing involvement in the Middle East, on the part both of consuming nations and of international companies, occurred at a time when changing circumstances were bringing to an end the political stability of the inter-war period. The defeat of the French in 1940 lost them their influence and prestige, so that Syria and Lebanon quickly claimed their independence in the immediate post-war period. In the rest of the region there was significant Anglo-American pressure to keep the area under political control. As oil became more important, ran the British and American argument, so greater control became more necessary. But control could not always be maintained. Certain territories declared notional independence to be effective independence. In Iran, this declaration took the form of the nationalization of the Anglo-Iranian Oil Company's facilities. The international oil industry – wholeheartedly backed by Washington and London – retaliated by isolating Iranian oil from world markets and for three years little crude oil from Iran's producing fields and few products from its massive refinery at Abadan moved on to world markets. The cessation of Iran's exports did not cause the world supply difficulties that had been anticipated, for they were quickly made up by additional production and exports from neighbouring countries whose oil development thus received a considerable stimulus. Iran was thus eventually obliged to accept a compromise whereby the Anglo-Iranian Oil Company, later renamed British Petroleum, along with American and other interests, were allowed back to work the prolific Iranian fields on behalf of the National Iranian Oil Company. The formula appeared to concede the national ownership of resources that Iran claimed, but, in reality, the work of the Consortium at first differed little in essentials from the earlier situation.

## The Western World's Strategic Interests

Anglo-American efforts to secure a degree of direct or indirect political control over Middle East countries were only partly motivated by their desire to ensure the 'rights' of their companies to explore for and develop oil. In part they were dictated by the constant fear that the Middle East might become increasingly susceptible to external political intervention by other nations. It was thought that the USSR had not only a political, but also an economic, interest in securing control over the Middle East oil producing countries. Observers noted the rapidly increasing demand for energy within the Soviet Union itself and assumed that this would require more energy than the Soviet Union could produce within its own boundaries. From this assumption observers predicted that the USSR would seek access to Middle East oil not on commercial terms through purchases from the producing companies, but within the framework of a political attempt to capture the growing nationalist movement in these countries. Given the international situation at that time, the coupling of a Soviet interest in Middle East oil with the more general view of Moscow's expansionist tendencies led to a real fear of the Soviet Union's intentions in the region. Contemplating this 'threat', the US and UK governments forgot their earlier differences on the sharing of the Middle East's oil wealth and, emphasizing the need to present a united front in the area to meet the external threat, they attempted to sell to the Middle East countries the idea of the peripheral anti-Soviet alliance, similar to that recently developed in western Europe through NATO. It was from these efforts that the intergovernmental organisation of CENTO emerged. Amongst other countries in the region one of the most important oil producing countries of the Middle East – Iran – decided to participate in this body. Behind the screen thus provided by Turkey, Iran, and Pakistan, Anglo-American influence in the Middle East was felt to be more secure. Both diplomats and oil investors breathed a little more freely once the treaty had been signed.

In retrospect, one can see that this concern about Soviet intentions in the Middle East was exaggerated, at least on the oil side. The USSR proved to have more than enough oil and gas within its own borders to satisfy its own requirements. Indeed, the rapid development of the Soviet oil industry not only enabled the country to change from a coal-based to an oil- and gas-based economy, but also, by the mid-1950s, to become a potential rival to the Middle East in terms of oil exports. Middle East oil now faced a new competitor, particularly in western Europe, where the Soviet oil export agency, in order to break into the markets controlled in the main by the companies producing oil in the Middle East, cut prices where necessary. The USSR also sought oil outlets in the developing countries such as Brazil, Guinea, India, Ceylon, etc. The steady growth of both Soviet producing and oil export potential quickly dispelled the idea that the USSR needed the Middle East to provide fuel and power for its economic development. Its recent decision to purchase large quantities of natural gas from Iran confirms this view, for the gas can be used in the Soviet Union to replace domestic oil, which will then be available for export, thus giving the USSR a net gain in hard currencies.

Defence of Anglo-American oil interests in the Middle East,

however, was not restricted to warding off the 'threat' from the USSR. Elsewhere in the world, countries and companies looked enviously at the oil wealth of the region and viewed with disfavour the control exercised by a limited number of Anglo-American companies which, although producing oil cheaply, were able to sell it expensively because of their control over transport, refining, and marketing facilities. As a reaction against this, French, Italian, Japanese and other outside interests attempted to secure influence in the Middle East. Companies from these countries, both state and private, started to tender for the right to explore for oil in new concession areas offered by the Middle East governments. The degree of profit-sharing, state participation and guarantees of various other kinds the newly-interested companies were prepared to concede, was greater than that made by the international companies which had not bargained for being outbid by these outsiders. The formers' defence against the latter rested not only on the technical and commercial expertise of the Anglo-American companies; it was also backed by diplomatic and political pressure by the American and British governments. In certain instances the pressure was successful, but the growing hostility of some of the nationalist groups in the Middle East (encouraged by diplomatic counter-offensives by the French, the Japanese and the Italians), coupled with the greater commercial and economic attractiveness of the newcomers' proposals for oil acreage, eventually secured rights in the area for oil interests other than those of the United States and the United Kingdom.

## Nationalism Undermines the Traditional Oil Concession System

The first move in this direction emerged out of Iran's expropriation of the Anglo-Iranian Oil Company's interests in 1951. The 1954 agreement negotiated with the Consortium limited the latter's right to produce and explore for oil to those areas in which some work had already been done. The rest of the country's oil areas was vested in the National Iranian Oil Company, which since 1957 has signed contracts with several other foreign companies and other (sometimes state) entities for exploration and development work. In particular, the Italian state oil organisation, ENI, has secured acreage in areas formerly conceded to Anglo-Iranian. More recently, Kuwait has insisted that the Kuwait Oil Company (owned 50 per cent by BP and 50 per cent by Gulf Oil) give up its 'rights' in almost half the country. The acreage relinquished is now being allocated to other companies. Most recently, Iraq has taken the most extreme action in this respect by unilaterally declaring that the rights of the Iraq Petroleum Company – which had the whole of the country under concession – extend only to the areas it actually worked. As these amount to only about one per cent of the country's area, the remaining 99 per cent of Iraq is now 'on offer' to other parties. And in spite of the IPC's claim that the Iraqi government's action is illegal and that no other party has any right to work oil in the country, Iraq has not been unsuccessful in interesting other potential oil seekers. Though it is reported that strong diplomatic pressure from Britain and the US (whose oil companies are dominant in IPC) secured the withdrawal of an offer from ENI to develop acreage in Iraq, this has not prevented a French state oil company from concluding an arrangement to develop some of Iraq's oil resources hitherto controlled by IPC. Most recently, even Saudi Arabia ended the monopoly rights of the original concession holder – ARAMCO – and has accepted tenders for oil exploration and development from other companies. Thus by 1968 – in spite of the considerable commercial power of the traditional companies and the diplomatic pressure of their governments – the former simple pattern of oil concession arrangements in the Middle East, essentially countrywide arrangements within the hands of eight companies working either alone or, more frequently, in alliance with each other, had been replaced by a complex jig-saw pattern in which over 50 companies and institutions are involved.

These new concessions are important not only because they have brought in companies and interests other than British and American (even India now has an interest in Iran), but also because the Middle East governments have taken the opportunity, when agreeing to new concession arrangements, to insist on some degree of national control over the eventual exploitation of any resources discovered. Typically, the new-style agreements involve initial obligations by the companies concerned to undertake a rapid exploration of their areas, and to move equally rapidly into developing any fields discovered. The state then claims a share in the development on payment only of an appropriate share of the development costs. Governments thus secure the right not only to share in the profits, but also to participate in deciding how the field shall be operated: not an unimportant right in respect of decisions on the speed at which production shall be built up and off-take increased – for an international company can restrict development in order to protect its investments elsewhere. Through this approach to direct participation in oil ventures, state investment becomes virtually riskless, for it makes no contribution to the risky exploration expenditure. The new-style agreements often also insist that the companies agree to build a local refinery to process part of any crude oil eventually produced; and they also require that the national tanker fleet (the development of which is another recent development by some of the producing countries) be given first option on the transportation of some of the oil. All this, of course, adds up to much more than merely sharing the profits of an operation run entirely by the oil company. Instead it opens the possibility of achieving significantly enhanced economic status for the countries concerned, not only vis-à-vis the companies, but also in relation to the outside powers to which the companies owe their allegiance.

To date, however, the increasing number of concession arrangements and the establishment of new terms for newly negotiated agreements have been less significant than the changes which the Middle East governments have been able to secure in the terms of the old concessions, from which some 75 per cent of Middle East oil is still produced. The pre-1950 concession arrangements essentially reflected the semi-colonial status of the area. Supported by their governments, the companies secured concessions which offered a royalty of modest size on any oil that might eventually be produced, and received in return the exclusive right to search for oil – usually over an area covering the whole of the national territory concerned.

By 1950 the international oil companies elsewhere in the world had already accepted a changed status as a result of nationalist pressure. In 1938, Shell and Esso interests in Mexico had been nationalized because the companies had been unwilling to renegotiate their concession agreements on terms acceptable to the government. In 1943, the Venezuelan government passed new oil legislation that marked a turning point in government-company relationships. It introduced the concept of profit-sharing between government and companies, and of company participation in the domestic economy over and above that demanded by the mere production of petroleum. Given political change and increasing awareness in the Middle East in the immediate post-war period of the unfairness of the concessions granted, it was only a matter of time before similar nationalist reaction to the companies began to make themselves felt. Nationalization in Iran in 1951 was the most obvious manifestation of this, but the crunch effectively came a year earlier in Saudi Arabia which by then had become a very successful producing area with an even greater potential remaining to be developed.

Given the low-cost nature of Saudi Arabia's production – due not only to favourable geological conditions, but also to the proximity of the fields to the coast and hence to ocean transportation facilities – and the value of this crude to the American companies concerned for serving their rapidly growing markets of the Eastern hemisphere, the joint concessionary company, ARAMCO, was not unwilling to accede to the government's demand that the recently introduced 'posted price' for Saudi Arabian oil should be used for calculating total profits ( = posted price per barrel multiplied by number of barrels produced minus the costs of producing the oil and getting it to the export terminal). Total profits thus determined would then be divided equally between company and government. In this way Saudi Arabia achieved a profit-sharing status similar to that won by Venezuela some years earlier, and in addition established the principle of their 50:50 split between government and company. Once ARAMCO had accepted this new form of relationship with Saudi Arabia, there was nothing that oil companies elsewhere in the Middle East could do to prevent their own concession terms being renegotiated. They accepted the inevitable and sought only delay. Delays proved to be very short so that by 1954 all major oil producing countries in the area were on a 50:50 profit-sharing basis.

## Changing State/Company Relationships

Although profits to the companies remained high under the new formula – by virtue of low production costs and high posted prices – the effect of the agreements was to reduce significantly the gross returns made by the oil companies in their producing operations. This fundamental change in government/company relationships might not, however, have been achieved so easily had it not been for decisions by the United States – and later Britain – to allow their oil companies to offset all tax payments to the governments of the producing countries against their tax obligations to the US and British treasuries. Thus, the net additional tax burden on the oil companies was greatly reduced and led to a situation in which it was, in effect, the taxpayers of the US and Britain who suffered the main consequences of the new financial arrangements between producing countries and companies.

Since the Middle East countries took this action to secure a greater and more direct interest in oil operations in their territories, government/company relationships have become a more or less continuing dialogue in which first one government and then another has sought to improve the terms of the oil concession arrangements. The manoeuvres have been mainly concerned with ways in which pricing and profit-sharing should be carried out.

Profit-sharing based on posted prices had guaranteed Middle East countries rising revenues from oil whilst the demand for oil continued to rise and whilst posted prices remained high. After 1957, however, the world energy situation changed, bringing increasing competition between energy sources. In their search for markets, oil companies started to cut prices. At first, price cutting was achieved by shaving profit margins, but soon, companies producing oil in the Middle East started to offer crude to independent and state-owned refineries at less than the officially posted prices. As this practice grew, it produced an incentive for the companies to bring down the posted prices for oil in order to reduce their tax obligations to host governments. The companies assumed they had an unfettered right to change prices, as they had done for a quarter of a century in the US and the Caribbean, and indeed in the Middle East from time to time in the 1950s. Until then, however, the price changes had been essentially upwards in sympathy with changes in the western hemisphere. The companies were thus surprised when small reductions in posted prices in 1958 and 1959 produced some indications of disapproval from certain Middle East governments. It was not, however, until they announced major cuts in prices – of the order of 10 to 15 per cent – in 1960 that a storm broke over the heads of the companies whose decisions threatened to reduce the oil revenues of the countries concerned by 5 to 7.5 percent. The companies, at first, insisted on their right to take such action when necessitated by commercial consideration, but they were eventually obliged to reappraise their position under the combined onslaught of all the oil producing countries, where the strident nationalist reaction threatened not only the interests of the companies, but also the standing of Britain and the US in the difficult post-Suez period.

The posted price issue turned out, however, to have much broader ramifications. Indeed, it was a sufficient spur to the producing countries for them to act collectively by forming the Organization of Petroleum Exporting Countries. One of its aims was to secure the cancellation the price reductions. Though this was not achieved, OPEC's stand made it quite clear to the companies that future unilateral action in this direction was impossible. Since 1960, therefore, the use of posted prices for calculating the companies' notional profits has been sacrosanct. Thus, governments have retained their revenues per barrel in spite of a continuing fall in the actual price levels at which crude oil has been sold. Thus the division of profits between Middle East governments and companies has moved from the agreed 50:50 basis towards an effective 70:30 split in favour of the former. This development has effectively removed most of the 'fat' from oil company profits, which on a world-wide basis no longer provide a return on capital out of line with that in other industries.

## The Impact of the New Relationships

Thus in 1968 the way in which the oil industry is organized in the Middle East has little in common, politically or economically, with the position up to the early years after the second World War. Vestiges of economic and political colonialism still remain in the traditional concession arrangements with international companies, but the days of these seem to be numbered as they are certain to be converted to more acceptable forms of agreement within the next few years. Change has been achieved more as a result of the constant pressures brought to bear on the companies by the increasing number of Arab oil professionals in government service, than from the political calls for the nationalization of oil resources emerging from time to time from the Arab League and other organizations. The most recent outstanding example of 'professional pressure' is seen in the continuing negotiations between the Iranian government and the Iranian oil consortium whereby the former is attempting to secure the right to a voice in decisions on the future planning of the latter's business – and, in particular, on the volume of future production levels. Hitherto, such decisions have been a function of secret agreements between the member companies of the consortium – in spite of the fact that such decisions are the main determinants of Iranian government revenues from oil operations and thus crucial to the economic planning of the country. The pressure for change will almost certainly succeed and few companies now operating in the Middle East expect anything other than moves in this direction in the future. They will not like the changes, but they will acquiesce in them not only because the alternative is expropriation – enthusiasm for this can always be whipped up at short notice in most countries – but also because they recognize that such changes do, in fact, make commercial sense by guaranteeing the continuity of oil operations that would otherwise be interrupted by occasional, but recurrent, political crises, as in the last 15 years. The participation by governments in decision making will reduce their propensity to use oil as a political weapon: hence the flow is likely to be smoother and more regular, thus ensuring continuing returns on company investment, albeit at a lower rate of profit per barrel of oil produced.

## External Moves to curb the Oil Producer's Power

If the changing relationship between governments and companies succeeds, as suggested, in diminishing the significance of Middle East oil as a direct political weapon, then one of the main potentials for creating instability in the region will have been removed. In the 1956 crisis, the dangers in the situation were exacerbated by the knowledge of all parties that a decision to close down oil production and oil transport facilities would affect the consuming countries, owning countries, and international companies more seriously than it would the producing countries. But with the changes in the organization and control of the industry, this interpretation no longer holds good, for the producing countries are now moving into a position in which their own benefits from oil far exceed those of any other party, and in which their immediate and longer-term economic (and political) viability depends increasingly on keeping their oil flowing.

Consumers of Middle East oil, on the other hand, decreasingly depend on keeping this particular oil flowing, for they have not been content merely to await the next unilateral decision by the 'oil sheikhs' to turn off their essential energy requirements, but have sought means of reducing their dependence on Middle East oil. In 1959 the United States introduced mandatory quotas on crude oil imports, restricting them to some 12 per cent of the country's total needs. Part of the reason for this decision was concern for security of supplies from the Middle East whose crude oils were moving to the US in increasing quantities by the late 1950s. In Japan – currently more dependent for energy on Middle East oil than any other major industrial nation – companies are being persuaded to develop oil resources elsewhere, notably in the Far East and especially in Indonesia, whose oil potential has remained undeveloped as a result of post-war difficulties with the previous concession holders, *viz*. some of the international oil companies. Japan in also looking to the possibility of investing in oil developments in Australia and Alaska, and is also about to look for oil in its own backyard – the Sea of Japan. Even countries in the developing world have advanced the insecurity-of-oil-supplies-from-the-Middle-East argument as one reason for spending scarce domestic capital on initiating an oil search at home.

In the meantime, the countries of west Europe, which collectively still take the bulk of Middle East oil and which at the time of Suez depended on it for 75 per cent of their total supplies, have noted the dangers to their economic progress of too rapid a change to imported oil as the basis of their energy economies. Without the fear of interruptions to these supplies, most countries in west Europe would have had fewer inhibitions in pursuing cheap energy policies based on imported Middle East oil. Instead, indigenous coal has been given a greater degree of protection, whilst alternative supplies of imported energy – including coal from the US and oil from the USSR – have not been unwelcome as a means of reducing the risks. European attitudes towards the dangers of excessive dependence on Middle East oil have also helped to accelerate interest in the continent's own natural gas potential. The discovery of resources of gas under the North Sea – indicated by the earlier discovery of the world's largest gasfield at Groningen in The Netherlands – was quickly followed by an international agreement on its exploitation by all the nations bordering on it. As a result very large quantities of natural gas have already been located in the British section and the search is under way in Norwegian, Danish, German and Dutch waters. The great potential of North Sea gas (plus possibilities of similar developments in the Adriatic and Baltic Seas), coupled with the 120 million tons of oil equivalent per year from the Groningen field, seem likely to eliminate much of the annual rate of growth in the market for oil in west Europe within the next five to seven years.

These reactions by major consumers of Middle East oil to the threats and realities of supply dislocations arising from political action have, moreover, been accompanied by action on the part of the oil companies themselves to limit their own degree of dependence on the Middle East. Iranian nationalization of the Abadan refinery in 1951 finally persuaded the companies that refineries were best located away from the Middle East. There were already increasingly significant economic reasons for doing this, and nationalization clearly demonstrated that capital in refineries should not be put at risk in the Middle East, when the refineries could be located in consuming countries. No major refinery has since been built in the Middle East producing countries. The Suez crisis of 1956–7 had a serious effect on the international oil business and the companies concerned were stimulated to further diversify their areas of producing, particularly into those areas, such as Nigeria and North Africa, which could serve the west European market without depending on the Suez Canal. Simultaneously, the tanker situation was reappraised and decisions taken to enlarge company fleets – particularly in respect of tankers of 80,000 tons upwards, which can use the Cape route to Europe without much of a cost disadvantage over the Suez Canal route. It is already apparent that the closure of the Canal since the Arab-Israel war has acted as a final consideration in persuading the companies to switch their crude oil transport to Europe to 250,000 to 300,000 ton tankers, which cannot use the Suez route even in ballast. These tankers – the first orders for which have already been placed – will eventually realign a significant part of the main flow of oil from the Middle East to Europe (and America) around the Cape of Good Hope to the detriment not only of Egypt, but also of

Lebanon, Jordan, and Syria, all of which secure revenues from the movement of oil by pipelines to the eastern Mediterranean. The giant tankers will also undercut the cost of moving oil by these pipelines.

## Reactions by the Exporting Countries

These developments have not gone unnoticed in the Middle East, even though in public and the media they receive little attention compared with references to the industrialized world's dependence on Middle East oil. The fact that realists are now largely in control, however, is demonstrated by the relatively limited support, in terms of oil sanctions, given to the Arab cause in the recent crisis. It is surely not coincidental that only two producers – Libya and Algeria, less sophisticated in the oil business than the longer-term producers of the Middle East proper – persisted for more than a few days in boycotting certain markets. For Iran, the dislocations of the crisis presented an opportunity for expanding production immediately and for pursuing with renewed vigour its claim to participation in decisions on future levels of oil off-take by the Consortium. For Saudi Arabia, Kuwait and Iraq, the choice was more difficult, but in 1967, in contrast to the destruction of pipelines that took place in 1956, nothing was permitted to occur which would have meant a long-term interruption of supplies. The boycott was restricted to certain destinations and/or certain ships (depending on nationality). Nor do any serious steps appear to have been taken to ensure that the companies did not get away merely with observing the letter, but not the spirit of the boycott decisions. The companies reorganized their supply arrangements so that all bans on particular oils to particular destinations could be observed, but switched tankers and cargoes so that no nation went short of oil. In other words, the Middle East oil producing nations finally made up their minds that, in both the short and the longer term, only they would suffer from 'holding the West to ransom'.

Since the war, however, at the series of Arab conferences held to determine future policy, a new philosophy towards the use of oil as a means of achieving desired ends has begun to emerge. In brief, the idea is to make deliberate use of revenues from exported oil to assist in the achievement of Arab political aims. In this, some support can be expected from the Soviet Union which, as a first step, has already agreed to offer finance and technical help in increasing Iraq's oil export potential. The scope for such an idea is clear and it has, of course, emerged not only from the growth in total exports, but also from the growth in per barrel revenue that has been secured, from 30 cents in

1946 to 80 cents in 1966. In 1966 Middle East countries (including Libya and Algeria but excluding Iran) exported about 3500 million barrels of oil and thus earned over $2500 million in government revenues. The monthly payments of $5 million that are being made to Egypt while the Canal is closed and to Jordan for an indefinite period by the major Arab oil producing nations is an indication of the possibilities which exist in using oil revenues for political ends – but one notes that these payments amount to only about 3 per cent of total monthly oil revenues received by the Arab states. Whether or not there is a reasonable chance that oil revenues will be used to achieve both political and economic aims in the regional context (and it should also be remembered that Kuwait is also using oil revenues to finance a Development Bank which underwrites economic development projects in the Arab world), rather than in a series of separate national contexts as in the past twenty years, is beyond the competence of an observer of the oil scene to judge. However, if the chance is taken, then it will be somewhat ironic to find that the commodity which originally helped to divide the region into the spheres of influence of different outside powers, and which later emphasized national rivalries within the region and created 'have' and 'have not' nations in the Arab world, is now providing the means whereby greater regional cohesion can be achieved – albeit through the use of an unfriendly, but commercially necessary, outside world. In aiming to sell as much oil as possible, under arrangements which gives the overwhelming part of the profits to the producing nations, through international companies which started off as 'agents of imperialism' and which will soon be little more than brokers and transportation agents matching up supply with demand, the Middle East oil producing nations, carrying their less fortunate neighbours along with them, would seem intent on guaranteeing the future availability of all the oil that can be sold out of the area (in competition with increasingly significant developments elsewhere). In return, the economic and political advantages that the oil revenues can buy could produce an Arab Middle East with much greater internal cohesion and a more significant power potential *vis a vis* the world's other power blocs in the latter part of the twentieth century.

## Bibliography

There are a large number of books and other publications on oil in the Middle East. The following selection provides more detailed description and analyses of points made in this essay.

H. Cotton, *The Evolution of Oil Concessions in the Middle East and North Africa*,
      New York, 1967.

D. Hirst, *Oil and Public Opinion in the Middle East*, London, 1966.

C. Issawi and M. Yeganeh, *The Economics of Middle Eastern Oil*, London, 1962.

W. A. Leeman, *The Price of Middle Eastern Oil*, Cornell University Press, 1962.

G. Lenczowski, *Oil and State in the Middle East*, Cornell University Press, 1960.

S. H. Longrigg, *Oil in the Middle East*, Oxford University Press, 3rd ed., 1967.

H. Lubell, *Middle East Oil Crises and Western Europe's Energy Supplies*, Baltimore,
      1963.

M. A. Mughraby, *Permanent Sovereignty over Oil Resources*, Beirut, 1966.

P. R. Odell, *An Economic Geography of Oil*, London, 1963.

E. Penrose, *The Large International Firm in Developing Countries: a Case Study of
      the International Petroleum Industry*, London, 1968.

B. Shwadran, *The Middle East, Oil and the Great Powers*, New York, 1959.

# Chapter II – 3

# International Oil and National Interests in Latin America*

## 1. Introduction

The fact that this seminar is sponsored and arranged by OPEC perhaps indicates that the participants will, in the main, be more familiar with the problems of oil company/state relationships in oil producing and exporting countries than with the problems which have emerged for countries whose main interest in oil lies in its significance as an import required to fuel the economy. Countries in this latter category, however, far outnumber those in the first and may basically be divided into two groups:

> *i) the countries of industrialised western Europe and Japan whose attitudes to oil and its relations with other sources of energy have been, or will be, discussed in other sessions of this seminar and with whose problems OPEC and OPEC member countries are familiar through their continuing studies.*

> *ii) the countries of the developing world which, as far as I could judge from the contents of the programme, are not scheduled to have their oil policies discussed elsewhere in this meeting. In the absence of speakers or representatives from*

---

* **Reprinted from the Proceedings of the OPEC Seminar on** *International Oil and Energy Policies in Producing and Consuming Countries*, **Vienna, 1970, pp.131–140.**

*important or potentially important oil consuming nations like
Brazil, India or many others in Latin America, Asia and
Africa, I would, therefore, appear to have a formidable
responsibility in speaking of their problems and attitudes
towards oil. I hope that I can adequately discharge this
responsibility in view of the importance of the issues not only
to the member countries of OPEC, but also to the consuming
countries themselves.*

These countries have interests in common arising from their
under-development and low living standards, whilst participating in the
economic system of the non-communist world – although often only
marginally so, through the export of one or more primary commodities
to the industrialised nations in return for which foreign manufactured
goods are imported for use by the minority of their populations in the
market for such things. At the same time, many – or even most – of the
people of these countries pursue their own subsistence, or largely
subsistence, way of life, largely unaffected by such external trading
relationships in light of their very rudimentary wants.

There are a very large number of such countries – over 100 in
Africa, Latin America and Asia and, from the point of view of our
deliberations, it is very significant that all except six of them use oil as
their most important energy source – excluding that collected or
gathered within the framework of a subsistence way of life. Their
position in this respect differs markedly for that of the industrialised
nations where coal has, in most cases, been the main source of energy
and where it is only in quite recent years that there has been a significant
switch to oil. And in that almost all the developing countries are now
pursuing policies designed to secure more rapid economic development,
they are using energy in much greater quantities and thus facing the need
to increase their oil supplies very quickly. Therefore, they often have
problems connected with oil which are even more critical than those
faced by the countries of western Europe and Japan. Many of them thus
have to give a great deal of attention to their relationships with the world
of oil.

Incidentally my characterisation of these countries matches, in
some respects, features of many of the important oil producing nations
but in that these latter countries have unlimited quantities of oil and gas
available for their own use and, through its production and export,
generally enjoy favourable balance of trade positions and buoyant
government revenues, they thus have advantages not enjoyed by the
countries we are concerned with here. The importing countries look

with some degree of envy on the good fortune of the oil producing and exporting nations and have even sometimes argued that they are the means, in part at least, whereby the oil producing countries have secured their wealth! To the governments of many under-developed oil consuming nations it has often appeared that oil prices internationally have been so arranged as to transfer income from themselves to the wealthier oil producing countries! They are thus also characterised by a desire to change this situation – particularly in light of the increasing gap in income per capita that has opened up between most of the developing oil producing nations, on the one hand, and the developing oil consuming nations on the other; and also because of the belief that, given the continuation of the old trading patterns in oil, the discrepancies will continue to grow.

To return, however, to the characterisation of the developing, oil consuming nations. Their initial steps along the pathway towards economic development often produce energy bottlenecks. As societies emerge from a largely immobile, subsistence economy into one in which transport, both of goods and people, becomes more significant; in which urbanisation becomes a marked phenomenon; and in which industries are established to make use of the country's resources or to substitute imports, so there emerges a situation in which the demand for energy starts to grow quickly. Rates of increase in annual demand for energy of the order of 15 to 20% can be expected.

This process of a rapidly increasing energy use in developing countries is, of course, not a new phenomenon. The developing countries of the 19th century – Britain, Germany, the United States etc. – went through much the same process. But for these countries the development of their coalfields provided the energy basis for their economic growth and the growing economic strength of particular nations and regions was closely associated with the local availability of this energy resource. Even into the post World War II period it was still fairly central to ideas on economic development that countries or areas wishing to follow in the economic footsteps of Europe or of North America would have to rely similarly on the expansion of indigenous coal resources. However, partly as a result of wartime dislocation in coal supplies and subsequent significant increases in its price level, and partly from the increasing ability and willingness of oil companies to serve any markets in the world, it was gradually accepted that economic development in general, and industrialisation in particular, could be based on oil, rather than on coal. The resultant increasingly widespread use of oil as the basic fuel for such processes was assisted, moreover, both by the inherent physical characteristics of oil and by the organisation of the industry internationally.

As a liquid – easy and generally safe to handle, even with only limited technical resources and expertise – oil could be relatively cheaply shipped around the world in small, as well as in large, quantities and could thus be made available at all points at which energy was required in the development process. Moreover, the relatively unsophisticated and divisible nature of oil transport and use ensures continued advantage for oil over more recent competitors – notably natural gas, whose distribution and use requires an infra-structure only viable economically with large scale demands, and atomic power whose technological and capital requirements are beyond the capabilities of most developing countries.

Moreover, the international organisation of the oil industry produced a situation in which the opportunities created by the technical suitability of oil as a source of energy in the developing countries could be exploited. Ready availability of capital within the framework of the international oil groups, no matter where the opportunity for investment arose, meant that local financing difficulties did not stand in the way of providing energy. Thus, wherever they have been permitted to operate, the international oil companies, with personnel used to and willing to work anywhere, have both sought and created opportunities for selling oil products by establishing their terminals, depots and other infra-structural requirements for their operations. It is difficult to find a case anywhere in the developing world where (politics permitting) the international companies have failed to respond to an energy 'gap', or to produce a profit out of their activities.

Thus, the developing world has become increasingly dependent on supplies of oil for its economic advance. Looking, for example, at Latin America as a whole one finds that between 1939 and 1966 the total demand for energy increased by over 3 times – from under 70 to over 210 mmtce. Over this period the use of coal increased only from just under 10 million tons to a little over 11 – its share thus falling from some 15% of total energy to only 5%. The bulk of the increase in the total energy demand was provided by an increase in the supply of oil – from under 50 to 150 mmtce. As previously indicated, in the whole of the developing world only six countries – Taiwan, India, South Korea, Turkey, Zambia and Mozambique – still consume more coal than oil. This is clear evidence of the way in which oil has become the dominant energy source in the developing countries, almost all of which have missed out the coal-based stage of economic growth which Western

Europe, North America and Japan experienced. Thus, energy in the developing world increasingly means oil and is likely so to remain for the foreseeable future.

But this fact of economic life for the developing world has given rise to characteristic problems and reaction – with political and economic aspects often difficult to disentangle. These arise out of the conflict between the increasingly intense nationalism of the developing countries, on the one hand, and the foreign control exercised over their oil sectors, on the other. In the main, it is the international oil companies that have been responsible for the establishment of the complex international infra-structure whereby the oil is produced, refined, delivered, and marketed in the right quantities and qualities to markets in the developing countries. They have almost invariably achieved reliable operations. Indeed, in many developing countries the local affiliates of the major international oil companies present an enclave of efficiency in a generally less than efficient public and private administration. For example, their employees are usually smartly turned out; their service stations and depots never seem to run out of the required products; their transport is clean and well-maintained; they often produce the only easily available maps of a country; and they sometimes finance or sponsor facilities ranging from technical training centres to young people's sports competitions. But, ironically, this essentially 'American' image which they present to the societies in which they operate is often a contributory cause of difficulties for them. For them to be so wealthy, to afford to be so efficient and to be such 'good citizens' must mean, in the view of some of the local populations, that they are 'exploiting' the local economy and that the poverty of many is a function of the wealth of the oil companies! Thus, oil companies in developing oil consuming countries become the targets of attack as the 'agents' of neo-colonialism – the force whereby nominally independent nations remain subjected to the control of the metropolitan powers through economic considerations.

Aside from the visual manifestations of this phenomenon – as mentioned above – the most general neo-colonialist charge levelled against the oil companies is concerned with their pricing policies in selling overseas-produced oil to the consuming nations. In essence, the countries concerned object to the international oil posted price system. They see this as a device used by the oil companies to keep prices above what is considered a reasonable level by eliminating competition. The system in effect conjures up visions of an 'International Oil Cartel' making monopoly profits out of the poor, oil importing nations. Though the developing nations are not the only ones to have considered themselves unjustly treated in this way, the problem for them has been greater, not only because the system has lasted longer and worked more effectively in their cases, but also because oil for them is generally relatively more important in their economies and has a larger proportionate effect on their balance of payments. The oil exporting nations which in the last decade have insisted on the validity of the posted price system, ought perhaps to be somewhat thankful that it is the international oil companies that have had to bear the brunt of the attacks on the system by their customers in the developing world! But, in this respect, as well as in others, one notes how the international companies have served to keep producing countries at arms length from consuming countries − even in the case of Venezuela and the rest of Latin America to whose experiences in particular we now turn, having set the general framework for our investigation of oil-short, developing nations.

## 2. Latin American Experience

In many ways Latin American countries have led the way for the developing world in tackling the problems associated with a growing need for oil in a situation in which its supply was largely controlled by a small group of international companies. Certainly, in most Latin American countries, oil has long been viewed as one of the 'commanding heights' of the national economy: its importance in the energy sector gives it the same significance that coal had in countries of western Europe and where, in many cases, such industries were brought under national control. But oil in Latin America has had another attribute which increasingly made it even more susceptible to political involvement, *viz.* the fact that foreign companies, almost all American, were centrally involved in the development of the industry. It was this consideration which made the oil industry subject to critical attention from a wide spectrum of political forces in Latin America for it thus became involved in the major issue of nationalism − and its associated phenomenon of anti-Americanism. This nationalistic reaction against foreign control of a major sector of the economy has also had the effect of making oil a foreign policy issue for many Latin American countries − as in 1938 in the case of the Mexican expropriation of oil companies' interests in the country and, most recently, in the case of Peru's nationalisation of Esso's local subsidiary, the International Petroleum Company. Thus, throughout almost the whole continent there is an area of broad general agreement between parties and groups deeply divided on other issues, *viz.* that oil should be brought under effective national control. Disagreements on this arise over 'means' rather than 'ends'. An examination of the various 'means' employed in pursuing such policies illustrate clearly the ways in which relationships between the international oil industry and national interests have developed in Latin America.

## 3. Nationalisation, State Monopoly and State Control

### a. Mexico

Resolution of the conflict between the interests of the international companies and national interests has taken the extreme form of total expropriation in two cases – Mexico in 1938 and Cuba in 1960. In the case of Mexico, expropriation marked the culmination of over 15 years of difficulties and friction between the companies and successive governments following the Mexican revolution in 1910. In this period the companies were less practised in the art of achieving a *modus vivendi* with a system of government that they did not care for, and expropriation followed their unwillingness to accept the implications of the Mexican revolution and to renegotiate their concessionary arrangements and their largely extra-territorial and statutory positions. However, the reasons for Mexico's action need not detain us for long – the conditions and attitudes under which it took place are no longer very relevant. What is worth evaluating, however, is the technical and economic performance of PEMEX which for over 30 years has had responsibility as a state oil entity for all aspects of oil operations from exploration through to final sales in a rapidly developing and industrialising economy largely dependent on oil in its energy sector.

PEMEX's ability to attract foreign private loans to finance its expansion is often used as evidence of the organisation's success. But in accepting this as a valid criteria for judgement, one should also note the length of time that PEMEX has taken to achieve 'financial respectability'. Prior to 1961 it was able to borrow only on a maximum 3 year basis and at unfavourable rates of interest. Its A1 credit rating and its ability to negotiate long-term loans at favourable rates of interest is a recent development. Moreover, success in this direction has, in part, been a function of PEMEX's recent ability to generate investment funds out of sales income – and this has been achieved as it obtained the right to charge reasonable prices for its products and thus the right to act much more as a commercial enterprise, rather than as an instrument of state!

PEMEX has also successfully supplied the country's energy requirements in a period of 30 years of rapid economic growth. But one should note that it had a favourable start in this respect as it was, for 10 years, able to supply the expanding domestic sector with the oil that had previously been exported by the expropriated companies. It was only in 1947 that domestic demand finally rose to the level of Mexico's productive capacity in 1937 – the last year before nationalisation. By 1947 PEMEX was in a position to be able to undertake an appropriate exploration and development programme. Since then, with the help more recently of contractual arrangements with American and other companies, it has had the logistical problems of meeting domestic needs well under control, with the construction of facilities as and when required. The charge sometimes made that PEMEX 'let the country down' as far as its energy requirements were concerned appears to lack validity. In fact, in as far as one can evaluate the evidence available, it seems possible that through PEMEX the country's energy requirements have been met as effectively and at lower direct cost to the consumer than through any other alternative. This conclusion seems to be the only one possible for it is difficult to visualise private foreign companies selling oil products in Mexico at the prices at which PEMEX has had to sell; or meeting the specially low price requirements of public transport operators in Mexico City; or accepting the social obligations to sell oil products at all and any locations in the country at the government controlled prices. PEMEX has not necessarily liked having to do these things – but it had no option. Foreign private companies would have had more freedom to argue for commercially acceptable pricing policies and could probably have persuaded the government to accept their views – particularly as they would have had great bargaining power in light of the significance of petroleum in the country's energy economy.

In one other important respect there is a large question mark over the role of PEMEX in assisting Mexico's economic development. This is in respect of attitudes, abilities and policies towards exports. Nationalisation eliminated exports overnight – because the expropriated companies were able to exclude them from virtually all world markets, either by their direct control over outlets or by threat of legal action. Even after these problems were cleared, PEMEX was not technically in a position to take production to a level at which exports were possible, given the top priority of satisfying domestic markets. By the post-war period Mexico's ability to produce oil barely exceeded domestic demand and thus only small exports were possible – in spite of the existence of a 'world-wide sellers' market which could have produced significant foreign exchange earnings had oil been available. For this period one can certainly argue that had the international companies remained in Mexico they would have created an oil potential from which exports would have earned many millions of dollars a year. Some of the massive investment which the companies put into Venezuela would have gone to Mexico, with a view to achieving exports of both crude and products. Later in the post-war period, PEMEX succeeded in increasing reserves and production capacity to levels which could have made crude oil available over and above domestic requirements: but now Mexico chose to forgo the opportunity to sell such oil abroad on the grounds that crude oil is an inappropriate export commodity as it can be converted into refined products which are 'worth' several times as much and the manufacture of which produces jobs and income within the Mexican economy. But this doctrine of the 'inherent worth' ignored the reality of the world market situation. There were very few possible outlets for Mexican oil products and so upwards of 5 million tons of oil a year – which could have been exported as crude oil – stayed in the ground bringing no benefits to the Mexican economy, so limiting the ability of PEMEX to earn the revenues it required for expansion and necessitating a greater reliance on foreign borrowing. Herein lies one nationalist reaction to the international petroleum system which appears unworthy of emulation by other developing countries.

## b. Cuba

Cuba's nationalisation of the local oil industry was part and parcel of that country's move from the American to the Soviet system. It goes without saying, therefore, that the results of the expropriation have completely severed Cuba's connections with the international oil industry as represented both by companies and governments. Soviet supplies from over 6000 miles away have now kept the Cuban economy going for almost a decade and though there is some evidence that the Soviet Union would not be unwilling to negotiate an exchange agreement with one of the 'majors', so that Venezuelan oil could again meet Cuba's requirements, such action could only be contemplated as part of a political solution to the problem of relationships between Cuba, the United States and other countries in the Caribbean. The issue is no longer an oil industry problem.

Apart from Cuba and Mexico, no other Latin American country has resolved the conflict of interests between national aspirations and the activities of the international oil companies by total nationalisation – though several have moved a long way in this direction. Indeed, Bolivia, did nationalise existing company interests a year of so earlier than Mexico, following national suspicions that the companies were working against Bolivian interests in its territorial disputes with neighbouring countries. New legislation in 1964, however, permitted foreign companies to return to Bolivia and one, Gulf Oil, has since successfully developed fields which are now producing oil for export. Even there, however, Y.P.F.B. – the state oil entity – has had its monopoly over domestic oil interests completely maintained.

Within the framework of the general acceptance of the idea of the desirability of state intervention in the oil sector of the economy, three other countries – Chile, Brazil and Argentina – demonstrate varied responses to the problems involved.

## c. Chile

Chilean intervention goes back as far as 1927 when Shell and Esso sought concessions to develop potential oil areas in Tierra del Fuego. Chile chose not to grant the concessions on the grounds that they would infringe Chilean sovereignty. Instead, Congress passed a monopoly law which had the effect of confining almost all oil industry activities to the state. Since then a series of legislative measures have confirmed the exploration, production and refining functions of the industry in state hands – a position which has never been seriously challenged in political terms. To a large extent Chile has thus divorced itself from the activities of the international oil industry and sought self-sufficiency in oil through national efforts – a policy which is justified on the following grounds.

Chile is a country poor in oil opportunities and any production can only be for local consumption. If this were to be developed by the international companies their costs per barrel of output would almost certainly be higher than in their main producing areas such that they would be inclined to import, rather than to produce locally. By way of contrast, Chile's state oil monopoly has as its first consideration the overall effects of production or imports on the Chilean economy so that, even if it could import somewhat more cheaply, local production could still be justified because the saving in foreign exchange would mean its availability for furthering Chilean industrialisation and modernisation policies. Herein lies the essence of the nationalistic argument against allowing the international companies to dominate the oil industries of economically weak and developing countries. Paradoxically, it has had little to do with securing oil products at prices lower than those which would necessarily apply if the international companies were responsible for the country's oil supply. E.N.A.P. – the state oil entity – in fact, makes use of the Caribbean posted prices system for valuing and selling its own products: and, incidentally, thereby ensuring its own profitability which is also helped by the inclusion in the price of a national tariff component. But without local production it is highly unlikely that tariffs would have been charged on imported products and thus prices for products in Chile are well above the levels at which they could have been imported, even at posted prices let alone at the discounted prices at which they could have been obtained over the last decade or so. Further costs of the 'Chileanisation' of oil also arise from the state-intervened marketing system which has been evolved in order to curb the earlier dominance of Shell and Esso. A local Chilean marketing company – COPEC – was formed with government assistance and was guaranteed 50% of the market – the rest being left to Shell and Esso. Though the prices of products have been controlled by successive governments, the system could still have been costly for the economy, in that it eliminates competition and thus allows the possibility of excessive distribution and retailing margins. Overall, then, Chile's comprehensive attempt to secure its independence from the international oil industry has been successful in political terms – and appears to give a high degree of national satisfaction in the country. In economic terms, however, there is room for differences of opinion on the effects of the policy.

## d. Brazil

Brazil's policy has evolved rather differently – and in one main respect has been centrally concerned with the costs of oil. This is in connection with the foreign exchange costs of imports but, before looking at that aspect, we should also note that, as in Chile, exploration and production were designated as, and remain, a state monopoly. Petrobras, the state oil entity, has been at work since 1950 but for much of the period to date has been vexed with politically orientated control and government imposed policies. This ensured high costs and a relative lack of success in its activities such that, until recently, its output of Brazilian oil met less than one-third of the rapidly rising national requirements. Here again it has been argued that had the international oil companies been allowed to search for oil in Brazil then they might have been more successful. On the other hand, they might have preferred only to 'pretend' to search for oil in Brazil knowing that greater profits could be earned by continuing to sell oil from their low cost fields elsewhere in the world!

Petrobras' release, since 1964, from involvement in political in-fighting and its designation as a technical entity with a clear-cut responsibility for making Brazil as self-sufficient as possible in oil production, has radically changed the situation as far as oil production is concerned and it is now covering about half national requirements. Though this still begs the questions of: 'at what cost?' and 'how does this compare with the cost of alternative supplies from abroad?' these are, in fact, barely relevant for a country with a more or less permanent balance of payments crisis and thus permanently short of sufficient foreign exchange with which to import goods required in the industrialisation process. The benefits of substituting domestic crude oil for imported supplies in terms of the foreign exchange that can thus be made available for other requirements are considered to outweigh any considerations of comparative costs of domestic and imported oil. This attitude appears unlikely to change.

It has been concern for the balance of payments, added to elements of nationalism, which have motivated Brazil in its attitude to oil imports (necessary because of a short-fall in domestic production) and oil refining. Until 1950 the latter was in private hands – nominally owned by independent companies, but which, in practice, were 'sponsored' by some of the international oil companies which saw them, therefore, as 'integrated' outlets for their crude. In 1950 refining was brought into the state sector – for political and nationalistic reasons – and since then most refinery development has been the responsibility of Petrobras, though permission has been given from time to time for the private refineries to be expanded.

Once it was in business as a major refiner, Petrobras quickly appreciated the opportunities which existed for it as a third-party buyer of international crude and, in a global situation of 'surplus' oil production, it started to negotiate its purchases at discounted prices. Its negotiating abilities steadily improved and in 1966 (the last year before the Suez War upset transport costs) it paid an average of only $1.96 per barrel – an apparent discount of 25% off the formal delivered prices using posted prices and assessed freight rates. Even by 1964 the contrast between Petrobras' purchase prices of imported crude oil and those of the private refining companies had become so significant (over 12%) that the Brazilian government eliminated the right of these private companies to buy crude oil abroad from their international associates which were, of course, charging prices as near as possible to usual transfer prices in the vertically integrated global system. Instead, Petrobras was given the responsibility for obtaining the crude oil requirements of these private refineries and in the 3 years from 1963 to 1966 brought down the average delivered price of their crude from $2.48 to $1.95 per barrel – a reduction of more than 25%. In small part this price reduction was a function of economies of scale, as oil for the private refineries was moved with the greater quantities required for the state refineries: but in larger part, it was a function of the divorce of the prices from the international system of the majors.

In total, Petrobras' activities as the monopsonistic purchaser of crude oil (and certain other products) for Brazil have achieved a foreign exchange saving of the order of $40 to 50 million per year (equal to over 3% of Brazil's annual availability of foreign exchange). Though the foreign exchange component in the costs of building the refineries in the first place must be offset against these savings, one can see that Brazil's independence of the integrated international oil system has not been an insignificant advantage. One also notes that Brazil is not interested in the slightest in the alleged validity of posted prices as the 'fair' prices for oil: so much so in fact that Brazil has not hesitated to stop buying oil from neighbouring Venezuela. In 1962 Venezuelan oil accounted for 70% of Brazil's imports: it is now only about 20% and its share is still falling. Companies selling oil out of Venezuela have found it virtually impossible to secure Brazilian contracts in a situation in which the discount off posted prices they were able to quote, in part as a result of pressures from the Venezuelan government that prices should be kept up, failed to match discounts from other supply sources – even after the higher transport costs from more distant supply points had been taken into account! Brazil sees its national interest over oil as differing markedly from the interests of both the international companies and the oil exporting nations – and its attitude is not untypical of much thought – and associated actions – in the world of developing, oil importing nations.

## e. Argentina

Up to ten years ago Argentina's attitudes and policies towards oil were similar to that described above for Brazil. In fact, Brazil's approach owed much to the earlier experiences of Argentina which first established a state oil entity as long ago as 1910 and in 1927 formed a national petroleum authority – Y.P.F. – which was given a monopoly over all future petroleum exploitation and production and some responsibilities for refining and marketing. But at that time, Shell and Esso – already operating oil fields in Argentina – were allowed to continue their operations and they, and other private companies, still ran the bulk of other oil activities. Thus, Argentina was very far from being entirely divorced from the international oil system and as Y.P.F. was not very successful in developing Argentina's known oil reserves, Argentina gradually became an important market for imported crude oil and products. These, in general, were allowed to enter the country at the transfer prices of the international oil system and as these prices were used as a basis for determining retail prices and for the valuation of Y.P.F.'s oil production, Argentina remained strongly connected to and dependent on the international system. One result of this policy was a very high foreign exchange cost of oil – by 1956 oil accounted for 25% of total exchange availability.

This high and unexpected cost of nationalism towards the oil sector emerged, in part, from the fact that known domestic resources remained in the ground because Y.P.F. was unable to get them out (owing to lack of capital resources and technical expertise); and, in part, because no foreign companies were permitted to help! In 1957, therefore, this intensely politically nationalistic policy was replaced by one intended to be economically nationalistic. Argentina's oil resources were opened up for exploitation by overseas companies which were guaranteed a market for every barrel of oil which they produced at prices based on the existing landed prices of foreign oil. Needless to say, the policy was technically successful for, within three years, production was expanded threefold and Argentina became self-sufficient in oil. However, the flow of profits generated for the producing companies appear to have cost Argentina almost as much foreign exchange as oil imports had cost it previously! This, coupled with political responses to the new situation, helped to cause the over-throw of the government of President Frondizi. The new government declared the oil contracts be made invalid – and Y.P.F. regained its monopoly position. Even more recently, with another change of government, the policy has again been reversed. Now, under a new petroleum law, foreign companies are being invited to take concessions in oil potential areas, whilst Y.P.F. concentrates its activities on the production of oil from the existing fields. The results so far of this two year-old policy suggest that Argentina will again secure self-sufficiency in crude oil. The government is also encouraging the expansion of domestic refining – both state-owned and private – to ensure that refinery capacity keeps up with rising demand for oil products so as to eliminate imports.

In the Argentine response to the 'oil problem' we see something of the dilemma into which many developing countries get themselves over this commodity. They dislike the international oil companies on principle – but in trying to manage without them as far as the exploitation of domestic reserves is concerned, they find themselves increasingly dependent upon them for imports out of the international system! But this they cannot stand for balance of payments reasons and thus have to seek help in order to curb the foreign exchange cost of oil imports. Paradoxically for the major oil producing and exporting countries, it is the 'rejection' of the international companies by the developing oil importing nations that is more to their benefit than the alternative – 'rejection' has created the need for greater imports from the oil exporting countries.

## 4. External Effects of Latin American Oil Attitude and Policies

The time allocated to this paper has been sufficient to examine only five Latin American countries' oil policies. The ones chosen, however, do illustrate attitudes and policies concerning oil which are representative of the whole of the continent. They also demonstrate how attitudes and policies towards oil will develop in most of the rest of the developing world as countries elsewhere start to accumulate the necessary expertise and experience, as independent nations, for dealing with the international oil industry – though one notes in this respect that no Latin American country has so far taken action similar to that of India in its search for oil independence – *viz.* direct investment in one of the world's major producing nations. As indicated below, however, this could be a solution to Venezuela's problems if other Latin American nations could be persuaded, and were allowed to, invest in oil exploration and development there!

To conclude I would like to indicate in brief the external implications of the evolving relationships between the international oil industry and the national interests of Latin American countries. These might serve as the basis for discussion of this theme.

> *a) National insistence on domestic production to achieve as large a degree of self-sufficiency as possible has had the effect of isolating Latin America from the international industry in terms of market openings for crude oil from the major producing nations. Such policies seem certain to persist and are likely to be increasingly successful, particularly as almost all Latin American countries have favourable territories for oil exploration and as the incremental markets for oil are likely to be limited by the growth of natural gas use – once the intensity of energy demand justifies the investment in gas transmission facilities.*

> *b) The insistence on domestic refining is also going to continue such that markets for internationally traded products will be*

available only on a short-term basis. This is particularly significant for the Venezuelan refining industry.

c) The insistence on lowest possible prices for any oil imports which may be necessary will be accentuated. This makes any important flows of oil to Latin America within the framework of the internationally posted price system virtually impossible. Latin American countries view the posted price system as a device for overcharging them for their oil and it cannot be envisaged that they will come to accept it as a 'fair and equitable' system. In this respect note how Venezuela's insistence on the maintenance of price levels has already resulted in its virtual exclusion from the main Latin American markets.

d) As a result of the combined effects of points a) to c) above, almost the whole of Latin America has become a government-controlled market for oil. This implies a limited degree of freedom for the international oil companies in their efforts to work within the continent. They will be increasingly hard put to resolve the problems arising from conflicting government pressures – arising, on the one hand, from governments of the producing countries; and, on the other, from governments of the developing, oil importing nations.

e) It remains an open question as to the degree and speed with which this situation will be developed further into a Latin American regional policy rather than a series of national policies for oil. The moves so far towards regionalisation in Latin American oil, through ARPEL – Asistencia Reciprocal Petrolera Estatal Latino Americano – have been political only and have had no effect in economic or even technical terms. ARPEL is a body for talking, not for action. But, out of the talk, there may well eventually emerge positive and practical proposals for the regionalisation of Latin American oil divorced entirely from the international system.

f) In any such development, the role of Venezuela is crucial. As it loses its international oil role, is its policy going to change fundamentally to embrace the idea of Venezuelan oil fulfilling *a Latin American role? The development certainly cannot be envisaged in the short term because the 'sacrifices' it would involve, both by Venezuela and by the Latin American oil importing countries, are too great. For Venezuela, its oil would have to be sold at lower prices to its Latin American partners: for the partners it would mean foregoing access to even cheaper Middle Eastern and African oil. But sacrifices in the interest of the longer term good of all parties are not unknown and one can visualise such developments in oil as part of a wider political settlement in Latin America. This might include reciprocal investments by Latin American countries in each other's potential resources – for Venezuela this would mean investments in Latin American oil developments.*

*g) Overall, therefore, one sees the Latin American way of ensuring that national interests in oil are fully met as implying the region's continuing withdrawal from the international oil system; the continued functioning and well-being of which is presumably the main motivation for the Organisation of Petroleum Exporting Countries. Much diplomacy and a greater degree of mutual understanding is going to be required to ensure that this clash of interests between the major oil exporting and the poor oil importing countries does not become an irreversible process.*

## Bibliography

This, unfortunately, was not published in the Proceedings. The original manuscript is no longer available.

# Chapter II – 4

# The First International Oil Price Shock

## A. Against an Oil Cartel*

## Introduction

"The battle for oil in Europe is all over bar the shouting." This was the view I expressed less than two years ago – in my book, *Oil and World Power* – in an evaluation of the success of west European oil consumers in getting their supplies at prices well below those of the United States, whose oil prices had traditionally determined those everywhere else. This had been achieved, moreover, in spite of the tremendous growth in oil demand – as oil had been substituted for coal – and in the face of periodic political and/or military crises affecting the Middle East, from which most of Europe's supplies originated. Europe's success arose not only out of division between the oil-producing countries and the oil-production companies, but also out of disputations within these two groups.

## The Oil Companies

For the companies the disputes emerged out of their growing number. Until the middle 1950s the only companies of any importance were the so-called international majors – for example, Esso, Shell and BP. There were seven of these firms and they controlled well over 80 per

---

* **Reprinted from *New Society*, No.437, February 1971, pp.230–232.**

cent of world trade in oil. Their influence extended throughout the non-communist world. They understood each other's ways; talked the same language; had been through the same baptism of post-war fire when they had been establishing new-style relationships with politically more independent producing countries; and had fought a joint battle over their oil pricing policies in Europe against the administrators of the Marshall Plan and other American government agencies.

They had, moreover, run a successful oil cartel in the years before 1939 and had not forgotten the advantages they gained from such a form of organization, even though they now had to respect the reinforced American anti-trust legislation, whose control extended to wherever American companies operated. They thus hankered after "orderly" marketing arrangements – that is, arrangements which would have guaranteed their profits. But, from the mid-1950s onwards, they were effectively inhibited from making such arrangements by their need to defend their interests against the increasingly important activities of additional companies that were then entering the world oil scene.

These newcomers were American domestic oil companies (like Philips, Amoco, Conoco and Murphy) which had ventured outside the protected world of North America for the first time. They were all now aggressively seeking to justify their geographically wider set of activities. They faced difficulties early on, because their original motivation to seek oil abroad – namely, to find lower-cost oil to substitute for the higher-cost oil they had to buy or produce in the United States – had been undermined by the introduction, in 1959, of American quota barriers against foreign oil. The quotas kept out most of the foreign oil they were now capable of marketing.

In order to achieve revenues which would reimburse them for their considerable exploration and development expenditure, they had to seek markets elsewhere. Deep price-cutting secured them the markets they sought – principally in western Europe. These "new" companies paid much the same revenues to the governments of the producing countries as the international majors did. So their rapid growth endeared them to the producing countries because the latter's total revenues from oil got an unexpected boost.

Fighting this new competition kept the majors busy. They saw no point in any fundamental bargaining with the producing countries over the organization of the industry to the mutual benefit of both parties. Instead, they concentrated their efforts on keeping to a minimum the extra revenue they were now under pressure to pay to the governments.

Over the years, however, the new companies grew to a size at which their combined contribution to the total world oil industry made them a force to be accommodated. They, in their turn, came to learn the type of behaviour, over pricing policies and negotiating attitudes, expected in the world of oil outside their small backyards in Texas, Indiana, California or wherever. They "earned" their inclusion in a possible united front which all the companies could present, to both the producing and the consuming nations, in order to stop the nonsense of increasing taxes in the producer countries and decreasing sales prices in the consumer ones.

## The Oil Exporting Countries and OPEC

By the late 1960s, the companies were prepared to discuss world oil arrangements with the main producing countries. But the producer governments were most certainly not ready for discussions. Most of them – excluding only Nigeria, whose oil fortunes had been decimated by the civil war – had spent nearly ten long, hard years setting up the organization of Petroleum Exporting Countries (OPEC) and trying to learn something about the world of oil away from the producing wells and the crude oil export terminals. They were in no mood even to concede the idea of a possible deal with the oil companies. They viewed the companies as "enemies" in spite of the wealth the ever-growing rate of oil activities of the companies had generated for them.

As political and strategic problems – like the renewal of Arab-Israeli fighting and the second closure of the Suez Canal – started to limit the amount of oil that could be delivered to the main markets of western Europe, so they sensed that unilateral action rather than collusion with the companies, would be a better short-term bet for improving their share of the profits to be made from oil. By last year, an aggressive new government was firmly in power in Libya, trans-Arabian pipeline closures were adding to the transport problems over Suez, and there was an unexpectedly rapid increase in the demand for oil. The OPEC countries were then in a position to strike hard for their short-term aim of higher revenue per barrel of oil exported.

## Company/Government Negotiations

They have taken advantage of this favourable situation. In the last year they have successfully renegotiated the terms of the oil concessions with the companies – picking them off one by one, in much the same way that the companies used to do with them. For a time, the companies were not particularly unhappy. They successfully called "crisis", and "crisis" again, and put up the delivered prices for their oil in western Europe by a percentage several times higher than was justified by the increased taxes in the producing countries and the higher costs they were having to pay for their ocean transportation. The oil companies' financial results in 1970 reflected this basic improvement in their position.

Thus, both the producing countries and the companies which produced, transported and distributed the bulk of the world's oil, outside the communist bloc and north America, had now achieved their short-term aim of higher revenue and improved profits respectively. The world oil power game could turn to maintaining this over the longer term. The producer countries concerned co-ordinated their strategies at the OPEC meeting in Caracas in December.

In the meantime, the dozen or so American-based international oil companies decided to seek permission from the Nixon administration for collective action. This meant removing the anti-trust restraints on such action. (Previous anti-trust legislation had broken up the oil empire of Rockefeller early in the century and the oil cartel of the 1930s.) This time, US administrative decisions, as opposed to Congressional decisions, enabled the companies to work together. They, too, plus the few non-American companies involved – including Shell and BP, could now produce a co-ordinated strategy for consolidating their higher level of profits.

## ...and Collusion...

The pattern of development both parties were now consciously, or accidentally, aiming for was one similar to that which has long obtained in the United States. This is a system of collusion between the main producing states and the producing companies, whereby the two parties effectively hold the consumers to ransom. In the United States, the ransom is an estimated minimum of $350 million per year. The immense sum paid by American consumers in higher than justified oil prices seems likely to pale into insignificance besides the prospects that now emerge for both oil-producing countries and the oil companies to earn monopoly profits out of the world oil market. The apparent "failure" of negotiations between the two parties last week in Teheran and the subsequent unilateral decision of the producing countries to increase the posted price of oil, seems unlikely to worry the companies very much for, acting together, they will obviously pass on the increases – and more – to their customers, and so enhance their own profitability.

The reported likely increase in the price of oil per barrel is equal to under one penny per gallon: but the talk is of 3d per gallon increase in the oil company price to customers

In economic terms, collusion would mean a redistribution of the wealth earned from oil production and use towards the shareholders of a few large (mainly American) companies and towards the national exchequers of even fewer oil exporting countries. Neither element in the shift seems justified.

## ...for Mutual Advantage

The companies already rival many countries in the size of the resources they control. Any increase here is inherently "unfair" and possibly dangerous, unless good evidence is produced to show otherwise. Many of these companies already have more than enough resources to finance their oil activities, so they are busily diversifying into activities like building chains of motels across Europe (Esso) and producing commodities like salt and aluminium (Shell) – to say nothing of their rising dominance in the world chemical industry. Any further increase in the flow of cash they achieve from oil seems likely to produce their faster growth into industrial conglomerates with influence in many different sectors of many different national economies.

On the other hand, most of the oil producing countries do not stand out as poor nations whose wealth almost everyone would wish to see increased. Some of the largest oil producers – like Libya – have such small populations, in fact, that their per capita incomes – mainly arising from the oil industry's activities – exceed those of the United States or Sweden. All other members of OPEC – except Indonesia, the smallest oil producer by far – have levels of income which very comfortably exceed those enjoyed in the rest of the developing world: and, as we shall see later, the rest of the developing world is likely to be seriously affected by the collusion now being planned in the oil industry. Only Iran and Venezuela, apart from Indonesia, really need more capital for development than is now available to them from oil revenues. But their problems could be dealt with apart from the bloc of oil-producing countries.

Realistically, the idea of "need" as a determinant of the way the world's wealth should be distributed is of zero significance. Neither the oil-producing countries nor the companies are likely to restrain their efforts to improve their positions because they do not "need" the money. We require appropriate counter-action to make sure that the collusion between producer companies and producer countries does not work.

## Consumers' Reactions

It is very odd that the governments of the major oil-consuming nations of the industrialised world should apparently – through discussions in OECD – have "approved" (in effect) the steps that are being taken to establish an oil cartel, whose first result must be higher oil prices for consumers. The consumer governments have shown a very singular lack of responsibility to those sectors of their national economies which will not enjoy any benefits from the cartelization of the oil industry. And this means all sectors, except the oil industry itself and possible shipping.

The only explanation for such irrational behaviour lies in the ability of the sheer weight of the propaganda of the oil companies to bamboozle people who should know better. The propaganda line of the oil companies has been that only their collective action will "save" the consumer from the perverse behaviour of the "baddies" in the oil-producing world – notably Libya, which is bearing the brunt of the propaganda output in this respect.

In brief, we have a ludicrous situation in which the most sophisticated oil-consuming countries have not only willingly put their heads through the noose that has been carefully prepared for them, but before doing so have thankfully patted one of the hangmen on the back for saving them from the death from oil starvation that he persuaded them they really faced. And all this in a world so full of oil-producing potential as to make the development into a very sick joke.

European countries' support for the companies appears to stem from their being at a complete loss as to what else they should do. They have permitted their indigenous coal industries to be all but killed off by oil, and have even allowed their indigenous oil and gas reserves to fall largely under the control of the same companies who will now find it in their best interests to get their increasing requirements of oil from the producing countries with which they are in collusion. Moreover, governments in Europe seem to be more worried about conserving their oil and gas resources already discovered for hypothetical use well into the 21st century than they are about finding ways in which indigenous output can be maximised in the next critical five years.

## Higher prices from the Cartel

In the short term, Europe seems ready to accept the higher prices arising out of the establishment of an international cartel and not ready to accept the idea of constraints on the use of oil through, if necessary, a system of rationing and the elimination of non-essential uses to conserve accumulated stocks as long as possible. Why the strategy of a consumers' strike should be so readily discounted is not apparent. Nonsense among the producing nations and companies – for automatic adjustments of oil price levels to take account of inflation – should surely be rejected, particularly by a government like the present British one which is resisting the idea that wage rates among its own people should rise automatically with inflation.

Even an acceptance of the inevitability of the oil cartel prices in the short term is no excuse for inaction in the longer term. We require a fundamental reappraisal of the decision, taken almost a decade ago, to make Europe a more or less open energy economy. Neither the United States nor the Soviet Union accepts other than a highly protected energy sector. Western Europe would now seem to have just cause for following suit, particularly in view of the oil and gas finds that are being made across the continent.

## European Policy Requirements

Such a change in policy would be most quickly marked by a decision to maintain the coal industry at least at its present level. More important than this is the need for positive action to make Europe self-sufficient in oil and gas – both through the rapid exploitation of indigenous resources and by direct European state or private involvement in oil and gas resource developments in areas (like Canada or Australia), where development will remain free of the restraints introduced on development elsewhere by the collusion between traditional producing countries and the international oil industry.

Is it beyond the vision and capability of Europe's statesman to encourage, sponsor and exhort their capital-owing institutions and individuals to create a European oil and gas industry entirely independent of the embryonic oil cartel? Where companies involved in this cartel already have interests in European oil and gas production, then the decisions about how fast these resources are exploited must lie with an authority other than the companies themselves. The latter will restrain the rates of development of these resources by their obligations to governments elsewhere.

Other companies not involved in the cartel should be given preferential tax treatment in their efforts to develop Europe's own resources. Consideration must be given to diverting more state investments into such activities (say, out of the funds set aside for building roads as these will not be much use in the absence of adequate oil supplies).

Finally, how about an unorthodox state-sponsored or supported attempt to mobilise individually small amounts of risk capital from millions of Europeans for investment in major oil and gas development project in the North Sea and elsewhere? It seems not beyond the bounds of possibility that at least ten million Europeans would each take between £10 and £100 out of their traditional savings for such a high risk, but exciting, investment – if the opportunity were to be persuasively presented to them. The resultant multi-hundred million pound fund – for investing quickly and exclusively in the development of Europe's oil and gas – would buy the most expert advice and advanced equipment available and make a generation's difference to the build-up of production, and considerably enhance Europe's possibilities of achieving self-sufficiency.

My arguments have been almost exclusively concerned here with the relationships of the oil producers (countries and companies) to the oil consumers in western Europe. These consumers and Japan stand quantitatively most exposed to the dangers of an oil producer cartel. But, for some Europeans, there will also be a moral issue involved.

## Impact on the Third World

Europe's politico-economic power will be mobilised to win it cheap energy. Is this not somewhat unfair on the rest of the generally poorer world? As I have already indicated, not too many tears ought to be shed for most of the major oil-producing nations. They have been doing very nicely and, with the establishment of an oil cartel, they will do even better. But the few countries in this group are outnumbered by the really poor countries whose economies depend on imported oil – and who cannot afford to pay more for it.

These oil-consuming nations of Asia, Africa and Latin America have interests which are not going to be served in any way whatsoever by the emergence of an international oil cartel. Many of them will thus further strengthen their own resolve to take action to secure their independence of the world oil industry run by uncaring outsiders.

As Europe moves towards its effective reaction against the cartel, might there not be time and energy enough to spare for advice and help to be given to these other energy-importing nations whose undeveloped economies depend on the continued flow of oil at prices which do not reflect the impact of collusion between oil supplying nations and a cartel of international oil companies.

## B. The World of Oil Power in 1975*

### Introduction

In a situation in which the world political and economic crisis over oil is still deepening, it would be normal to expect governmental and media activity which sought to inform us about the reality and complexity of the issues involved and which aimed to prepare us for the consequences of the worsening crisis. Instead, we are beguiled by the results of wishful thinking in the Western world. In this there are two main components: first, the exaggeration of incidents which purport to indicate the imminence of the collapse of OPEC, and, secondly, the idea that the problem is amenable to solution through 'co-operation'. As a result, the Western world is having its power in the international economic and political system gradually whittled away and is choosing to take no effective action to stop this happening.

The following elements appear to be behind the emerging pattern of events and developments, *viz.* solidarity within OPEC; wishful thinking and disarray in the West; the watching and waiting game of the Soviet Union; changing views in the Third World consequent upon the revolution in the world of oil power; and, finally, a new international outlook arising from the interaction of these elements.

### The Solidarity of OPEC

Over the last few months the media have had a field-day in reporting the unimportant details of what is happening in the OPEC world, whilst at the same time ignoring – or, at best, misunderstanding – the fundamentals of that organization's strategy. Thus, Abu Dhabi's reduction of its oil prices by a few cents a barrel to eliminate the disadvantage of its high-sulphur crude compared with oil from other OPEC member countries, has been magnified in headlines as heralding the break-up of the oil cartel. Similarly, the fact that a few heads of state of OPEC countries did not attend the organization's summit conference in Algeria last March is cited as evidence of internal dissension strong enough to break the organization. But these minor events stand in marked contrast to the continued success of OPEC in fulfilling its aims and objectives to a higher degree than its leaders could ever have imagined. For, in spite of the reduced demand for oil in the importing

* Reprinted from *The World Today*, vol.31, No.7, July 1975, pp.273–282.

world (mainly as a result, incidentally, of industrial depression and warm winter weather, rather than of deliberate conservation), prices are being kept up, so that the revenues earned by the producing nations are now higher per barrel of crude produced than they were a year ago, or even at the beginning of 1975. Though prices may not have kept pace with inflation, they have even gone some distance towards this objective – particularly as OPEC countries are getting their own imports at lower real prices than a year ago, given the intense competition for their business between Western suppliers. Oil production is being held without much difficulty to the diminished demand level, in spite of the increasing potential to produce oil, which, at a marginal cost of a few US cents a barrel, would make handsome profits for the producing nations which went out to seek special low-price sales to importing nationals. And whilst Abu Dhabi and Libya may express their concern at the high degree to which their production has fallen away (and, in doing so, receive the effective support of other OPEC members), other producing countries are still cutting back on production levels. Venezuela will only produce 130 million tons of oil in 1975 and 120 million tons next year, compared with 150 million tons in 1974 and a peak annual production of 170 million tons. Kuwait is enforcing yet another cutback, a policy decision which will be made easier by the final nationalization of BP and Gulf Oil. And even Saudi Arabia has announced a lower production target.

There can thus be no doubt now that the OPEC countries have found the key to their prosperity and development potential; no doubt that they wish to continue to use the strategy which has brought them this success; and almost no doubt that they are developing the information system and the institutions which will enable them to implement the strategy as a permanent feature of their relations with the rest of the world. If there remains any doubt about their success, their motivations, and their continuing intentions, then it centres solely on the role of Saudi Arabia whose wish to retain the friendship of the United States relates to the 'anti-Communist' views of its government. But even these views are always tempered by the higher priority which Saudi Arabia gives to a solution of the Jewish problem, particularly the future of Jerusalem. Whilst this problem exists – and who would be optimistic enough to see it disappearing in the foreseeable future – Saudi Arabia is going to go along with OPEC and its strategy, to the cost of the Western world.

## Illusions and Disarray in the West

The solidarity of OPEC is thus marked by firm views on price maintenance and on production controls whereby the power of the cartel can be maintained. We might also bear in mind that the OPEC countries have, over the last eighteen months, further consolidated their power in the international oil system by taking over from the oil companies the ownership of the producing facilities and the oil reserves which the companies developed through their exploration efforts over the last 25 years. Whilst this might have been expected in the case of OPEC member countries with radical governments, such as Libya and Iraq, it is much more surprising as far as other member countries of the organization are concerned. The oil companies in Venezuela, for example, have survived 25 years of threats in this direction and now most of their concessions have only less than 10 years to run. Nevertheless, it is the relatively mild left-wing government of President Carlos Andres Perez which proposes that the oil industry be nationalized this year. In the Middle East, the negotiations in Saudi Arabia for the take-over of ARAMCO, the consortium of American companies which has hitherto exploited that country's massive reserves, are almost complete and the United States appears to have resigned itself to securing little more than a slight preferential treatment for its purchases of Saudi Arabian oil and the continuation of American management of the oil installations. And even Kuwait has announced the imminent nationalization of the joint BP/Gulf Oil company responsible for most oil production in the country.

Such a wholesale take-over of Western world assets – in return for compensation which is usually little more than nominal – when added to OPEC's strength in terms of the control over the supply and the price of oil, reflects a revolutionary change in the West's relationships with its raw material suppliers. Yet we have persisted in the same sort of wishful thinking that has marked our entire reaction to the oil crisis. The Western line is still essentially that 'co-operation' will guarantee the oil we need and ensure that the producers do the right things in their industrialization, monetary, and general international policies. But co-operation, as a meaningful concept for regulating relations between sovereign states, implies, firstly, a willingness of both sides to co-operate, and, secondly, the possession by each side of sufficient power to 'force' the other to co-operate. At present, the West is willing to co-operate because the oil producers have the control and ownership of a resource which is required for the survival of its socio-economic system. But the OPEC world lacks any motivation to cooperate because it is not obliged to seek formal co-operation with the West. It is sometimes argued that OPEC countries *must* have our capital and consumer goods, our expertise and know-how, and that because of this they have no alternative but to co-operate. This is wishful thinking with a vengeance, for two reasons: firstly, most OPEC countries can, in the final analysis, survive without the goods and services which the West has to offer, because the great majority of the people in these countries has not yet come to participate in the revolution of rising expectations – and what has not yet been enjoyed can be foregone, without such sacrifices leading to immediate dangers of social and political unrest; secondly, formal co-operation with the West is not needed by the OPEC countries because they can get the goods and services they want quite easily and cheaply by having Western countries compete against each other for the business. The Western world in turn enhances this ability of the OPEC countries by playing the competitive game, in the absence of any effective common front in its relationships with these wealthy buyers. If individual Western countries are prepared aggressively to seek markets in OPEC countries without any concern to ensure that their willingness to supply goods and services is related to OPEC changing its supply and pricing strategy over oil, then the oil producers simply have no motivation to co-operate and can act without restraint in devising their oil policy.

Viewed from the OPEC side, the West is in disarray over its response to the challenge of the new world of oil power. Compare the essential solidarity of the OPEC summit meeting in Algeria (at which even long-standing and deep-seated problems such as the frontier dispute between two member countries, Iraq and Iran, were solved within the unifying atmosphere of the successful oil cartel), with the continued dispute in the OECD over Western attitudes towards the oil problem. Or compare the willingness of the OPEC countries to put the essential needs of their joint oil strategy before individual national interests, with the unilateral decisions of Western countries to give highest priority to domestic problems (of unemployment and of reflation especially), irrespective of the effect which such policies have upon the demand for oil and hence upon the West's relationship with OPEC.

In such a situation, the proposed (but not yet agreed) meeting of oil-producing and oil-consuming nations seems unlikely to yield more than further appeasement by the West in response to the enhanced demand by OPEC for more rights in the international system. This is due to the West's continuing failure to recognise the need for a fundamental restructuring of both national and international economic policies, as a prelude to regaining an ability to constrain OPEC and so 'save' the Western system. Further delay in this respect makes the ultimate solution not only more difficult, but also one which is increasingly likely to be found only in the ultimate alternative to appeasement, i.e. armed conflict.

## The Soviet Watching and Waiting Game

In its attitude and reactions to the oil crisis since 1973 the Soviet Union has come to resemble the sleeping giant. She has indulged in very little by way of 'huffing and puffing' while relationships between the oil producers and the Western industrial nations changed radically; and, unusually for the USSR's normally brisk commercial reactions with respect to oil, there was a delay of well over a year before its oil-export prices for its East European consumers were adjusted even part of the way towards the quintupled price level of international oil.

One senses here an inability on the part of the Soviet Union to make up its mind as to its best strategic response. It was probably hoping that the industrialised nations would not be too badly affected by the oil crisis and so took no action to exacerbate the situation, knowing that its continued development depends on the prosperity of the industrial world, especially Western Europe. This may still be interpreted as its preferred option – as reflected in the continued lack of its intervention on the international scene. However, the Soviet Union must be becoming increasingly sceptical about the ability and willingness of the West to take the counteraction against OPEC which is required to safeguard continued economic development and progress in Western Europe, Japan, etc. Thus it could well be preparing a second-line strategic option, designed to ensure that its own interests are not too adversely affected by a radical change in the economic order of the Western world. And in so far as this second-line option implies an aggressive policy against the unity of Western Europe, it may well have devotees in the Soviet Union who would wish it to be made operational without delay – and ultimately incorporated into the détente with the United States.

The Soviet Union needs to ensure a supply of capital and consumer goods for its development.[1] A depressed – and hence a politically and socially unstable – Western Europe would no longer offer such a guaranteed supply. The Soviet response to this possibility could well be to seek a complete and wholehearted rapprochement with West Germany, based on an offer of unconditional reunification linked to the

Soviet Union's agreement to provide a guaranteed market for all the capital and consumer goods which Germany would like to sell. In a situation in which West German industry and commerce was depressed because of the lack of sufficient demand in the oil crisis-affected Western world, this package could be very attractive indeed. It would, of course, be in exchange for the 'Finlandization' of the reunited Germany, with all that this implies for the Western alliance. In addition, the Soviet Union could and would – for the short to medium term – offer Germany guaranteed supplies of oil and natural gas to substitute for the latter's dependence on OPEC.

In this context another European element enters Soviet strategic thinking – this time in respect of Norway, whose proven and potential reserves of oil and gas offer a longer-term solution to possible Soviet difficulties in meeting both West German and East European needs for energy. It will not have passed unnoticed in the Soviet Union that Norway is pursuing a highly conservationist and nationalistic policy towards the exploitation of these reserves – and that this is creating tensions between Norway and its energy-hungry sister nations in Western Europe. The Soviet Union and Norway are, moreover, beginning to negotiate over the division of their off-shore waters in the Barents Sea where there are high expectations of very large oil and gas resources being found (the Soviet Union, indeed, considers it to be the most highly prospective area in the world). All this provides the basis for a Soviet offer to Norway whereby favourable consideration of its claim in the Barents Sea would be extended in return for acceptance of Soviet protection for Norwegian resources against all outside powers. This would amount to a kind of Norwegian/Soviet condominium over these important oil and gas resources, in the framework of which Norway's continued membership of the Atlantic-style organizations would no longer be appropriate. It should be recalled that Norway has not become a full member of the Western International Energy Agency (IEA) for fear of having to produce its indigenous oil and gas resources beyond the level which Norwegian policy-makers consider desirable.

Finally, consolidating the separation of both Western Germany and Norway from the rest of Western Europe, in a new situation which offers either political or economic advantages to the two countries concerned, the Soviet union would become a sort of 'honest broker' introducing the oil resources which Norway wishes to produce to the markets of Germany and Eastern Europe, thereby eliminating the problem which the USSR could have in the medium to longer term of meeting the energy requirements of these countries as well as its own growing demand.

The Soviet Union's second-choice strategy is thus aggressive – but very sensible from its own point of view. It also has much to attract those parts of Western Europe which could be divided from the north Atlantic system. And, in the final analysis, if Western Europe insists on thwarting US policy with respect to the containment of OPEC, and hence creates mortal dangers for America as well as for the Western system as a whole, then the US may well decide that the Rhine and the North Sea are just as good dividing lines between East and West as the lines of the last 30 years, and subsequently have the reality of the new division built into the détente which it still hopes to achieve with the USSR.

## Indecision by the Poor, Oil-Importing Countries

For many non-oil-exporting countries of the Third World, the impact of the initial traumatic increase in oil prices was moderated by the concurrent high prices on world markets for their own product exports – both mineral and agricultural. These high prices were a function of a strong demand at that time by the rapidly expanding economies in the industrialised nations of the world for those commodities. Now that the latter have moved from expansion into recession, engendered in part by the oil crisis, their demands for primary materials have fallen away and, as a consequence, so have prices. Thus, the exporting countries are doubly affected, by both a reduction in the quantities they can sell and the reduced unit price obtainable in the changed market conditions. Unlike the oil producers, the exporters of other commodities have, of course, no effective oligopolistic arrangement whereby prices – and hence export earnings and government revenues – can be maintained.

As a result, many developing countries are getting into increasingly severe economic difficulties, with consequential internal problems in both welfare and political terms. The key question for these countries is to which group of the world's powerful nations should they ally themselves for purposes of rescue, the Western advanced industrial countries or the new wealthy oil-producing and-exporting nations? The former, it is argued by some Third World leaders, have provided and will continue to provide the markets for the commodities that the Third World countries export. Hopefully, they speculate, the Western world's economy will soon be stimulated sufficiently by a series of policies designed to reflate national economies and reduce unemployment, thus ensuring that the demand for primary commodities will once again improve. After the time-lag caused by the need to reduce stocks, there will then be favourable results for the primary producing countries owing to higher sales levels and hardening prices. Meanwhile, the argument continues, many advanced industrial nations will have done their best to maintain their programmes of aid and other assistance, in spite of economic difficulties at home, partly as a consequence of the European Common Market's step in 1974 to strengthen and broaden the bases of its economic relationships with Third World countries. In the light of all this, the argument concludes, it might be better for the developing countries to stick with the proved economic system of the West – much though this might be deplored politically by many leaders in the third World. This is an attitude which shows a great deal of faith in the flexibility of the Western system and its ability to overcome even the oil crisis.

However, for other Third World leaders, the oil-exporting nations clearly provide a more effective basis both for short-term help over the immediate economic problems and for the development of a newly orientated economic power group in world economic and political affairs. A few poor, oil-importing nations have already received short-term help from the oil-producers – for example, the small nations of Central America from oil-rich Venezuela, and Pakistan and some other Islamic countries from the oil producers of the Middle East. But to date the OPEC nations have not played their short-term potential in this direction very strongly, and many countries – notably in Africa – are still waiting for aid and assistance promised by oil-exporting countries up to more than a year ago (for example, the $200m. loan offered by the Arab League to the Organization of African Unity in late 1973 has not yet been paid into the African Development Bank as requested by the OAU).

Though this is undoubtedly frustrating – and even incomprehensible for many of the world's poor nations, which have seen their development efforts thwarted by the high oil prices determined by the exporting nations (rather than by the impact of expected profiteering by the international oil companies) – many Third World countries are increasingly putting their faith in the growing political and purchasing power likely to be exercised by the OPEC countries within a few years. After all, it is argued, the OPEC countries will soon be involved in all the international monetary and other decisions as they force the International Monetary Fund, the World Bank, and other international organizations to heed their views. Moreover, if their solidarity holds and oil prices are thus maintained, they will soon be in the position of having access to more funds for international help purposes than the weakened member nations of the traditional Western system. Is it not Algeria, a member country of OPEC, which is leading the struggle for the New International Economic Order? And is not the former Algerian

Secretary-General of OPEC, Dr El Rahman Khane, now the Executive Director of the United Nations Industrial Development Organization, the UN Special Agency which the Third World created against the opposition of the West and which is seen as an important instrument for bringing in the new economic order? And are not Algeria, Venezuela, and other oil-exporting nations taking the initiative in the Third World's "Club of 77" – a device for evolving a strategy whereby the poor shall inherit the earth? And is it not the case that there are rich, or potentially rich, oil exporting countries in all the continents of the Third World (Venezuela and Ecuador in Latin America; Nigeria and Algeria in Africa; the Persian Gulf countries in the Middle East; Iran and Indonesia in Asia) on the basis of which the problems of poverty can be put into a regional context and then reduced by means of mutual help within the regions? In this emerging framework, the advanced industrial world becomes more than an irrelevance; it becomes an impediment to the development of the new order – especially in the next few years in which it still has a chance to break the power of OPEC.

For many OPEC nations this option is attractive – not to say inspiring – and it ought to be encouraged. And perhaps the best way of encouraging it is not to do too much too quickly for the poor Third World countries, for if their economic prospects are improved they might revert to traditional relationships with the Western world before the new system has had a chance to become effectively established – and, in so doing, help the Western system to survive! Instead, a further squeezing of Third World countries, together with many advanced industrial countries, will soon create conditions in which the poor, oil-importing countries have no option but to ally themselves on a long-term basis with the oil-producing nations and so provide the framework for a revolutionary change in the world's economic and political order.

## International Implications

To sum up; the system of advanced industrial nations appears to stand in the greatest danger of falling apart as a result of the failure of its leaders and peoples to recognise that it is under attack from a powerful external force which, in part at least, sees the destruction of the Western system as a desirable end. This failure has been accompanied by the resurgence of the forces of nationalism, so that to the external enemy there is added the shorter-term danger of internal dissension between the Western countries. In as much as the United States appears to recognise this and may soon despair of persuading its European allies to forget their minor, local differences and to stop 'knocking' the Americans in monetary terms, it is not inconceivable that she may re-evaluate her expensive protection of Western Europe, retire to a 'fortress America' strategy, and leave Europe to its fate. This is a fate which Western Europe certainly recognises as being impossible and unacceptable, but one against which is has so far chosen to make little effort to protect itself – in either strategic or economic terms – in relation to the power of the OPEC countries.

It is within the context of a weakened, or even abandoned, Western Europe that the Soviet Union could make its expansionist effort to secure the adhesion of Germany and of Scandinavia to her system, and so bring the frontier of the Soviet-dominated Europe up to the North Sea and the Rhine – a valuable strategic and economic gain for the USSR. It would give it not only open access to the world's oceans (their long-time geopolitical aim), but also access to the world's most efficient industrial base in a re-united Germany, together with the bonus of control over Norway's oil reserves, which now appear to be potentially larger than those of any other country, apart from the Soviet Union itself, Saudi Arabia, and possibly China. As for the rest of Western Europe, it may well get what it has been seeking under the inspiration of French policy, i.e., co-operation with the Arab world, but one fundamentally different from that to which French policy has aspired. The leadership and the control would flow from the Middle East countries, with their oil wealth and their justifiable claim to lead the Third World, and the rump of Western Europe would have to do the bidding of those who for so many generations were controlled by the French and the British. Revenge, indeed, for the exclusion of Islam from Europe many centuries ago.

In the Far East, Japan would be out on a similar limb. In an effort to avoid dependence on OPEC countries, Japan would be forced to link up with China in a system which would initially be mutually advantageous to both parties (with Japanese industrial goods and expertise finding limitless markets in China, in return for a flow of oil and other raw material which would enable Japan to do without the outside world). In the longer term, however, China's strength and its massive population would reverse the power relationship that developed between the two countries in the 1930s and the 1940s, and the best that Japan could hope for might be a relationship similar to that of a restructured Germany with the Soviet Union – internal independence, but with no opportunity to act independently in relations with other parts of the world.

Elsewhere, the 'new internationalism' could take the form of large regional groupings, with each one orientated towards the locally important oil-producing and -exporting nations: Venezuela and Latin America; Iran and the Indian subcontinent; Indonesia and South East Asia; and, of course, the greater Middle East itself, with the unity of the Arab world stretching from the Indian ocean to the Mediterranean Sea and the Atlantic.

The basic assumption in all this is that the revolutionary change in the world of oil power – and the failure of the Western world to react to the revolution except by appeasement – provides the catalyst on which the changes can be predicated. This is an assumption which makes oil and oil power more central to the shaping of world events than any other single factor; an assumption which would have been dismissed as more than unrealistic before recent events in the world of oil brought home to statesmen and others, the fundamental significance of this commodity in determining world economic and political developments.

The revolution implicit in the changed world that could be brought about by the new oil system may yet persuade Western statesmen that something more fundamental has to be attempted than half-hearted efforts to stop the waste of energy, and policies which are based on little more than pious hopes that we can persuade the new oil masters to co-operate with us. The key to the right approach lies here in Western Europe – in the necessary restructuring of our economy and society to minimise *immediate* dependence on OPEC oil, and on the determination over the slightly longer term to become completely independent of it, by discovering and producing our own considerable oil and gas reserves. And so the ball comes right back to our court: for decisions, that is, to restrain energy demand and to ensure the highest possible rates of exploitation of our indigenous energy resources.

## References

1    See Michael Simmons, 'Western technology and the Soviet economy' *The World Today*, April 1975.

# Chapter II – 5

# The International Oil Companies in the New World Oil Market*

The high degree of solidarity of the member countries of the Organisation of Petroleum Exporting Countries (OPEC) has, since the autumn of 1973, steadily become increasingly obvious to and reluctantly accepted by the western industrialised world. Earlier pious hopes – largely based on wishful thinking – that strains and stresses amongst the world's oil producers over questions of price and of levels of production would produce a near future break-up in OPEC[1] have now been generally discarded, and the idea of the oil-producing countries as the dominant element in the international oil system is being accepted with all the implications that this has for future international relationships.

## I – The Role of the International Oil Companies

There is thus a much greater willingness on the part of most nations to *consider* ways and means of reducing dependence on the OPEC cartel. Unfortunately, however, this is not always accompanied by a general realisation that this also means either reducing the contribution which the international oil companies make to the world's energy supply or the need for steps to bring these companies under the effective control of the nations concerned. On the contrary, the companies are often still largely believed to be the essential elements needed to ensure the continuity of oil supply and especially for organising the necessary

* Reprinted from *The Yearbook of World Affairs*, Vol. 32, 1978, Stevens and Sons Ltd., London, pp.76–92

infrastructure of supply arrangements in the event of any further interruption of production and transportation for political and/or economic reasons by the OPEC countries. Such beliefs are moreover – at the level of analysis concerned with the technicalities of supply problems – justified, for it is a matter of fact that most of the world's oil which is internationally traded continues to flow at the behest of the international oil companies through facilities which they own and operate around most parts of the world.

It is, however, as a result of a recognition of these "facts", that the dangers to the oil-dependent western world are enhanced because this obscures the interdependence and the willing – indeed, the necessary – co-operation of the OPEC member countries and the international oil companies which thus jointly constitute a producers' cartel working *against* the real interests of the world's oil importing nations – rich and poor.

### (a) Oil Companies' Profits from OPEC Oil

The motivations of the companies in this respect are not difficult to determine. These are, simply, a recognition of what they have to do in order to ensure that their commercial interests are best served for, in the changed oil supply and price situation, they are able to achieve a higher cash flow by marketing OPEC oil than by seeking to secure alternative supplies from elsewhere. Indeed, throughout 1976, it seems that the companies, on average, achieved a weighted up-stream gross profit margin on crude oil delivered to the refineries in Rotterdam (the world's single most important market for OPEC crude oil) of about 85 cents per barrel.[2]

Table II-5.1, derived from official Dutch government foreign trade statistics and from oil industry publications, indicates the costs of crude oil purchases from OPEC countries and the transportation costs involved in getting the oil to Rotterdam. It thus shows how the up-stream profit margin was achieved in 1976 in respect of this market.

The profit per barrel made on importing OPEC oil to Europe has, moreover, special qualities which enhance its attractiveness. It is virtually an after-tax profit for two reasons. First, as it is earned "off-shore" it can, if appropriate, be accounted for in tax havens. Secondly, under existing legislation in the mother countries of the international oil companies (especially the United States but including also the United Kingdom), profits earned on operations in the traditional producing countries and which are remitted back to the home country are free of additional tax obligations. This is because the taxes already paid in the producing

## TABLE II-5.1

### The cost and price of OPEC oil and oil companies' margins at Rotterdam 1976

| Country | Assumed Average A.Pl. Gravity[a] | Imports (million metric tons)[b] | % of Total Imported to Rotterdam | Average Value – $ per metric ton[c] | Adjusted Value – $ per long ton[d] | Average Freight Cost – $ per L.T.[e] | Netback Value – $ per L.T. | Netback Value – $ per bbl.[f] | Average Crude Oil Cost to Companies – $ bbl.[g] | Company Profit Margin – $/bbl. |
|---|---|---|---|---|---|---|---|---|---|---|
| Libya | 40° | 1.21 | 2.40% | 98.20 | 99.80 | 2.28 | 97.52 | 12.57 | 12.48 | 0.09 |
| Nigeria | 34° | 8.80 | 17.30% | 99.28 | 100.90 | 3.61 | 97.29 | 13.00 | 12.24 | 0.76 |
| Iran | 32° | 14.49 | 28.50% | 96.96 | 98.54 | 8.05 | 88.91 | 12.23 | 11.31 | 0.92 |
| Saudi Arabia | 33° | 14.36 | 28.20% | 96.16 | 97.73 | 7.83 | 88.33 | 12.06 | 11.04 | 1.02 |
| Kuwait | 31° | 7.04 | 13.90% | 94.86 | 96.36 | 8.12 | 86.74 | 11.99 | 11.08 | 0.91 |
| United Arab Emirates | 38° | 4.90 | 9.70% | 99.87 | 101.50 | 7.80 | 92.07 | 12.08 | 11.56 | 0.52 |

Weighted Average of Company Margin[h]  0.85

a  Based on the range of crude oils listed for each country in the list of "representative crude oils" per country in the *Petroleum Economist*.

b  From the Dutch Department of Foreign Trade Statistics, Crude Oils Imports Section, Ministry of Economic Affairs, The Hague. The figures show the volume of crude oil imports for processing in the Netherlands. All such imports reach the Netherlands directly from the exporting countries via the port of Rotterdam.

c  Calculated for each month from the value of the oil in Dutch gulden and converted into U.S. Dollar values at the middle-rate used each month by the *Foreign Trade Statistical Department – ibid.*

d  Calculated for a ton of 1016.05 kilograms.

e  Based on 1976 World Scale rates from the various ports of origin to Rotterdam. For the purposes of the calculation the following assumptions were made on the cost of the VLCCs used for transport: 50 per cent. of volume moved in company-owned tankers operating at WS 55.4 (the rate needed in 1976 to cover the trading and capital costs of an owned tanker – see *Financial Times*, January 24, 1977): 25 per cent. of volume in time-chartered tankers at an average of WS rate of 50.0 – the average of AFRA freight rates in 1976: and 25 per cent of the volume in spot-chartered tankers at an average rate of WS 26 – the average spot rates for the first half of 1976 as given by Shell Nederland in its publication *Het Tankerovershot: wat doen we ermee?* (1977), p.3. This gives a weighted average of WS 47.5 for average freight costs. This is a cost allowance which significantly exceeds the price of tanker transport on the open market and which, to that degree, therefore, overstates the allowance that should be made for freight costs in bringing oil to Rotterdam when viewed from the strictly economic standpoint. See M.A. Adelman, *The World Petroleum Market* (1972), pp. 113–114.

f  This is calculated from the netback value per long ton by dividing by the appropriate factor for the assumed average API gravity of the crude oil imported from each country.

g  Calculated from the cost of "Buy Back Oil" and/or "Equity Oil" as given for each country each month – in the list of "Representative Crude Oil Prices" in the *Petroleum Economist*. Where oil at both prices is still available an assumption of a 50/50 split in oil company supplies has been made. Note that "Buy-Back" oil, where available, is now priced at the State selling price, i.e. the price of oil to all buyers so that it gives no advantage to the integrated oil companies.

h  Weighted by the share of each country in the total supply and allowing for losses in transit.

countries (amounting in 1976 to over 10 dollars per barrel) give the companies a substantial tax credit position. This "after-tax" position of the oil companies' earnings on handling OPEC oil makes the apparent per barrel profit "worth" up to twice as much as profit made on the production of alternatives to OPEC oil in all those countries where such profits are subject to corporation (and other) taxes at rates which are normally of the order of 50 per cent.

## (b) Companies' Downstream Investments Tied to OPEC Oil

The profit situation is, in itself, a powerful motivation for continued oil companies' collaboration with OPEC countries but, in addition, there is another – and perhaps an even more important – component which helps to determine oil company policies towards the new oil supply and price situation. This is because the companies do not *need* to make any new investment whatsoever to obtain and to handle oil from OPEC countries. As far as the production of the oil, together with the production development facilities required to maintain and/or increase future output is concerned, the OPEC countries themselves now largely take care of the investment required, given their nationalisation of the companies' assets over the period since 1973.[3] And in so far as technical and managerial help is required from the companies to keep production going and/or to develop new potential, then the companies collect a fee – either in cash or in the form of oil at specially discounted prices. These arrangements are in themselves highly profitable, so adding yet another element favourable to companies' decisions to maintain or even increase their use of OPEC oil in the world's markets.

A similar situation favouring such traditional flows of oil exists in respect of the capital investments made by the companies in downstream facilities in the oil importing countries – especially in Western Europe and Japan where almost 70 per cent of all OPEC oil is used. In these countries, as a result of the fivefold increase in oil prices since 1973, demand for oil is running at about 35 per cent below the level that was hitherto expected to be in demand by 1977. However, investment plans made by the oil companies in the early 1970s, in anticipation of a continued upward trend of over 7 per cent per annum in oil demand in these countries, led to projects for terminals, refineries, pipelines and distribution systems which are only now being completed. As they come on line they necessarily add further to the already existing over-capacity of the West European and Japanese oil industries to handle supplies. Moreover, not only is this additional capacity not really needed, in light of present and now expected future market demands, so increasing the unit costs of supplying oil products to the markets, but most of the new developments are also of a kind and in locations which are related directly to the earlier expectation of increased flows of crude oil from the OPEC countries.[4]

## (c) Oil Companies must favour the Status Quo

In light of these facts of oil life at the beginning of 1977, any greater than really necessary changes in the patterns of oil supply – such as would occur if there were too large and too rapid a rate of substitution of oil from traditional sources by oil or by alternative energies from elsewhere – would have a serious adverse effect on the economic viability of the oil companies' operations. This would, moreover, be in a situation in which the much reduced demand for oil has in itself already caused serious problems of retrenchment for the companies in their operations, with much increased unit costs in processing and in distribution and, hence, a consequential squeeze of their down-stream profit margins.

In brief, the oil demand situation and outlook in those areas of the world most heavily dependent on OPEC oil, and the fact that the industry's infrastructure in these areas has been developed for handling such OPEC oil, combine to counter any enthusiasm that there might have been on the part of the international oil companies for curtailing their demand for oil from the traditional exporting countries. In other words, oil companies' interests are again bound up with those of the member countries of OPEC. Thus, at the beginning of 1977 one can readily recognise the existence of a set of common interests between the oil producing countries, on the one hand, and the traditional international oil companies, on the other. This situation, however, represents a continuation of the communality of interests which the two parties have shared now since the late 1960s,[5] and thus constitutes a main stabilising element in the changed world of oil power. Its importance for the medium-term future lies particularly in respect of the effect that it must have on the degree to which major oil importing countries are able to develop alternatives to continued dependence of OPEC oil.

## II - Independence from OPEC Oil

For the purposes of our analysis we shall assume that a reduction in the degree of dependence on OPEC oil is, indeed, an accepted objective of the national policy-making of importing countries. This, indeed, appears to be the case in respect of countries as different in their economic and political structures as the United States, Brazil, India and the countries of the European Economic Community (EEC).

## (a) Independence in the Developing World

For developing countries like Brazil and India the battle for independence from imported oil is, paradoxically, least subject to the influences of the activities of the international oil companies. This is so in spite of the fact that all such countries have significant financial and technical problems to meet and to overcome in their efforts to develop alternative indigenous resources. This relative lack of influence by the oil companies arises because such countries are very strongly motivated, both politically and economically, to achieve energy independence. Such independence is, indeed, seen as an essential element in national policy-making for the immediate future.[6]

Both India and Brazil, which are leading examples of these countries, have, as it happens, a reasonable opportunity for achieving such a policy aim for both have recently had important off-shore discoveries of oil and/or gas along their extensive coastlines and are already in the process of developing them. Furthermore, given their national ownership of and/or control over the petroleum sector of their economies, there is no question but that all domestic supplies of oil which become available will automatically be given an absolute position of preference over imported oil. Thus, the important outstanding question is the degree to which such countries can successfully pursue policies of indigenous oil exploration and development, given their technological and financial limitation. It is not, however, that such limitations are anything like an absolute barrier to such developments because the off-shore waters of the lower-latitudes offer quite modest technical and cost challenges compared with, say, the North Sea or the Atlantic coast of the United States. In the cases of India and Brazil both the water depths involved and the nature of the climatic environments are much less unfavourable for the exploitation of hydrocarbon resources.

As a result both India's National Oil Commission and Brazil's Petrobras have already been successful in developing their countries' off-shore resources and they both have continuing and increasing capabilities in this respect. They have, moreover, not unimportant possibilities for the expansion of such nationally-based petroleum developments as a result of help which has been, or will be, made available to them from state oil entities in other part of the world. This includes the oil producing countries of the Communist world, but help may be sought and given from the growing number of State-owned or State-controlled oil companies amongst West European countries (such as ERAP of France, Statoil of Norway and Hispanoil of Spain, etc.). And one must not exclude the possibilities of financial and even technical assistance for such developing countries from the State oil companies of the OPEC countries.

Although these possibilities do not entirely exclude future projects and ventures in developing countries like Brazil and India by the international oil companies, they obviously limit the openings quite severely. Moreover, they make necessary a willingness on the part of the oil companies to negotiate terms of involvement which will tend to isolate their activities locally almost entirely from their otherwise internationally integrated methods of working. Thus, for example, although Brazil has signed agreements with B.P., Shell and other major the international oil companies for their involvement in off-shore exploration, any physical success on the part of the companies in finding oil and/or gas can only be turned into a commercial success within the context of what Petrobras and the Brazilian government wants.

In any such ventures the profitability of the international companies depends on national decisions, and the profitability can, moreover, exist only in Brazil itself which, like most other developing countries, has strong controls on the repatriation of profits and the export of capital assets. Such ventures on the part of the international oil companies are, thus, really high-risk ones – for not only do they have to accept the whole of the risk involved in the exploration and the development effort, but they also have to face the risk of a successful development then being effectively cut off from the rest of the company's international system. In light of this it seems highly unlikely that the international oil companies will be too anxious to get involved on such terms in too many developing countries – except in the case of the very few which, like Brazil, are large enough and with a great enough potential for development to more or less ensure a long-term future for Shell or B.P. ventures isolated from the rest of the companies' activities in the commercial and financial (if not the technical) sense. And, moreover, it must be remembered that the international companies are not even welcome in some countries which fall into this category – including, indeed, India itself.

## (b) Can the Oil Companies Trade OPEC Oil in the Third World?

To a significant degree, therefore, the international oil companies will become cut off from involvement in energising the Developing

World unless they are prepared to participate in ways which effectively eliminate most of the usually accepted benefits which flow from the internationally integrated actives of such companies. To make their position even worse, moreover, they also seem likely to be decreasingly involved in trade in oil with such countries – in so far, that is, as the Developing World's countries will have to import oil from the petroleum exporting lands. On the one hand, the latter have taken over, or are in the process of taking over, the production and associated facilities of the companies. And, on the other hand, the oil importing countries have themselves been nationalising or bringing under direct government control, the refining and distributing facilities developed in the most part to date by the international oil companies. As a result the necessary infrastructure for direct state-to-state trading between oil exporting and developing-world oil-importing countries is steadily being expanded. Moreover, and perhaps even more important, the political will required for the establishment and development of such activities is emerging – with an additional factor in this being the increasingly firmly expressed wishes of OPEC countries to serve the oil importing countries of the Third World in this way.[7] Before the end of the decade, therefore, it seems likely that the international oil companies will no longer have much opportunity to contribute to this particular part of the international movement of oil.

## (c) The United States Market and International Oil

This much, however, the international companies must already be largely anticipating in their forward planning. Although it means the loss of business which has traditionally been highly profitable (given the oligopolistic position which the companies have generally enjoyed in such international trading activities in most of the developing world), it does nevertheless account for rather less than 10 per cent of the companies' involvement with OPEC oil exports. Of the remaining 90 per cent the United States accounted for about one-fifth in 1976. However, because of the growing short-term imbalance between United States demand for oil and its own ability to produce the commodity, both the absolute amounts involved and their percentage contribution to the total quantity originating from OPEC countries, will grow: and this oil will continue to be handled mainly by the international oil companies and a number of other largely American-based enterprises. However, given the general level of acceptance already in the United States on the need to keep a close watch on the delivered prices of foreign oil because of suspicions that the companies otherwise generate oil-price rises for purposes of increasing profits; and, further, given the advent of a Democratic administration led by a President from a State which is heavily dependent on energy imports, then the companies may well expect their profit margins in respect of their role as suppliers of foreign oil to the United States to be subject to close surveillance.[8]

Moreover, the companies must also bear in mind that the United States government remains likely to insist on indigenous oil being given preference over imports even in situations where its production and distribution costs make it more expensive to deliver to markets than imported oil – in the interests of as high a degree of energy self-sufficiency as possible. This is a factor which is of particular importance in respect of Alaskan oil which, having been produced, must then be delivered to markets in the continental United States via the West coast. In terms not only of the quantities of oil involved and of its characteristics *vis-à-vis* patterns of oil product demand in California and other west-coast States, but also in terms of the supply/demand patterns of individual companies it seems as though there will be an infrastructure problem which prevents this oil being absorbed in United States markets at a cost below that of alternative patterns of supply based on imports. And this could possibly lead to a squeeze on the profit margins of the supplying companies concerned. The tentative indication by these companies that they might well prefer to sell Alaskan oil to Japan, whilst, at the same time, importing OPEC oil into the eastern/southern parts of the United States, in the interests of optimising their own profit opportunities, has been roundly squashed by the new United States Administration which is not prepared to accept the higher-than-necessary degree of dependence on foreign oil that such supply schedule proposals involve.

## (d) Western Europe and the International Oil Companies

Thus, the disappearance of opportunities for profitable activities by the international oil companies in the Developing World, plus the likelihood of increased constraints on their ability to make profits in selling foreign oil to the United States, make it all the more important that the companies concerned should try to defend and consolidate the greatly improved possibilities which they have had in recent years to generate revenues from their activities in exporting OPEC oil to Western Europe. Their opportunity in this respect arises, naturally, from the fact that these companies play a so much more dominant role in energising the Western European economy than in the case of the United States. Their contribution to the energy needs of Western Europe is more than

60 per cent, compared with a less than 40 per cent contribution to the United States' energy system.

We have already demonstrated above how the international oil companies are now able to supply Western Europe with imported oil at a profit which is higher than that which could be earned as a result of any alternative oil supply patterns – given the limited extent to which the companies concerned have to make investments either in production facilities or in transportation and refining infrastructure in order to enable them to continue to supply oil from OPEC countries. This, however, is *only* possible in the case of Western Europe where the companies continue to enjoy the right to determine the region's oil supply pattern – a right which, as shown, is one that the companies do not enjoy either in respect of developing countries or of the United States: or, for that matter in the case of Japan which, as the single most important importing country of OPEC oil, insists on a high degree of national involvement in the decision on oil supply patterns though MITI and other official entities associated with the Ministry of Trade.

In other words, it is essentially only in Western Europe that the international oil companies still – in 1977 – continue to enjoy the uninhibited right freely to trade the oil which they are committed to take, or which they prefer to take, from their OPEC suppliers at over 14 dollars per barrel (landed at a West European port). This is not only one of the main – or even perhaps the main – factor which ensures the maintenance of the OPEC system. It also has two other consequences which are specifically European in their importance. First, it means that Western Europe remains – and will continue to remain – unnecessarily over-dependent on insecure and expensive supplies of oil. Secondly, it means that there is an effective barrier to the rapid development of Western Europe's own indigenous energy sources – most notably those of North Sea oil and gas which constitute immediate, or near-immediate, and directly substitutable resources for imports of OPEC oil.

## III – North Sea Oil and Gas as Alternatives to OPEC Oil

The exploitation of North Sea oil and gas is placed in jeopardy because so many of the developmental decisions belong to exactly the same international oil companies which, as shown above, have a strong commercial interest in maintaining the level of oil imports to Western Europe. Or perhaps it would be more correct to say that the development of the North Sea's recoverable reserves has been placed in jeopardy because the companies are still allowed to take supply, production and refining decisions in Western Europe that are virtually free of relevant and meaningful government or inter-governmental control.

And matters must remain like this until action is taken by the governments, the EEC and other relevant institutions in Western Europe to secure the divorce of the indigenous oil and gas industry from this degree of external control over its affairs. Such a divorce can only become absolute as the international oil companies are required to take their decision on the provision of energy in Europe in the light of the best interests of the countries or the region, rather than in the context of maximising profits for the member companies of the international oil system together with the continued enrichment of their OPEC partners.

## (a) The Resource Base – and its Exploitation

Unfortunately, misunderstandings – arising partly out of deliberately misleading statements[9] – on the nature and probable size of the Western European off-shore oil and gas resource base, and on its possibilities for determining the medium-term outlook for the Western European energy economy, have detracted attention away from the fundamental issues involved in securing the full exploitation of the already known resources of the North Sea fields. Thus, whilst much effort has been directed by the United Kingdom, Norway and the Netherlands to ways of toughening up the conditions under which the companies have been allowed to operate in the North Sea (so that the countries concerned could secure a larger share of the economic rent arising from the recovery of each barrel of oil or gas that the operating company decided to produce), no European country has yet taken any action which obliges the international companies to produce all the recoverable oil and gas from each field as quickly as possible to substitute for imported oil. And neither have the three producing countries (the Netherlands, Norway and the United Kingdom – in order of importance at the beginning of 1977), nor, indeed, any West European oil importing country, yet offered to guarantee profitability on the oil that could be produced by the oil companies as a means of getting them to develop fields more intensively than they will do in the light of their own calculations of their optimal strategy in the open-market conditions of the West European energy economy.

Indeed, the very opposite has been happening – from two points of view. First, the main potential customers for North Sea oil – *viz.*, France and West Germany – appear to view with suspicion the idea that they should underwrite or guarantee the production of high-cost North Sea oil. Instead, they prefer to keep their energy supply options orientated into other directions: first, to the protection and the financing of extraordinarily expensive and very large programs of nuclear power development; and secondly, to special deals and/or greatly enhanced trading relationships with the Arab producing countries of the Middle East seen, in terms of such trading potential, "as a natural extension of Western Europe". Thus, special relationships with these countries to try to secure essential oil supplies seem to be preferred to the alternative of special arrangements with the oil and gas producing countries of Western Europe.

Secondly, the oil and gas producing countries of Western Europe have not themselves done very much to persuade their neighbouring countries to change their policies in this respect. Indeed, the Netherlands, Norway and the United Kingdom all appear to be convinced that their resources of oil and gas are so scarce that they need to be protected against the rapacious demands of their neighbours. Such resources, these countries argue, are better left unproduced now for the sake of future generations. The validity of such arguments of inevitable scarcity are very much open to question and appear to emerge mainly out of a too unquestioning acceptance of oil companies' sponsored views on the size of the reserves (ignoring the fact that the companies do have a good commercial motivation to minimise rather than maximise their estimates as to how much oil is recoverable) and partly out of a serious misinterpretation on the nature of the resource base. Here we are not concerned with the validity or otherwise of the scarcity hypothesis as such, but only with it in terms of the impact that it has on the way governments look at the development decisions on North Sea oil and gas reserves.

### (b) The Optimal Development of North Sea Fields – Conflict between Companies and Governments

A development decision which involves the installation of fewer platforms and/or fewer wells on a field than are needed to produce all the technically recoverable oil from a field in an economically relevant time-period is likely to be counter-productive in terms of national policy concern for the conservation of resources. In a multi-field province such as the North Sea there may be something to be said on grounds of conservation (if not of economics) for not producing some of the fields which have been discovered or, even better, for not discovering so many of the fields in the first instance. But, neither on grounds of conservation nor on grounds of national economic considerations is there any reason to allow, let alone encourage, a field which is in the process of development not to be developed to its maximum possible extent – to a degree of development that is, which lies beyond that which is optimal for the oil company.[10]

This is so because a decision which means the development of a field only up to the level of investments, and hence of production, which is optimal for the company concerned, instead of to the maximum possible level of recovery (at or approaching the level of the technically producible reserves), will have an adverse effect on the degree to which the exploitation of the oil benefits the economy of Western Europe. In the short-term, this will arise because jobs and profits which could be generated from oil development related activities will be less than they could be. In the medium- to the longer-term, the benefits to the country concerned will be reduced in terms of the size of the government revenues arising from oil production, of the contribution of the production of oil to the Gross National Product (GNP) and to the balance of payments. It also affects the degree to which oil imports to Western Europe can be substituted by North Sea production.

One must thus hypothesise an *inevitable* and *normal* divergence of international oil company and West European interests in the development decisions on North Sea oil production. On the one hand, one has to recognise the validity, from the oil company's point of view, of the strictly commercial methods used by a company to calculate the optimum recoverability of the oil in light of competing alternative opportunities *at the international level* for its available investment funds. On the other hand, the European government has an interest in maximising the returns to the country from a field's development by ensuring that it contributes as much as possible to the growth of the GNP, to the creation of employment and a local as well as regional multiplier effect, to the balance of payments' situation, to government revenues and to substituting insecure supplies of foreign oil.

## (c) Government Action to Secure
## Optimal Developments of Indigenous Oil

If a government's interest thus lies in maximising the recoverability of oil, then, in order to persuade the company concerned to do what the government believes to be necessary, flexibility in the arrangements for taxing the production of oil from the field and/or the possibility of the government contributing to the financing of the field's development will be necessary. Such a government contribution is necessary in order to create conditions in which the company concerned will be at least no worse off when it is obliged to install a production system which aims to maximise the degree to which the technically recoverable oil is recovered, than it will be if the company's preferred less complex system is employed; given that the latter, on the basis of the company's own evaluation of costs, prices and taxes, appears to provide it with its optimum investment strategy as far as the development of the field is concerned.

A full analysis by government of each of the several possibilities for developing each field is thus necessary in order to establish the "best" solution from the point of view of the government and country – and then there must be the means to "impose" this solution on the company concerned – though this means that the company has to be assured of a minimum acceptable level of return on its investment. Agreements between government and companies to produce this kind of result seem to be possible based on appropriate calculations of the contrasting risks and present values which are attached to the possibilities of the future production of oil from North Sea fields by the international companies, on the one hand, and by European governments' interests, on the other.

## IV – Reducing Uncertainty in Western Europe's Energy Outlook

It is only by divorcing decisions on the production of the North Sea's oil and gas reserves from the control mechanism of the international oil companies that there is a chance to reduce the uncertainties in the energy sector of the European economy to manageable proportions. This is because it is only in this way that it is possible to maximise the production of indigenous resources. The essential element in achieving this is an effective co-operative effort between the governments of European countries and the international oil companies. Ironically, such co-operation can, however, only be possible when the latter have been brought under control – in respect, that is, of eliminating their continuing opportunity to exercise, jointly with OPEC, the right to determine the supply and price of energy in Western Europe. Only when this has been done – within a necessarily interventionist framework of the kind that Europe has built up over the years in respect of other, and less-important, sectors of the economy and as other countries are, as shown, already building up in respect of their oil sectors – will it be possible for Europe effectively to realise the opportunities for independence from the international oil system which have been generated by the oil and gas resources of the North Sea province – now generally accepted to be a major one by world standards. This is development which is as important to the United Kingdom, Norway and the Netherlands – the important producing countries of Europe – as it is to the other nations in Western Europe, whose energy needs can also be more securely and more cheaply met out of the North Sea's resources than they can from their continued reliance on insecure and increasingly expensive oil from the OPEC countries.[11]

## A New Role for the Oil Companies in Western Europe

This analysis of the European future of the international oil companies may be construed as a criticism of them. But this is not so, for these companies justifiably, from the point of view of their own interests, seek to achieve the most profitable set of operations possible. And they have to do this in relation to governments and to inter-governmental organisations around the world. On the one hand, they have to deal with OPEC which, in its wisdom and as a result of many years of experience with the international oil companies, has created very effective constraints on the companies' activities. OPEC has, however, also ensured the continuity of highly profitable operations for the companies. In Western Europe, on the other hand, it is the lack of the *right* sort of governmental constraints on the companies which enable the latter to pursue policies which lead to the inevitable divergence of what they do from what they should be doing for Europe's well-being and security. And this occurs without the companies, in their turn, really knowing if and how they are going to be able to make adequate profits out of their activities in Western Europe.

In no way is this gap between governments and companies of more importance in 1977 than in respect of decisions which affect the speed and the degree of development of the North Sea's considerable resources of oil and gas. Too many of the province's known or expected recoverable reserves remain unnecessarily unexploited basically because the international oil companies and European governments have not yet found a *modus vivendi* which makes the development of them a top priority of common interest to both parties. As a result, the energy supply potential from the indigenous oil and gas resources of Western Europe remains much more limited than it need be, with the consequential dangers of a continued too high a degree of dependence on imported OPEC oil and/or the need for an economically unsustainable and environmentally unacceptable degree of development of nuclear power. The companies, in their turn, do not really know where they stand in relation to the medium-term development of the European energy market. Perhaps one way of clarifying the issues involved – such that appropriate solutions could then be sought – would be a European take-over of all the European-located subsidiaries of one of the American-based major international oil companies, say Exxon. Shares in the resultant large oil company – with producing, refining and marketing activities throughout the continent – could well be placed with Europeans (*en masse*) in order to form a kind of European Peoples' Oil Company (EPOC Ltd). The efforts of this could then be centrally directed to ensuring that its activities maximise returns in and for Europe rather than, as hitherto, being employed to achieve what is acceptable to OPEC's member countries and in order to ensure as high as possible a rate of return on the investments that Exxon's European subsidiaries have made for the benefit of the American parent company. For a sector as important as oil to the welfare of its inhabitants, West European policy towards the international oil companies and their activities cannot afford to be any the less interventionist and nationalistic than in the case of countries like Brazil – or the United States.

## References

1       These hopes were examined in P.R. Odell, "The World of Oil Power in 1975", *The World Today*, July 1975.

2       This overall up-stream profit margin for crude oil delivered to Rotterdam should be compared with the nominal 25 cents/bbl. profit margin which the companies claim they are allowed on purchases of crude oil in most OPEC countries. It seems that the methods of company operations permit these nominal f.o.b. profits to be multiplied several times before the crude oil is delivered to importing countries. There are, however, weaknesses in some oil-products markets in many oil consuming countries and these necessitate some products being sold below "cost", so reducing the final overall profit per barrel to less than the almost one dollar per barrel figure calculated in Table II-5.1. These calculations have, incidentally, been discussed with representatives of several of the companies involved. They do not accept the results, yet, on the other hand, they cannot find any major component in the assumptions on which the calculations are based which would give an upward bias to the per barrel rate of profit.

3       The one remaining exception to this – at the beginning of 1977 – was in the case of Saudi Arabia, where the proposed nationalisation of ARAMCO had not by then been effected. In this case, therefore ARAMCO remained responsible for investment in new producing facilities. This, however, seems unlikely still to be the case by the time this paper is published.

4       One example is Shell's £60 million crude oil import terminal on Anglesey. This was planned in the late 1960's in the expectation of a continued growth in oil demand in Northern England and a consequential need for a new deep water terminal to handle oil from

500,000 ton tankers. It was only opened in 1976 by which time it was clear that the new capacity was not needed because oil demand had declined. Moreover, the refinery which it was built to serve will increasingly run on oil from the North Sea. This will either be brought to the refinery by pipeline or by small tankers which cannot or which do not need to use the terminal. The terminal could thus become an expensive white elephant.

5    Co-operation, amounting even to collusion, between the international oil companies and OPEC in the period since 1968 is described in P.R. Odell, *Oil and World Power* (1975). See especially Chapter 9, "The World of Oil Power since 1973." See also J.M. Blair, *The Control of Oil* (1977), for further discussion on this point.

6    See, for example, *National Energy Balance*, Brazilian Ministry of Mines and Energy (1976).

7    See, for example, L.Valenilla, *Oil, the Making of a New International Economic Order* (1976), for a clear statement on this development in respect of Venezuela and its neighbouring countries of the Caribbean region.

8    This has been made clear in President Carter's National Energy Plan (April 1977).

9    It is, for example, inconceivable that the October 1973 statement of Sir F.S. McFadzean, the Chairman of Shell, that the North Sea oil and gas could not possibly provide more than 15 per cent. of Western Europe's energy was made for other than political reasons; *viz.* to awake Europe from its lethargy over its energy supplies in the light of the deteriorating situation in the Middle East. Similarly, with the comment in his Chairman's speech in April 1976 that Britain will not be energy self-sufficient for more than 10 years "because all the big North Sea fields have already been found." This seems to be a McFadzean effort to stop the United Kingdom relying on North Sea oil to solve its economic problems and is as unlikely as his 1973 statement to be based on the knowledge that Shell has about the probability of North Sea oil and gas production potential. The major oil companies are thus non-credible in their pronouncements about the North Sea, indicating that they are motivated by other considerations – perhaps those related to their role in the international oil system.

10   The technico-economic background to this is presented in full in P.R. Odell and K.E. Rosing, *Optimal Development of the North Sea Oil Fields* (1976).

11   A possibility finally recognised by the EEC at the very end of 1976 when the then out-going Commissioner for Energy, Mr H. Simonet, at his final press conference on December 22, 1976, spoke in terms of the Community developing a guaranteed market for British and other nations' oil in return for EEC participation in the decisions on the exploitation of the resources. See *Europe*, Nr. 2120 (New Series),

December 23, 1976, pp. 6–7. More recently (July 1977), the new Energy Commissioner, Dr. G. Brunner, has suggested that a fixed upper limit should be set on the EEC's oil imports from OPEC, so implying guarantees for indigenous production.

# Chapter II – 6

# The Pressures of Oil*

## a. The Consequences of the Revolution in Oil Power

### Introduction

We have attempted elsewhere in this book to specify the way in which reactions against the international oil companies finally produced a revolutionary change in the world oil system. This change has been one in which power has passed from the oil companies to the member countries of OPEC. Today, OPEC's members not only control the supply and price of oil, but they also own the bulk – over 60% – of the world's proven oil reserves. The consequences of this are, as with any revolution, far reaching. In this case they are particularly difficult to visualise and interpret as they have technical, economic, and political components, the inter-relationships of which create the complexity of the international oil system.

### The Technical Component

This emerges from the dominant role that oil had come to play by 1973 in almost all countries of the non-communist world. The world ran on oil, so that concern for security of supply was not simply concern for the supply of a commodity which helped to make life easier or more pleasant, but for a commodity without which organisational

* Reprinted from the book with this title (Harper and Row, London, 1978, pp.37–50 and 79–101) jointly authored by Dr. L. Vallenilla.

systems essential to the maintenance of the modern way of life could not survive for very long – if, indeed, at all. One thinks here of systems of societal organisations which are found developed, to a greater or lesser degree, in all countries of the world – such as systems of cities and systems of transportation. During the 20 years prior to 1973, countries, as well as individual consumers, had become used to the idea of oil being 'on tap': available against all instant demands in exactly the quantities and qualities required. In most countries outside the communist world this meant familiarity with and dependence upon household names such as Shell, Esso (Exxon), BP, Mobil, Texaco, Gulf, and Chevron: dependence, that is, on major oil companies which, irrespective of their faults in other directions and the problems to which they gave rise in political and economic terms, could, nevertheless, be relied upon to deliver the goods as and when required. It would, indeed, be difficult around the world to find a single customer with much by way of a serious complaint on the effectiveness with which such companies met his energy needs. In other words, the oil companies had built up a formidable reputation for reliability and for customer service – in the best traditions of Western capitalist endeavour – such that the significant diminution of their control over the oil supply system since 1973 and their replacement by suppliers of unknown quality raised questions of uncertainty about the future.

In particular, these uncertainties, at both the national and the international (inter-governmental) levels, have been reflected in terms of concern for the security of supply. As a result, since 1973 action has been taken around the world to build up stocks of oil in order to give consumers some protection against the changed circumstances. This has become a general attempt by many of the world's nations, including all those in the OECD group, to do something which had hitherto only been necessary for countries like South Africa, Sweden and Switzerland.

In the case of South Africa, a strategic stockpile of oil was considered to be an important tool in the fight against the country's possible exclusion from normal international trade because of its domestic policies. Thus, it is now reported, South Africa currently stocks enough oil to cover more than two years' demands, even though it is a country which is not very dependent on oil, given its large availability of low-cost coal from which most of its electricity is made and on which its industries run. In the cases of Sweden and Switzerland, the traditional neutral nations of Europe, their need was to be prepared for emergencies which could arise because of conflict between their neighbours and into which they might be dragged if they were not able to isolate themselves from dependence on one or the other for a commodity as important as oil. Thus, they took their politico-economic decisions to build up strategic stocks of oil, representing 180 days or more supply, to be held as a reserve against such eventualities.

Since 1974 these hitherto rather exclusive arrangements have been generalised throughout the industrialised world. As a result, an elaborate inter-governmental agreement has been created by the OECD countries (working together under the banner of the OECD's newly created daughter organisation, the International Energy Agency (IEA) ) to monitor the developing supply of oil, to supervise the build-up of strategic reserves in the organisation's member nations and to control and allocate the flow of oil to and between the nations of the group should this ever become necessary. The conditions in which such actions should be taken have, moreover, been defined. The control system becomes operational should the supply of oil from the oil exporting nations fall to 93 per cent or less of the total amount needed to meet evaluated demand, while the power to allocate supplies would be taken by the IEA if supply of OPEC oil fell to 86% or less of expectations. In brief, the world's most important oil consuming counties (accounting for about 75 per cent of total world consumption) have collectively decided that power in the new world oil system is now so distributed that they need these contingency plans to meet a potential supply crisis. This clearly reflects the high degree of uncertainty which they see in the new situation – especially when this is compared with the 'faith' which they had in the system when it was run by the traditional international oil companies.

This sort of reaction has not, moreover, been limited to the industrialised countries of the world. Some of the more developed countries of the Third World, particularly those in which oil consuming industry and road transportation have become important components in the economies, have also tried to build up stocks of oil to protect themselves against potential supply crisis conditions. This motivation is not, however, the same as in the case of the member countries of the IEA where there is the feeling that they might again be subjected to supply embargoes for political and/or economic reasons. Instead, the Third World countries' motivations arise from fears that, in the event of oil supplies being limited to less than the total world market demand, they may simply be outbid by the richer nations for the oil that is available and/or they may be given low priority for supplies by the international oil companies which still own and/or operate most of the facilities required to get oil to the consumers. As the oil companies are institutions which "belong" to the member states of the IEA, many Third World nations fear that IEA members may be given priority for scarce supplies. Again, these are attitudes and actions reflecting the uncertainty which now exists concerning the technicalities of the changed oil supply position of the post-1973 period – seen from the point of view of the oil importing nations.

There has, however, been an equally important technical component in the consequences of the change in the world system as far as the oil producing and exploring nations are concerned. For this group of nations the problem has been a much more immediate and a relatively much more important one, namely how to take over not only the oil operations, but also the managerial role within their oil industries, most of which continued to be run basically by foreigners right up to 1973. Thus, a take-over of the companies posed difficulties, even in terms of keeping the domestic producing and refining operations going. Even more so, it involved difficulties for the oil exporting countries at the level of international relationships and organisation, given the fact that, under the global regime of the international oil companies, domestic operations in each country were not independent, but rather completely integrated into the international activities of the one or more companies that had been involved in marketing the oil. This meant, of course, that decisions on questions as important as the level of production at any given time had been based on information systems centred on New York and London.

The degree to which the technical and managerial challenges of this kind could be accepted, let alone overcome, by the oil exporting countries in the short term was limited, as the required infrastructure just did not exist. The establishment and the manning of a national oil company was a challenge in itself and, to start with, some such entities could do little more than exist in name except where, as in the case of Venezuela, the staff of the former local subsidiaries of the international oil companies – 97 per cent of whom were already Venezuelan – could be taken over *en bloc*. Even then, such taken-over entities lacked some essential groups of personnel, such as those able to undertake activities like planning and research. Such functions had hitherto been almost entirely centralised in the international oil companies' head offices outside the producing country.

Thus, in order to keep the oil flowing from the wells and to the refineries and the export terminals, it was still necessary to depend on technical help from the companies. For this help significant fees were negotiated by the oil companies, sometimes in terms of a discount off the published price of oil for the companies concerned, and sometimes as a per barrel fee on all oil produced. This payment to the companies could then be increased further as a result of an agreement between the producing country and the companies which had been nationalised. As part of such agreements the latter agreed to continue to take certain minimum quantities of the state's oil in their international marketing systems. Fees for managerial and technical assistance negotiated by the oil companies lie in the range of 20 to 35 cents per barrel. Discounts off official selling prices for agreeing to market some of the state's oil production appear to rise as high as 50 cents per barrel. The companies are doing 'quite nicely' out of such arrangements. Of course, the state entities would have liked to be responsible for their own sales abroad, but what they could have achieved initially was somewhat limited. Even Petroleos de Venezuela, the Venezuelan state oil corporation, which was better placed than most of the other state companies to operate abroad, was only able to manage about 20 per cent of the country's total sales in its first year of operation: partly by sales on the open market and partly as a result of direct government to government deals.

In order to achieve total success in producing and selling their oil, the new state owned entities require to build up their so called 'downstream' investments and operations – in terms of tankers, refineries, and distribution systems. The purchase or the creation or these is, however, a long-term, rather than an instant, process and this automatically places a technical constraint on the degree to which the changes in the oil system looked for by the oil producing countries can be implemented. Or else, of course, it means that the new system may fail to deliver the goods required by the consuming world – a danger against which the International Energy Agency has, as shown above, attempted to prepare. It is, moreover, a danger which, if it became a reality rather than simply a potential development, would then lead to external pressures on OPEC and its members from both industrialised and Third World oil importing nations. This would generate even more serious consequences for continuity in the supply of oil, given that OPEC's members hardly seem likely to stand by and accept such pressures without retaliating by using some of their own power. In brief, it is clear that even the technical component in the overall pattern of consequences of the change in the oil power system is a potentially disturbing influence in the evolution of international relationships – for at least the relatively short-term future.

## The Economic Component

This is the component to which one turns for the most significant actual disturbances to the international system arising from the revolution in oil power. The initial traumatic impact of the oil price rises on the economic systems of the OECD nations has been well documented and often explained. The impact has been expressed in terms of the degree of inflation which it generated and in terms of the disequilibrium caused in the world monetary system by the flow of petro-dollars. Certainly the Western world reaction was, at first, one of consternation and of disbelief in the very idea that a group of countries like the petroleum exporters could unilaterally take such effective action to upset the world economy. Thus, initially, the general Western expectation was that OPEC could not hold together, so that the much higher prices for oil would be short lived and there would then be a return to the *status quo ante*. It was only after some delay caused by this sort of wishful thinking in the part of the OEDC countries that a serious appraisal was initiated on how to cope with the new situation of OPEC control over the supply and price of oil: and by this time the Western world was in its most serious recession since the 1930s.

Thereafter, however, increasing efforts were devoted to 'managing' inflation and the petro-dollar recycling problem within the framework of existing national and international institutions – and as a result of which some success was achieved. However, the situation has not been restored to the industrialised nations' norm of the more or less interrupted growth of the 1950s and 1960s. Indeed, there still remains a high degree of uncertainty as to the future shape of economic developments which are likely to be achievable by the Western industrialised nations in a situation of continuing OPEC control over the supply and price of oil.

To date, therefore, the consequence of the oil revolution for the Western industrialised world has been a much lower economic growth rate in the period since 1973 than the rate to which these countries had become used over the previous 20 years. Most important in the set of economic consequences arising out of this basic change in the outlook has been the rapidly rising level of unemployment in all the industrialised nations. Full employment (in practice meaning that less than 3 per cent of the working population would be without jobs) had become accepted as a central aim of economic policy making in all these countries and had given rise to the expectation on the part of their working populations that a job was indeed a permanent right which all should enjoy. The basic ability of the Western-style economic system to deliver permanent full employment has, however, now been severely undermined. Rates of unemployment have risen in general to about 6 per cent and in some countries to almost 10 per cent. Even more significant, one finds amongst specific groups – for example, minorities such as the blacks in the United States and school leavers in most countries – that unemployment rates are of the order of 20 per cent.

As a consequence, there has already been some deterioration in the degree of social cohesion in many industrial Western countries – particularly in those countries which, because of their generally inferior economic performances and/or because of their particular historic economic problems, were already relatively weaker than other countries in the same group. This deteriorating social situation and outlook has, moreover, sometimes led to increasing political extremism with the most important element in this being the recent rise of so called Euro-communism as a force to be reckoned with in the political outlook for, say, Italy or Spain. In other countries the same sort of political uncertainties have been expressed in political divisiveness which has led to parliamentary situations so finely balanced between 'left' and 'right' that the instability of governments has become a marked feature of the present political outlook.

It is, of course, inappropriate to say that the revolution in the international oil system has 'caused' all these developments. Obviously there were already pre-existing conflicts and tensions in most Western industrial nations which threatened the national successes achieved since 1950. The 'oil crisis', however, as defined that is from the viewpoint of the OECD countries, so exacerbated the conditions that the ability of the existing institutions to control the situation was undermined. Consequently, the whole Western industrial system took a serious turn for the worse; as yet, there is no certainty of recovery.

However, what has happened so far in the industrial nations cannot be described as disastrous for the short term. Most people still have jobs; real incomes are, in most countries, still rising, albeit slowly; and the ability of most Western countries to mitigate the effects of unemployment, etc., through generous social benefits has limited what might otherwise have been the socially very divisive consequences of stagnant or declining economic sectors and regions.

The same is not true of the other group of nations in the non-communist world, namely the poor, oil-importing countries of the Third World. These have suffered very much more severe consequences as a result of the oil crisis, and in most cases the situation continues to deteriorate without the nations concerned, moreover, being able to do very much about it themselves. The results of the changes in oil supply and price have been both direct and indirect and can be briefly described as follows.

Direct consequences arose because of the high degree of dependence of most of the world's poor countries on imports of oil. In the process of the early stages of development, involving industrialisation, urbanisation, and the motorization of their economies, through which these countries had generally started to proceed in the 1950s and 1960s, there is an inevitable rapid growth in the demand for energy. This demand could not be met locally for two reasons: first, because the lack of knowledge and/or of capital severely limited the exploitation of indigenous energy resources; and second, because of the long lead times that are inevitably involved in developing an energy production capacity. Thus, imported oil was the only short-term solution to the energy supply problem (note that no other sort of imported energy was available, except for limited quantities of coal or coke). However, as foreign oil appeared to be readily available at low and decreasing prices, through the infrastructure of the international oil companies which were ever ready, willing and able to oblige, it was a highly practical method of ensuring that the growing energy needs of a nation were effectively met.

Significantly, local alternatives to imported oil were sometimes stopped from being developed because international and national aid giving or loaning agencies, like the World Bank and the U.S. Export/Import Bank, would not extend their help in respect of oil developments – on the grounds that oil was readily available from the international oil industry. The oil industry's member companies were, it was also claimed, always prepared to put private-risk capital into new ventures providing, of course, that acceptable conditions for such ventures on their part could be agreed. Such conditions very often could not be agreed, because of national fears of getting involved with these companies.

Thus, by 1973, most developing nations of Latin America, Africa, and Asia (outside communist countries) were 70 or 80 per cent dependent on imported oil for their energy needs. These imports, moreover, could only be reduced at the expense of halting the newly developing sectors of their economies and/or at the cost of great hardship for the newly urbanised populations. These, by definition, are dependent on commercial energy (rather than previously, when the populations were still much more rural, on non-commercial energy – energy, that is, that is collected from the biological environment including wood, dung, waste material from agriculture, etc.). Thus their demand response to the greatly increased cost of imported oil was necessarily inelastic. They were not able to cut back imports without adversely affecting their basic development and, thus, their balance of payments.

It is in the context of balance of payments difficulties that the indirect consequences of the international oil crisis for the Third World oil importing countries can also be best seen. This arises because of the trigger effect that the oil crisis had on the level of economic activities in the industrialised part of the world. As the level of economic activities diminished, so demands for the non-oil exports of the Third World were in general adversely affected, leading, in many cases, not only to a decline in the amounts demanded, but also in the going prices of the commodities as their somewhat – or even very – inelastic supply failed to respond quickly enough to the reduced demand. Thus, for the Third World non-oil exporting countries, their much higher import costs coincided with the reduced incomes (in general) from the goods which they exported and out of which, of course, in the short term they could not diversify.

Additionally, to make their doubly bad situation even worse, they also had to face higher import bills for the goods and services they bought from the industrialised world. This occurred as the industrial countries, at the centre of the world economy, succeeded in 'exporting' the inflation which had been generated in their own economies, partly as a result of the oil system revolution. In other words, the lesser-developed countries' role in the international economic system – that of dependency on the other more powerful countries and institutions – had the effect on them of magnifying the economic consequences in the international oil system. They were left more dependent and vulnerable than ever and, in order to survive, had to resort to excessive borrowing against their all too limited assets. By 1978 many lesser-developed countries (LDCs) are at the very frontier of the borrowing abilities, so that even this 'solution' to the consequence of the oil crisis seems to have been played out.

Finally, in terms of the economic component in the consequences of the revolution in the world of oil power, it is necessary to put the vastly improved economic position of the OPEC countries themselves in perspective. The fourfold to fivefold increase in their oil prices, together with the take-over of the oil companies' concessions and other oil industry infrastructure at 'book' prices, rather than at replacement cost, has changed the economic outlook for most OPEC countries quite fundamentally. They became numbered – almost overnight – amongst the richest nations of the world in nominal per capita income terms, and so opened up the possibility of a reshaped international economic system in which their demands for goods and services would be able to sustain developments elsewhere. But for most OPEC countries there are difficulties in instantaneously creating the right mix of conditions for growth and development. There are two essential components to this problem: first, the countries' inability adequately to substitute for the Western world's industrialised nations in respect of their demand for exports from the lesser-developed countries; and second, the difficulties they have had – for social, political, and infrastructural reasons – in expanding their demand for the Western World's capital and consumer goods.

In brief, OPEC countries could not – as a matter of short-term economic reality – do much to create demand in the international system in spite of their new-found ability to earn much more foreign exchange from their exports of oil. Meanwhile, internally, the absence of infrastructure, expertise – both technical and managerial – and of instant absorptive capacity, led to serious economic management problems, most notably that of inflation as too many demands chased too limited a supply of most goods and services. Such inflation produced consequential pressures for further increases in the price of their oil exports in order that their real value could be maintained and/or as a means whereby governments could continue to ensure increasing revenues and foreign exchange holdings. In other words, even for the OPEC countries themselves there were unforeseen and sometimes well-nigh uncontrollable economic consequences which flowed, more or less immediately, from the changes in the international oil system. Thus, most of these countries were also unable to contribute to reducing the adverse international ramifications of the new situation. This situation remains unchanged in 1978.

## The Political Component

The consequence of the oil power revolution in political terms is also a very fundamental one. It is related essentially to the fact that the take-over of power in the international oil supply system by the OPEC countries constitutes the first successful challenge to the Western imperial system for well over 400 years.

The success of OPEC was not at first seen in this way – not even by OPEC's members, let alone by the Western industrialised nations. What successes OPEC had, in increasing oil prices and in 'cocking a snook' at the international oil companies, were initially seen as representing simply a temporary and an unsustainable phenomenon.

Success would soon expend itself and, thereafter, order and equilibrium would be reintroduced as the western world's advanced industrial powers reasserted their authority. In such circumstances OPEC would inevitably collapse, or, at least, be effectively constrained in its temporary success except, perhaps, that oil prices would remain somewhat higher than pre-1973 in real terms.

Politically, in other words, OPEC was supposed and expected to go away. It has, however, not gone away. Indeed, by 1978 it is significantly stronger than it has been previously. So much so, in fact, that the Western world now waits on its decisions on changes in the supply and price of oil with something closely akin to bated breath, while its 'experts' stand poised to make their instant calculations as to the implications of whatever decisions that OPEC takes. In essence, OPEC has quickly become an effective decision taking organisation which lies outside the framework of the traditional decision taking authorities in the western world and it is one, moreover, which traditional western politicking and diplomacy is able to influence only to a small degree.

The significance of this lies in the impact it has had on the small, select group of western industrial nations that – over the centuries and, most especially, over the post-1945 period – learned how to work together, basically for their own common good. Of course, major difficulties and differences continued to exist and to influence the relationships between the countries concerned. Nevertheless, by 1973 there was sufficient understanding, and enough effective organisations such as the International Monetary Fund, to ensure that major problems affecting relations between members of the group could be solved and that a generally common front could be presented to the rest of the world. Needless to say, the outside world was effectively excluded from any real degree of involvement in such decisions on the world's economic and political order.

This comfortable club-like existence evolved over the post-war years to the mutual benefit of all its members. Membership, moreover, was expanded only very occasionally in order to take in acceptable nations – such as the ex-enemy powers, notably West Germany and Japan – of World War II. However, no nation belonging to the Third World was admitted to the group. Additionally, in international organisations which nominally involved countries outside the western industrial nations, Third World representation was inevitably of a minority character, such that there could be no real effective influence by them on the majority views of the industrial powers. This situation extended, as far as was necessary, even to the United Nations

Organisation which was, however, very much more concerned with the East-West situation than with ways of involving the majority of the world's nations in helping to determine the future of the world system.

The success of OPEC has thrown a very large proverbial 'spanner' into these well-oiled works and has thus created conditions of uncertainty for the OECD nations. The latter, indeed, can no longer validly assume that the decisions acceptable to them, and carefully worked out between them so as to achieve such acceptability, will also be acceptable to a group of previous outsider nations which, if necessary, have the power given to them by their control over the supply and price of oil to undermine Western world decisions.

These uncertainties have now come to underlie the analysis of potential political options open for the reorganisation of the world system: particularly as the OPEC group of countries, being relatively new to each other in terms of their relationships and even newer in terms of their role in the world, cannot yet know either exactly what to do or, even if they did, then how to go about getting it in the most efficacious way. In the meantime, of course, uncertainty breeds suspicion and fear of the unknown – with consequential dangers arising out of hardening of attitudes. If, therefore, one accepts the basic premise that OPEC will not go away, then its creation and continued existence and the way in which it is treated by the Western industrial nations, could well be critical for the medium-term future of the world. As we shall try to show later, the consequences may be for 'good' or for 'evil', such that careful analysis and appropriate prescription becomes a must for an acceptable evolution of world conditions.

## b. On the Communality of Interests between the Industrialized and the Petroleum Exporting Countries

The revolution in oil power has led to a confrontational situation between the industrialised nations (the member countries of the organisation for Economic Co-operation and Development) on the one hand, and the petroleum exporting nations (the member countries of the Organisation of Petroleum Exporting Countries), on the other. Given this state of relationships between the two groups of countries a set of risks exists threatening not only the continued progress of the member nations of the groups, but also the peace and security of the world. Nevertheless, one can demonstrate that co-operation between the two groups of countries is possible. This would then set the stage for a possible integration of the interests of the two groups of countries in a joint effort to meet the challenge to the non-communist world system posed by the continued existence of so many poor nations in the developing world.

In order to examine this possibility it is necessary first to establish a basic point: namely that the industrialised nations and the petroleum exporting countries, despite the present confrontation over the oil issue, are all members of the same economic system – the Western system. These two groups are not, therefore, basically in conflict with each other, as in the case of the West's conflict with the Soviet system with its fundamentally different economic and political patterns of values and organisation. Given, however, that we have previously noted 'the evolution of a confrontational situation' between the Organisation for Economic Co-operation and Development and the Organisation of Petroleum Exporting Countries, it is necessary to explain this apparent dichotomy in our argument.

Within the Western system there now exist two wealthy blocs. On the one hand, a traditional, developed industrial group of nations which possesses technological, administrative and managerial and diversified industrial capacity, enjoying a system which is generally democratic and with an internally acceptable distribution of wealth and effectively diffused cultural institutions – including institutions of economic, social, and political significance. These attributes have combined to permit sustained progress.

On the other hand, there is the OPEC group of nations the members of which lack most of the organisational and traditional institutions referred to above, but which possess the oil resources which the OECD countries cannot do without. OPEC is well organised – partly as a result of the technical/managerial arrangements it has been able to make with the now generally dispossessed oil companies – from the oil supply point of view. This has permitted an ordered arrangement of the conditions they sought for the sale of their oil to the OECD countries within which there is now an awareness of the power which has been created and which can be sustained by the control exercised over the supply and price of oil. At the same time, however, OPEC countries depend on the OECD group of nations in order to be able adequately to utilize, for their own development, the great amount of money they receive from their oil exports.

Together, the members of the two groups generate about 90 percent of the gross economic product of the Western World, while accounting for under 40 per cent of its total population. In relation to these two groups, therefore, the remaining underdeveloped, oil importing countries find themselves in a difficult and even a dangerous economic situation. They have to pay not only high prices for the petroleum they need, but similarly high prices for the products manufactured in OECD countries. These Third World nations thus need the economic and financial support of the two rich groups of countries (they do, of course, also need other things like technological transfer and institutional changes, etc.), in order to ensure that their situations do not further deteriorate or even collapse because of the pressures which are put on them by the existence of so much poverty.

We would, therefore, argue that both the OECE and OPEC groups have a common interest in preserving the Western capitalist system. Indeed, the preservation of the system can be accepted as a positive, worthwhile objective because of the opportunities it gives for utilizing its productivity and its organisation to bring material and other benefits to an increasing number of people in an increasing number of countries. Its sustenance, let alone its survival, should, therefore, be an acceptable aim of the policies of both OECD and OPEC groups. Within the framework of such a fundamental common interest, the contemporary conflict situation between groups can be interpreted as representing the process of adjustment necessary following the revolution in which the latter group fought its way out of one-sided dependency into a mutually inter-dependent status with the former.

Moreover, part of this inter-dependence is in respect of the Third World to which the success of OPEC has given so much encouragement for possible future massive improvements in societal conditions. OPEC, in this respect, offers a hope and it is not in the political, nor, indeed, the economic interests of the Western system to see these hopes dashed – as well they might be if OPEC and OECD continue to conflict rather than co-operate with each other. Partly for these political reasons and partly for equally good economic reasons – most notably that of the continued expansion of the markets needed to absorb the increasing productivity of Western industry – the OECD countries also need the Third World in order to survive. It is in the context of such an important communality of best interests that OECD/OPEC co-operation is a *sine qua non* for the survival of the system.

In order to make such co-operation feasible as a matter of course, there are certain things which must happen. To begin with we would emphasize that, in exactly the same way as organisations like the OECD itself, or the European Economic Community as an example of a smaller grouping of industrial nations, are formally recognised as valid and desirable entities for the social, economic and political system of the industrialised countries, so it must be with OPEC in respect of its membership. OPEC must also be recognised not only as a valid grouping of countries within the Western capitalist system, but also as an organisation which provides an essential stabilising element in that system. And if this is so, then the protection and survival of OPEC should be an essential objective of the Western system.

This objective could be considered self-evident given that OPEC, as a Western capitalist system organisation, serves to protect and enhance the interests of its members in exactly the same way as does, for example, the European Economic Community, the break-up of which would generally be seen in other parts of the Western world to involve a weakening of the system as a whole. It is possible to evaluate OPEC in precisely the same sort of way, but there are also other special reasons that suggest that its continuation is not merely desirable, but also necessary at this particular juncture in the Western system's history. As a reason apart from all others, it is needed to complement OECD in the establishment of new relationships with the countries of the Third World.

Meanwhile, there is another set of reasons why OPEC should be defended. These we would summarize as follows:

> *(a) The break-up of OPEC will inevitably produce serious problems for its members leading to a high degree of unpredictability in the supply of oil to the Western system. We have already demonstrated the dependence of the West – including most of the poor (Third World), oil importing nations – on oil from OPEC countries and there is no need further to labour the point. The 1973–1974 oil embargo provided a real-world demonstration of the truth of this situation. If the embargo had been maintained for just a few months longer, many of the industrialised nations would have suffered serve economic failures, or even collapse: and to this must be added the serious political, strategic, and even military difficulties that could have occurred almost as quickly.*

> *(b) A rupture of OPEC could easily be the spark that set off a political and/or military explosion in the Middle East of incalculable proportions. It could, for example, open up the way for the Soviet Union to penetrate into this strategic zone which is so vital to the Western world. We do not think it worthwhile to take time in speculating on the probable or*

*possible consequences of such an event. We would simply note that, today, equilibrium in that area is precarious and that the countries concerned are moving only very slowly along the road towards an accommodation which will assure – or at least improve – the possibilities of peace in that conflict-ridden region. Indeed, the slow movement in the right direction might well be related to the additional major interests that have been created for the Arab nations concerned as a consequence of the significant economic and political results from OPEC's success. Perhaps the fact that they now have such an amount of wealth to protect has motivated them towards a basic understanding of the need for a peaceful and orderly solution to the problems of the region in a way similar to developments elsewhere in the world. For example, one notes that agreements have been achieved in Europe in the recent past, in spite of earlier conflicts between the countries concerned which had seemed likely to tear apart the continent – and which sometimes nearly did! Out of such agreement the nations concerned have become rich and have achieved prominence in world trade, etc. The opportunity now presented to its Middle East members by the success of OPEC suggests a similar possibility for them and has thus created the attitude and the atmosphere required for compromise. The failure or, even worse, the destruction of OPEC would negate this development and thus be a backward step, not only for the Middle East itself, but also for the rest of the world.*

*(c) More prosaic, but, nevertheless, still of immense importance, is the likelihood that the break-up of OPEC will not, as some observers and even statesmen have suggested, necessarily bring a lowering of oil prices, but more likely an abrupt increase within the context of a consequential chaotic supply and marketing situation. This could arise because oil buyers – especially the big and wealthy ones such as, say, Japan or Western Germany – would be very tempted to compete amongst themselves to assure a secure supply. The chances of this happening are enhanced when we take into consideration the fact that the international oil corporations now own very little of the potential production of the oilfields of all the big producing countries (except, and this only for the moment, those of Saudi Arabia) and so would be unable*

*effectively to intervene as a moderating influence in a possibly chaotic oil market.*

*Whilst it is true that oil consumption has not developed as hitherto expected (a situation made more acute from time to time by seasonal and other factors) and that there exists in total an oil producing capacity which is in excess of demand at present prices – so that there has to be regulation of production by the members of OPEC – it is by no means self-evident that the break-up of OPEC would lead to the former member countries of the organization simply opening all the wellhead valves and so cause ruinous price consequences. Individually, by choice or as a consequence of infrastructural problems (such as would be caused by the withdrawal of technical assistance by the oil companies fearing for the safety of their personnel), the petroleum producers may well cut back production as Libya did in 1973 and Venezuela in 1974 – or even cease to produce oil, in spite of the immediate consequences that sort of action would have on the flow of their revenues from the commodity. By now, however, most OPEC countries have a nest-egg of considerable proportions tucked away for use in such circumstances and there can be little doubt that they would choose to use it in the circumstances of a break-up of OPEC – especially if there was any suspicion that this was the result of foreign intervention.*

*(d) The failure of OPEC could be interpreted as evidence of a general malaise of the Western system and in particular as evidence of the lack of confidence which the member countries have in each other. Earlier in the book we noted the special relationship which appears to have developed between the United States and Saudi Arabia. An intensification of this could assure the United States of all the oil it needs and, in return, Saudi Arabia could receive guarantees of security with all the peaceful and military technology that it thinks it wants. Such a development could, for both parties, theoretically be better than OPEC. But this conclusion omits to take into account the fact that the development could lead to serious stresses between the United States and its associates in the OECD, given that European nations and Japan might then be left 'in the cold' for both petroleum supplies and for a*

*sufficient flow of petro-dollars. Thus, if the Untied States ceases to play fair with its allies, will the other members of OECD continue to play fair with the United States? And vice-versa? What will the attitude of Iran be if faced with this sort of situation? And what of the attitude of the other members of OPEC – especially the more radical members – towards the industrial nations? These, and many other uncertainties arising from the failure of OPEC, as a result of pressure or manipulation on the part of one or more OECD countries or because of other factors, constitute very difficult questions to answer and confirm the great uncertainty to which the demise of OPEC would undoubtedly give rise.*

Faced with the alternative of a dangerous game of bilateral agreements between oil exporting countries – many of which would need, but probably fail, to be kept secret – with all the high risks that this would carry for all the participants in both the OECD and the OPEC groups, the mutual benefits of a real understanding between the OECD group of nations and the member countries of OPEC appear to be considerable.

It is, of course, necessary to try to specify these considerable mutual benefits of OECD/OPEC cooperation. In our view the full acceptance of OPEC as a beneficial and stabilizing organism of the Western capitalist system, and its effective cooperation with the OECD on a basis of equality, would create the real possibility of the following desirable results being achieved:

*(a) The mutual guarantees of OECD and OPEC would ensure an adequate flow of petroleum for the world's anticipated needs in return, needless to say, for mutually acceptable levels of prices. This constitutes the bread on which the butter of additional advantages can be spread. This is self-evident as the basis for the revamped Western system of international organisation as to require no further elucidation.*

*(b) OPEC countries have had recent, first-hand experience of underdevelopment and its problems, and this may be of great value in the joint OECD/OPEC efforts to deal with underdevelopment elsewhere in the world. Hitherto, though OECD nations have had the task of assuming responsibility for helping the Third World nations and have achieved good results in some cases, much of the assistance going via the*

*international organizations or through various national agencies has been based on evaluations of a somewhat theoretical nature. The aid givers have not, as it were, felt underdevelopment in their own bones. Most of the aid givers concerned have not lived with the problems of underdevelopment – as in the case of a rich-born young man who, having been educated in the best schools and universities and then having been involved with the top echelons of his family's business, finally decides to dictate some effort and money to help the numerous poor in his country. His comprehension, without direct experience of the problems, is inadequate; and, likewise, his solutions will be inadequate. This has, up till now, been the case with much of the rich countries' aid to the poor countries.*

*For most of the OPEC countries, in contrast, it is now, in early 1978, little more than four years since they began to derive enormous financial returns from their petroleum and there still remains much direct experience of underdevelopment. They also have much first-hand knowledge about the process of assimilating wealth into poor systems. Their knowledge and direct experience in these respects could be useful for ensuring that a higher degree of success attends the OECD effort in trying to understand underdevelopment – and, indeed, in indicating what the underdeveloped think about underdevelopment and how they perceive its cause. The knowledge and experience would also be valuable in the formation of a more adequate strategy whereby the problems of development can be tackled.*

*(c) For their part, the countries of OPEC would have the opportunity to learn more rapidly about economic diversification and its associated problems. Such diversification constitutes a fundamental political objective for most OPEC countries, given the fact that their wealth is based on finite, non-renewable resources which can, moreover, to a large degree in the medium term (say, over 30–50 years) be replaced by alternative energy sources. Diversification sounds easy, but Venezuela's experience suggests otherwise. This is important because Venezuela is the relatively most developed – and the most diversified – member of OPEC. It*

*was as long ago as the 1920s when Venezuela was already exporting oil that Dr Uslar Pietri launched the idea of 'sowing the oil; that is, of using the revenues and income from it to 'grow' other economic sectors. The idea was well received and many understood its deep and dramatic message. Thus the message has been repeated many times over the years but, in contrast with the well designed advertisements which by reputation induce massive consumption, the idea of 'sowing the oil' has not been very effectively implemented. Thus, even now, in spite of 60 years of petroleum exploitation – and of petroleum income – in Venezuela, true diversification of the economy has not been achieved; in comparison, that is, with what has happened in the developed industrial countries.*

*Close collaboration between OECD and OPEC nations may help the former to understand the objectives and the mechanism of diversification. It could lead to the generalization, perhaps, of a phenomenon that has hitherto been very limited in its application in OPEC countries, namely technological and managerial transfers. The limited application arose from the experience that a few Venezuelans, for example, gained in management and in technical abilities and experience by working for private oil companies operating in the country. Their gradually cumulating knowledge of management and technology has permitted some of them to launch themselves, often with much success, into other private industrial and commercial activities in the country and so to form an important group for beginning the diversification of the economy. OECD/OPEC co-operation would seem to give a possibility to generalize this experience and so help to achieve benefits for both parties.*

*(d) OPEC/OECD cooperation can give credibility and permanence to OPEC countries as important contributors to the development of the world economic system. Thus, on the one hand, they can become a set of powerful partners for the OECD group of countries and, in sharing the responsibilities for the system, so create new opportunities for its development and improvement. The OPEC countries would, on the other hand, also become more credible in the eyes of the Third World countries. This is important because they are a familiar*

*group of peer nations which overnight became rich and are now in a position to help commit the total western system to Third World needs.*

*(e) Finally, the effort to maintain – and, indeed, to sustain – OPEC would permit OECD and OPEC jointly to proceed to create an agency for development. Such an agency, to be jointly administered and financed by the two groups of nations on a basis of equity, would provide a means whereby the pressing needs of the poor petroleum importing nations of the Third World could be met. This we see as the best option for greatly minimizing the risks to the stability of the western system and for contributing to its future well-being and development. We envisage it as an institution critical to the furtherance of worldwide economic development for the mutual benefit of both OECD and OPEC countries and, even more so, for the gradual elimination of poverty and gross underdevelopment in the rest of the non-communist world.*

# Chapter II – 7

# The Second Oil Price Shock

## A. The Great Oil Shortage Mystery*

### Introduction

For more than 20 years up to 1973 there was one essential component in the evolution of modern life styles in the west about which it seemed unnecessary to worry; namely, the supply and price of energy. The real price fell each year whilst the rapidly increasing quantities demanded seemed to cause no difficulties other than problems of pollution and environmental damage – about which there was little understanding and even less concern in the 1950s and the 1960s.

Householders could thus heat and power their homes, to the extent to which they were encouraged both by the energy supply industries and by the manufacturers of durable household goods, without giving too much thought to the cost of the greater quantities of energy they would need. On the contrary, tariffs were generally organised to favour the large rather than the small user so that more use produced a falling average price. Transport developments were devoted more or less exclusively to the expansion of energy intensive ways of going places – by air and by private motor vehicle – so that other more energy-conserving modes of travel such as trains, buses and ships, became uncompetitive leading to services being phased out or severely curtailed.

* Reprinted from *The Chelwood Review*, No.7, January 1980, pp.18–23.

The changing pattern of the location of homes – at increasing distances from work places and from the services provided by city centres – similarly encouraged high energy use. So little attention was given to this aspect of planning that facilities at locations which could not conveniently be reached by car were closed down or relocated to places where more energy had to be used to reach them. Opinion even swung round to the view that factories and offices should be located in dispersed and distant places so that the journey to work became dependent on the motor-car, used on average by less than 1.3 people, at a consequential high energy use factor per employee.

And within the factories and other places of employment, the declining real cost of energy and its ready availability encouraged energy intensive processes and patterns of work organisation, together with a tendency for energy to be wasted in inefficient systems because it was hardly worth anyone's while to avoid waste. Managers had to spend their time worrying about inputs which were increasing in price, rather than about energy, about which they could be sure that this year's supply would cost less than last year's and that next year's would cost still less.

Out of this combination of circumstances we have moved to our energy intensive patterns of development – of working, of moving and of living – to patterns which are, indeed, the very essence of the material western way of life for the greater part of the populations of the industrialised countries and to which much of the rest of the world's population aspires. The fundamentally changed situation over the supply and price of energy since 1973 now puts in question the validity of the patterns that have already emerged and casts doubt on the aspirations in terms of the likely availability of sufficient energy, given a continued development of the present patterns of societal organisation. If an appropriately increased supply of suitable forms of energy cannot be maintained, then the outlook for the stability of the economic, social and political system of the western world must be in question because the likelihood of our societies accepting radical changes in the short term is small.

The essential questions in respect of the development of energy policies are thus, first, what are 'appropriate increases' in the supply; and, second, what are 'suitable forms of energy'. We can dispose of the second question rather briefly in that it necessarily implies a situation of very little change from the present position. All nations in the non-communist world have, since the early 1950s, developed systems – both technical and societal – which run essentially on oil and they have thus committed their futures to oil-based economies and societies. Neither individuals nor communities can afford to change other than slowly in respect of this dependence. This is partly because of the costs that would be involved in re-equipping homes, offices and factories to run to alternative sources of energy and the even greater costs which would occur if energy delivery systems had to be changed from, say, low cost oil pipelines or bulk oil transport by water and rail to high capital cost and relatively energy ineffective high-tension electricity lines, as the only means presently available for making energy from nuclear power and most energy from coal available to consumers.

It is also partly because of the political and social 'withdrawal symptoms' which can be expected in the event of oil having to be replaced by alternatives incapable of sustaining the complete range of activities based on the use of oil products. These problems greatly reduce the possibility of the otherwise apparently more attractive use of alternative energy sources in the short term. Collectively all of the above arguments exacerbate the current problems of the adequate availability of oil to satisfy the needs of the world's most important nations.

## Demand for oil exaggerated

This, fortunately, does not apply in respect of the other essential question – that of defining the likely size of the 'appropriate increases' in oil supplies for which we have to plan. There still appears to be a widely held belief that the demand for energy in general and for oil in particular is still evolving very rapidly – based on the extrapolation of the extraordinary experience over the two decades prior to 1973, when the annual average rate of increase in oil demand was of the order of 7½%: a rate which gives a doubling time of under ten years. The use worldwide of 500 million tons of oil in 1950 thus doubled to over 1000 million tons in 1969 and then doubled again to well over 2000 million tons by 1970 when there was a confidently held belief in the oil industry that it would be required – and able – to deliver 4000 million tons in 1980.

But this experience and its associated extrapolation are now well out of date and even more out of line with what has actually been happening in the non-communist world since 1973. Indeed, in 1979 non-communist world oil demand is only a little more than 5% above its 1973 level. Had the growth trend up to 1973 continued then demand by now would have been well over 50% higher than six years ago. In other words, the world is now using only just over two barrels of oil for every three that were expected to be used by 1979. Instead of a rate of increase in demand of more than 7%, the last six years have seen an annual rate of increase of under 1%. Given the now expected continuation, or even the exacerbation, of the adverse conditions which have faced the world economy since 1973, the present outlook for the demand for oil seems unlikely to be very different from recent experience. The prospect of an annual increase in demand which gets back to anything like that of the 1950–73 period is virtually non-existent. Yet it is that out-dated experience which still appears to dominate contemporary thought on the size to which the oil industry will have to grow in future.

Against the now most likely outlook for the future demand for oil there is little prospect of oil running out – as has often been suggested – or of demand reaching some level at which the industry's capacity to sustain the production facilities necessary to meet it cannot be stretched. Indeed, at present rates of increase in demand it will be well into the second decade of the 21st century before production is 50% higher than the present level and almost 2040 before the industry needs to be twice its present size.

## But politics could constrain supply

As there is a general degree of acceptance that the oil industry is capable of being expanded to at least twice its present size, no technical restraint on the development of oil production for at least another two generations thus appears to be necessary. This eliminates the spectre of a physical/technical restraint on its use, excluding, of course, any constraints which may become necessary because of local and/or general atmospheric effects of burning hydrocarbons – including oil.

Though this conclusion may appear to be a relaxing one, compared with many alternative scenarios presented recently, anything more than a very modest degree of relaxation over the future of oil would be most inappropriate, given that there is an immediate and serious challenge to the western world's well-being posed by uncertainties over the supply and price of oil arising from international economic and political conditions. As the use of oil in the short term is not substitutable by alternatives and as demand is not actually declining, the power that the Organisation of Petroleum Exporting Countries (OPEC) has to determine the supply and price is very clear. The dominant oil exporters are in a firm position to lever up the price either when they choose to do so, or when an event affecting the level of production in one or more of the member countries of the Organisation makes it inevitable. These countries are, moreover, able to argue, in the light of widely accepted views that oil is inherently scarce, that such price developments are appropriate in order, first, to curb the growth of oil demand on the grounds that too high a growth rate is not 'good' for the system and, second, to protect their own resources from too rapid a rate of depletion in order to save some of their oil for longer term future needs.

Moreover, minor demand restraints need not have any adverse effect on OPEC's members as they can, if necessary, introduce yet further price increases and even more supply constraints. The essential solidarity of OPEC as an effective international organisation makes this possible. In addition, in periods of particular supply difficulties in the market (as in the first half of 1979), its members can not only take advantage of sales opportunities at prices standing well above the official levels, but can then also argue that their official prices are not high enough to clear the market and so escalate those as well.

## No effective challenge to OPEC

These conditions represent the reality of the present situation. The latter shows no sign of changing for the better as there is no effective challenge to OPEC power by the world's rich industrialised countries. Such a challenge is impossible because of a combination of energy/oil policy elements. First, those Organisation for Economic Co-operation and Development (OECD) countries with indigenous energy resources of significance see their value rising as a result of OPEC policies and thus judge their short term interests to be well served by the developments in the international oil situation. Second, most OECD countries still argue that policies which aim severely to restrict energy use as such are economically and/or politically unacceptable. In the short term costs of such policies do, it is generally thought, exceed the short term benefits. Thus reductions in energy use can only be sought through exhortation and by policies which do not affect the accepted energy intensive ways of doing things and of going places. This often restrains even the use of the pricing mechanism as a means of constraining demand, let alone the application of more interventionist, regulatory policies which, in the short term at least, appear to be necessary in order to oblige our energy wasteful and/or careless populations to change their energy using habits. As a consequence the chances of a marked reduction in the use of oil, as opposed to a reduction in the growth rate, are very limited. Third, many OECD countries have decided that their own oil – and/or natural gas resources – are so valuable that they ought not to be produced in the short term to anything like the maximum technical, or even the optimum economic, level.

Thus there are restraints not only on production from discovered resources – as in the case of Dutch and Australian gas – but also restraints on investments in the search for new resources, as in the case of Norway and the UK, where the licensing of potential oil bearing areas has been deliberately slowed down. Such policies, which, in effect, serve to discourage the maximum possible production of indigenous oil and gas, are pursued not because it would be uneconomic to increase exploration and exploitation. On the contrary, the supply price of oil is now well above the costs of production even of the most expensive continental shelf oil now in prospect. The restraints arise because of the mistaken fears for a scarcity of oil by the late 1980s/early 1990s and also because of the lack of recognition of the seriousness of the short term oil supply/demand relationship and the failure to recognise the influence that decisions for higher production levels in the short term could have on the world energy situation and outlook.

Overall, there appears to be a kind of fatalism in the west – namely, a belief that nothing can be done to offset the basic domination of the world energy market by OPEC suppliers in the foreseeable future. However, in terms of both demand and supply components for the 10–15 years outlook, such fatalism is, as we have shown, misplaced. It is, nevertheless, effective in undermining the OECD world's ability to open up negotiations with OPEC over a longer term strategy for the supply and price of oil: and, in the absence of such an agreement, the difficulties in the west's economic system created by the energy/oil problem must continue.

In the six years since 1973 we have been made painfully aware of the significance of oil and of the power of the oil exporting countries in determining progress in the western economic system. A continuation of the rate of price increases which we have had in recent years (the real price of oil is already an order of magnitude higher than it was in 1970) and, even worse, a further period of continuing uncertainty about the availability of oil, as a consequence of the inherent insecurity of OPEC oil supplies in the present situation, together appear to pose a serious threat to the economic and political system of the west for the next few years at least. In this context, effective national programmes for reducing oil use by very much more than the modest 5% agreed at the Tokyo summit in July 1979 and for effective national efforts in all appropriate cases to increase indigenous production of oil, need to be judged in terms of their potential contribution towards ensuring a higher degree of stability in the western system. All member countries of the OECD can, of course, contribute to the demand constraint element in this effort;

though some more than others as a result of particular national circumstances. In addition, however, those member OECD nations which have the opportunity to increase supplies of oil and of gas from their own resources have a particularly grave responsibility.

## An option to disaster

These countries' continuing failure to produce as much oil and gas as possible may satisfy their objective successfully to conserve their reserves so as to ensure for themselves some residual light and warmth in the 21st century world which, they fear, may be bereft of other accessible resources. Quite apart from the absence of any real evidence for such a physical scarcity of resources at that time or, indeed, at any other time, any immediate satisfaction from a decision to produce less than they could produce – or, indeed, as much as market forces indicate ought to be produced – will be fleeting indeed if, as a consequence of too few alternatives to OPEC oil and gas being available in the meantime, the whole western system goes into a downward spiral of diminishing confidence and of increasing economic, political and social difficulties. It is against this prospect for the 1980s – emerging out of the continuation of present policies being pursued by OECD nations in the context of the political control exercised by the small group of nations whose oil resources happened to have been discovered in the period up to 1970 and on which the world came unwisely to depend – that the alternative possibility of maximum indigenous production and severely regulated consumption must be judged. As an option, that is, that creates, first, a real chance of our avoiding the worst dangers of the near-term future and, second, of opening up the prospect for a longer-term global energy strategy based on a more rational use of the world's preferred, but finite, resources of oil and gas and on the scientific, technical and commercial development of acceptable alternatives for the 21st century.

## B. Oil – An Overpriced Abundance*

OPEC has just celebrated its 20th birthday. It is also almost 10 years since its momentous decision in Caracas in December 1970 to control the supply and price of oil. Since then the real price of Saudi Arabian marker crude has increased by an order of magnitude (from the equivalent of $1.60/bbl to over $16.00/bbl in 1974 $). Also, since 1973 there has been an equally marked change in the growth rate of the non-

---

**\* Reprinted from *Energy Policy*, Vol.8, No.2, June 1980, pp.82–83.**

communist world's use of oil. This has fallen from more than 7% per year during the 25 years up to 1973 to under 1% since then.

These recent changes in the industry are dramatic. Even more dramatic, however, is the loss of confidence by the western world's oil industry in its future. Up to the early 1970s the oil companies wrote with enthusiasm and optimism about the future. Inglis and Jamieson of BP, for example, felt able to say, 'it can be asserted that there will always be enough petroleum for the world's needs',[1] whilst Hols of Shell wrote, 'By the year 2000 the non-communist world's total energy demand will be 300 million bbl/day of oil equivalent of which oil will supply 170 million bbl/day'.[2] Recent documents from the same companies indicate a complete *volte face*.[3] They now stress the impending end of the oil age and the need to reduce demand and to develop substitutes as a matter of urgency.

The changes are not unrelated. OPEC's domination of the international oil system has undermined the status of the oil companies which have reacted by clear expressions of their belief in the inability of the oil industry to survive their overthrow by more than a few years. OPEC, for its part, has used the oil companies' declaration of the very near future bankruptcy of the industry (in potential supply terms) as the main justification for putting up the price of oil and for cutting back the level of production.

Most of the rest of the non-communist world's nations have, meanwhile, not only priced the oil they produce at the OPEC-determined level, but have also increased taxes on its use so as to raise prices even higher: in the interests, so it is argued, of curbing the demand for a commodity which is believed to be inherently scarce and in the belief that such pricing policies are the appropriate market solution to the longer-term adverse supply-demand situation.

There is, however, no evidence of a shortage of oil – other than that created by political action (the reality of which is not in question, but to which sort of action scarcity pricing is hardly an appropriate reaction). Even with currently known proven reserves and the degree of appreciation of those reserves which can be expected from technological developments and higher prices, there is enough oil to meet demand for the next 30 years – even if one assumes an annual growth rate in its use which is three times that of the average of the last six years!

Beyond this known availability of conventional oil there are also the world's other oil resources. first, the conventional oil in the more than 50% of the world's petroliferous regions which have not yet been effectively explored; and second, the even greater volume of unconventional oil in, for example, the tar sands of Athabasca (Canada), the oil belt of the Orinoco (Venezuela), and the oil shales of the USA, Brazil, Malagasy and India. In the case of *at least* 3000 x $10^9$ bbl of these oil resources (equal to three times present proven and probable reserves), the technological and environmental questions associated with their production imply nothing more than the continued development of methods which are already known and the real costs of which will decline as production experience is gained.

Against the background of more than adequate reserves in relation to the now most likely rather slow development of demand, the imposition/acceptance of a price for oil which is well above its long-term supply price would be a sick joke were it not so serious in terms of the damage it is inflicting on the western world's economic system. The mechanisms for imposing the rigid discipline and fundamental changes required by the system in the light of a hypothesized scarcity are, unhappily, receiving well-nigh exclusive attention in energy policy making, whilst the steps required for the establishment of the institutions and conditions necessary to ensure the production of the world's remaining resources are not being taken up.

The exploitation of all other energy sources on the scale required to substitute oil and gas creates barely acceptable environmental and safety problems. They require direct investments to produce them and indirect investments to make them compatible with existing energy-using systems. This puts their resources costs well above the present oil price of around $30/bbl. By contrast, there are few conventional oil and gas reserves which cannot be very profitably produced at this price and there are more than enough unconventional oil and gas resources which will become economically attractive to exploit, given a decade of development experience – should they then be required to supplement conventional supplies.

The contemporary too-high price of oil is the unhappy result of inappropriate institutional arrangements for their development and of inept national and international politico-economic policies. A still further increase in the oil price will be the result of the failure to correct these man-made faults in the system, rather than the inevitable outcome of the depletion of the world's still very considerable oil resources.

## C. An Alternative View of the Outlook for Oil*

## Introduction

Since the changes in the world of oil power over the past decade there has been a general tendency to view the international oil system as comprising a powerful and permanent producers' cartel selling inherently scarce oil to a set of competing and increasingly voracious importers: with the conclusion that the price of oil must rise still higher than the $34 per barrel of 1982. Current price 'weakness' is, it is argued, inevitably a short term phenomenon, fathered by the deep economic recession since 1981. Unhappily for the member countries of OPEC – and surprisingly for most of the rest of the world – the reality of the international oil market situation is already very different from this generally presented view. Reality is diverging yet further from the myth of oil scarcity and inevitable high prices built up since 1973.

There are now over seventy oil producing countries in the world and, of these, only thirteen are members of OPEC. The OPEC members collectively accounted for barely a third of world oil production in 1982 and their share is still falling. Moreover, at least 25% of oil producing capacity in the non-communist world is currently unused and this, together with the equally significant under-utilization of producing capacity in the other main energy supply industries (coal and natural gas), is generating fundamental downwards pressure on prices. The consequential strains within OPEC have been temporarily masked by collusion over supply limitations and price fixing between OPEC and other countries with vested interests in keeping up the price of oil, but all this has produced is a situation of highly unstable equilibrium.

It is, indeed, undermined by the clearly emerging longer-term oil supply/demand outlook which is one of imbalance, with potential supply outstripping potential demand as long as prices remain significantly above $15–$18 per barrel (for the marker crude in 1983 dollars). World proven oil reserves, both absolutely and relative to the annual use rate, are at an all-time high. In terms of the nominal reserves/production ratio they stand at a level of more than 35 years. Even if oil use were to increase at an average rate as high as 2.5% per annum (compared with an annual average rate of *decline* of 0.6% over the last decade), then it would be well into the first decade of the 21st century before the production of these already known reserves had to stop growing.[4] Most of these

* Reprinted from the *Petroleum Economist*, Vol.L, No.10, October 1983, pp.392–394.

reserves are, moreover, very low cost oil, as shown in Table II-7.1. Indeed, over $600 \times 10^9$ barrels can be profitably produced at an oil price of $5 per barrel or less, so that modest declines from current prices need make no difference to the supply schedules for almost all the oil that the world will be using over the rest of the century (see Table II-7.2). Finally, it is important to note that reserves are now being added to the existing stock of the industry at almost twice the rate of use, and that the additions are also geographically widely distributed, so further enhancing the prospects for a more diffuse world pattern of production. There is no evidence that any significant proportion whatsoever of these additions to the world's already generous level of reserves will cost more than $10 per barrel to produce so that upward price pressure, from today's price levels, arising from the costs of exploration and exploitation (with 'normal' returns to risky investment thrown in) is essentially non-existent.

### TABLE II-7.1
### Deriving the availability of low-cost oil* in 1983 (in $10^9$ barrels)

| | | |
|---|---|---|
| 1. Remaining Proven Reserves in 1973 (by definition these were economic at the pre-crisis oil price) | | 635 |
| 2. Oil Use 1974 | 190 | |
| of which, higher cost oil | 40 | |
| Net use of Low-Cost oil | 150 | −150 |
| 3. Remaining Reserves in 1983 of Low-Cost 1973 Proven Reserves | | 485 |
| 4. Additional Reserves Proven 1974–82 | 245 | |
| of which higher cost reserves | 120 | |
| Net additional Low-Cost Reserves Proven since 1973 | 125 | +125 |
| 5. Total remaining Low-Cost Oil Reserves (at beginning of 1983) | | 610 |

* Oil which is economic to produce with a price of $4 per barrel or less in 1980 dollars.

**TABLE II-7.2**

**Future use prospects for currently available low-cost oil**

(Reserves 610 X $10^9$ barrels as derived in Table II-7.1)

| | | |
|---|---|---|
| 1. | World Oil Use, 1982 | about 19.5 x $10^9$ bbls |
| 2. | Low-Cost Oil Reserves to Production ratio | 31.3 years |
| 3. | Effective R/P ratio with a –0.59% per annum continued decline in oil use (= average decline rate in 1973–82 period) | 35.3 years |
| 4. | R/P ratio assuming an average 2.5% per annum growth in oil use | c.23 years |

5.   Assume:
i)   a 2.5% per annum oil use growth
ii)  that 75% of world oil demand will be supplied
     by low-cost known reserves
iii) a depletion curve restraint (*viz.* a minimum R/P ratio
     of 11 years) on the use of low cost oil.

Then, the year when production of currently proven
low-cost oil will peak is:                                           2001

## Too high demand expectations

On the demand side expectations have been, and remain, too high, partly because the forecasters have under-estimated the impact of the oil price shocks (and other factors) on the prospects for economic growth in the world economy, and partly because they have tended to extrapolate the evolution of demand based on the experience of the unique conditions which existed over the 15–20 years prior to 1973 – and which combination of conditions, with high growth rates in all parts of the world simultaneously, under the stimulus of low, and falling, prices can never be repeated. Given the new post-1973 conditions, there was – and, indeed, still is to a high degree (as, for example, in a recent IEA study[5]) an inexplicable unwillingness to consider both the price and income elasticities of demand, and the significance of the substitution of the use of oil by other sorts of energy. Thus, markedly lower rates of use, compared with the pre-1973 period, are seen as the aberration, rather than the return to the normality of long-term energy growth rates from which the experience in the 1948–73 period was the essential divergence. Given continued weak international economic performance and the demand effects of continued high oil prices, it is inappropriate to predict the use of oil to grow at a rate exceeding 2.5% per annum between now and the end of the century.

In the context of these supply and demand conditions, the widely accepted hypothesis that the twelvefold increase in (real) oil prices between 1973 and 1981 was the classic market signal of resource scarcity is clearly undermined. The oil price increase has been no such phenomenon: rather has it reflected a virulent new strain of oligopolistic pricing in the context of short-term supply constraints by producers exercising a temporary control over output levels, and with the effects seriously exacerbated by inappropriate user responses caused by the industrialized countries' misinterpretations of the developing market situation. The level to which the price of oil rose by 1981 under these circumstances thus provides a singularly inappropriate base from which to speculate on future price developments.

## Long-run supply price

The low long-run supply price (LRSP) of oil in the early 1970s, was defined at a little over $1 per barrel by Professor M.A. Adelman at the time.[6] His conclusions have now been confirmed in a recent analysis of the early 1970s situation made by Dr C. Fayat[7] – with the benefit of hindsight. Given the pre-1973 expectation of a high rate of continuing increase in oil use it was then reasonable to hypothesize that the supply price curve would have to turn up to enable successively higher cost resources to be profitably exploited. Company spokesmen at the time suggested $7 per barrel[8] (= about $14/bbl in 1980 dollars) as a reasonable level for the early 1980s. Meanwhile, however, the collapse of oil demand has stretched out the availability of low-cost oil (which amounts to more than $600 \times 10^9$ barrels as demonstrated earlier in this article), so that a more likely future LRSP curve now has a much flatter shape – rising to no more than $10 per barrel (in 1980 dollars) by the year 2000.

In other words, for the period up to the end of the century – and probably well beyond that – current rational expectations on supply/demand relationships indicate a supply price for oil which remains much lower than the highest price of oil in 1981/82 and even lower than a level to which the price has now fallen (of about $24.50 in 1980 dollars). There is thus no basic economic justification for the present price of oil, and for the efforts by OPEC to maintain it at this level – let alone for the apparent acceptance by governments and others in many non-OPEC countries that high prices for oil are an inevitable part of the current situation and outlook.

The oil price battle that remains to be determined is the degree and speed with which prices in the market will fall towards the very slowly rising long-run supply price in today's near stagnant demand conditions. A price collapse is by no means out of the question, but it is much less likely than a continued attrition of the price (with occasional upswings for political or cyclical economic reasons) under the influence of convergent OPEC/OECD policies which elevate moderation in the rate of change to the status of a principle. Success in applying this principle will produce a 1990 price which is 65–95% of its 1980 level of $28.50; a 2000 price at 55–80%, and a price in 2010 at 45–75% of the 1980 price level (in 1980 $). This price development would certainly give OPEC the best it can hope for in the medium term (in terms of both revenues from oil and the evolving status of the organization), and it is at least arguable that the rich, industrial countries will secure a net benefit from the economic stability which ought to ensue from such an evolution of the level of oil prices.

What is less certain is a favourable impact on the world's developing countries of a price for oil which remains relatively high for the next thirty years. Their prospects for expansion will, as a consequence, remain constrained, with potential political and economic repercussions for the rest of the western world's economy. Such continuing high oil prices also put the long-term survival of the oil industry in doubt. The motivation that such high prices give for not using oil and for developing lower priced (though higher cost) alternative sources of energy is a powerful one. In these circumstances the world could well end up early in the 21st century with a rapidly declining international oil industry, in a situation in which most of the world's potential oil resources remain unused or undiscovered – and unwanted.

## References

1      K. Inglis and W. Jamieson, *Factors in the availability of petroleum to meet the changing structure of future energy needs*, a paper presented to the World Power Conference, Tokyo, 1966.

2      A. Hols, *Oil in the world energy context*, a paper presented to the Economist Intelligence Unit's International Oil Symposium, London, 1972.

3    British Petroleum, *Oil Crisis... Again*, British Petroleum, London 1979 and Shell, *Energy Importance for the Coming Decades*, Shell, London, 1979.

4    This, incidentally, does not assume exhaustion of these reserves: there would still be sufficient left after peak production for a reasonable decline rate to be possible. See Tables II-7.1 and 7.2 for the derivation of this conclusion.

5    International Energy Agency. *The Energy Outlook to 2020*, OECD, Paris, 1981.

6    Adelman, M.A., *The World Petroleum Market*, The Johns Hopkins University Press, Baltimore, 1972.

7    Fayat, C., "The Theory of Exhaustible Resources: Was the Price of Oil before the Crisis close to the Competitive Price?" *Energy Exploration and Exploitation*, Vol.1, No.4, pp.287–308.

8    Chandler, G., "The Changed and Changing Energy Scene". A paper presented at the Summer Meeting of the Institute of Petroleum, Harrogate, June 1973.

# Chapter II – 8

# Post-Shock Reactions

## A. Outlook for the international oil market and options for OPEC*

Since the changes in the world of oil power there has been a general tendency to view the international oil system as comprising a powerful and permanent producers' cartel selling inherently scarce oil to a set of competing importers. The many proponents of this view have presented the latter group of countries as having no choice but to use as little oil as possible in the short term and, in the longer term, to turn to the use of energy alternatives. Unhappily for the member countries of OPEC and surprisingly for most of the rest of the world in light of the generally presented and accepted view of the future of oil, the reality of the international oil market situation is already very different – and is diverging yet further – from the myth built up over the last 10 years.

There are now 73 oil producing countries in the world, as shown in Figure II-8.1. Of these only 13 are OPEC members. These thirteen collectively accounted for less than 30% of world oil production in 1982 and their share is falling rapidly. Moreover, about 25% of oil producing capacity in the non-communist world is currently unused, so generating short-term downward pressure on prices. (It is important to note that there is also significant under-utilization of producing capacity in the other main energy supply industries – coal and natural gas – with, for example, about 30% of Dutch gas producing capacity shut-in because of

*Reprinted from *Energy Policy*, Vol.12, No.3, 1984, pp.27–34.

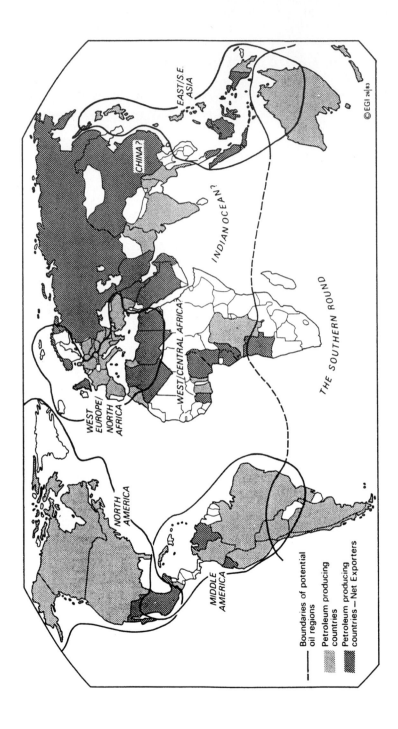

Figure II-8.1: Oil producing countries, net oil exporters and possible future world oil regions

lack of markets, so adding to the downward pressure on oil prices.) As far as the longer term is concerned it is also important to note that world proven oil reserves, in both absolute terms and relative to the annual use rate, are at an all time high. In terms of the reserves-production ratio they stand at a level of ±35 years so that, even if oil use were to increase at a rate as high as 2.5% per annum (compared with an annual average rate of *decline* of 0.6% over the last decade), these reserves – which simply represent the industry's 'shelf-stock' of the commodity – would sustain the increase in production until well into the first decade of the 21st century. Annual additions to reserves, moreover, are running well ahead to the rate of use. These additions to reserves are also geographically more widely distributed than even before, so further diffusing the prospective potential for production. These are the fundamental considerations which must shape the future of the world oil market for other than the very short term, when ephemeral, political and institutional factors will necessarily be important. They are considerations which offer opportunities for challenging the conventional views of the last decade on the international oil market. The need for a reappraisal of conventional wisdom has already been demonstrated by the rapidly falling prices and a number of other developments in the international oil market over the last year.

## The world oil market in 1983

It is, indeed, reasonable to speculate that the world oil market is on the brink of changes which are potentially even more astonishing than those which occurred in the early 1970s though, compared with that earlier period, there are now even greater uncertainties which face the analyst in his efforts to forecast the prospect of the industry over the next 10 years and through to the turn of the century.

In this article it will be argued that a combination of a set of inter-related factors has brought the hitherto powerful and expansionist oil industry to the verge of disaster. These factors are as follows:

a.  *An unsubstantiated but, nevertheless, widely accepted belief that oil is an inherently scarce energy source which the world has been depleting too rapidly.*

b.  *The overpricing of the commodity by OPEC in that organization's short period of politically and institutionally based control of the market. An increase in the fob price of oil (in real terms) in the first half of the decade by a factor of ±5, from its low in 1970, was just about acceptable, but a more than twelvefold increase in less than a decade (see Figure I-7.3) was certainly not.*

c.      *The loss of confidence by the international oil companies in their opportunity and ability to 'run' the industry in the aftermath of OPEC's success in unilaterally increasing prices so dramatically and, even more important, after the nationalization of the companies' assets in most of the oil producing countries in the second half of the decade.*

d.      *The 'greed' of non-OPEC oil (and gas) producing countries in using OPEC's control over the price to maximize their collection of economic rent from their own (relatively low-cost) production, coupled with the naivety of western economic policymakers when they took their decision to treat the rapidly escalating prices as providing the market signal for the competitive long-run marginal price of energy, instead of the short-term aberration from the latter that it really represented (see Figure II-8.2).*

e.      *A continued failure by the western world's policymakers to create and develop adequate institutional and financial arrangement to ensure the comprehensive exploration for, and the exploitation of, the world's oil resources, the greatest potential for which lies in the developing countries – the countries hardest hit by the economic crisis which has accompanied the difficulties over the supply and price of oil since 1973.*

As a result of the impact of these factors potential disaster now faces the international oil industry because of the consequential increasingly severe downward spiral of demand, a lack of interest in investment in oil and gas in many parts of the world, and increasing disequilibrium in the industry in respect of all its component parts (eg between crude oil producing capacity and production; refining capacity and use; tanker availability and demand; petrochemical facilities and their use, etc). These conditions not only create short-term problems, they also undermine the long-term prospects for the future of oil so that the world is in danger of failing to utilize more than 50% of even the

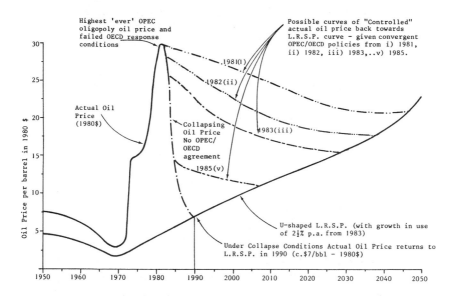

Figure II-8.2: The price of oil 1950–2050: alternative options

most conservative estimates of the industry's ultimate resources of $3000 \times 10^9$ barrels. There is a deliberate policy in many parts of the world of increasing the use of inherently more expensive and/or environmentally less safe alternative sources of energy (including coal, nuclear power and renewable energy forms). The survival of the world oil (and gas) industry – not excluding that important part of it which is in, or under the control of, the OPEC countries – thus now depends on the rapid implementation of policies which are radically different from those which have been pursued in recent years by various groups of countries. Major initiatives are required on the part of the industrialized nations with the objective of breaking the vicious circle of adverse conditions, whereby the prospects of the international oil industry is being undermined – with the hope that such action will also generate the prospect of an expanding world economy.

## The policy action required to change the situation

There are four separate but inter-related elements in the policy action which is now required as a matter of urgency. These are as follows:

> First, *the elimination of constraints – notably the penal fiscal regimes and the production limitation policies – which serve*

*to limit the level of investment in the oil and gas industries in the OECD countries. This is required in order to stimulate interest in, and the expansion of, the industry in the western industrialized nations, where about 85% of non-communist world use of oil and gas is concentrated and where the demand for oil is declining so dramatically under conservation and substitution policies.*

Second, *an end to state-subsidized investments in high-cost production of alternatives to conventional oil and gas – notably nuclear power, but also including oil from coal/shales, coal gasification, and renewable sources of energy where these involve high costs per unit of energy produced. Expenditure on such high cost alternatives to conventional oil (and gas) should be limited in the meantime to research and development investments.*

Third, *the creation of the necessary institutions and the appropriate conditions for the exploration and exploitation of the oil and gas resources of the Third World.*

Fourth, *the initiation of negotiations with OPEC for the establishment of an agreed global strategy on the future supply and price of oil. Prospects for the latter are illustrated in Figure II-8.2. OPEC's continued existence, which, as shown in the early 1983 negotiations to stabilize the falling oil price, now depends on OECD countries' attitudes and policies, is also a* sine qua non *for improved North-South relations which, in turn, are necessary for the continued development of the western economic system. In this context there is thus a communality of interests on which OPEC-OECD negotiations for convergent policies can be based.*

## The evolution of the world oil system – with appropriate policy action

This can be predicated in the context of the institutionalization of OECD-OPEC convergent relationships out of which a global energy strategy can be defined and implemented (albeit imperfectly, as with the results of all international agreements), so that uncertainty on the part of both producers and consumers is minimized. Success in this respect would have the following effects. *First*, low-cost oil and natural gas production would be re-established (relative, that is, to all other energy alternatives) as the central elements in both short and medium-term energy supply prospects (to the end of the second decade of the 21st century at the earliest). The renewed expansion of the oil industry which this outlook implies would, at best, be at a modest rate (of say 2–2.5% per annum). This would result from price-constrained demand, under conditions of a relatively slow fall in the price of oil, as shown in Figure II-8.2, given convergent OECD-OPEC policies and the continued impact of recently learned lessons on the potential for much more efficient use of energy.

*Second*, there would be resumption of growth in the exports of oil from the OPEC countries, the oil and gas reserves of which are already more than adequate in most cases to sustain an increase in production for at least two decades and, following further exploration and oil and gas field developments, for many decades thereafter. This would guarantee the OPEC countries a continued flow of resources on which to sustain their post-1973 development efforts.

The *third* effect of a successful global energy strategy would be the slow but, nevertheless, sure evolution of the exploitation of the relatively low cost conventional oil and gas reserves of those parts of the world where, to date, the process has hardly started. As with Mexico, Norway, the UK, Egypt, Malaysia and a number of other countries over the period since the oil crisis of 1973/4 9 (see Figure II-8.1), this would lead to the eventual establishment of new oil and/or gas exporting countries and, even more important, to a much more diffuse world pattern of oil and gas production and the increasing substitution of imports by indigenous output – so eliminating a main barrier to economic growth and development prospects in yet more countries.

The impact of these developments would have a relatively short-term revitalizing effect on the international oil industry, in contrast with the undermining process which has already gone on for some years and which is still becoming more intense. Such a change in the outlook for the world oil industry – from one of stagnation and even decline, to one of expansion in a geographically increasingly diffuse range of countries – would produce a favourable impact on the state of, and the prospects for, the international economy. This would emerge, indirectly, from the renewed confidence that would be generated by an OPEC-OECD rapprochement and its consequences, and, more directly, by the ready availability of energy at a lower real price than hitherto, as international oil was traded on the world market at a price closer to its long-term supply price, as shown in Figure II-8.2.

## Outlook for the world oil market and OPEC – without appropriate policy action

In the absence of policy action as outlined above, gross uncertainty and lack of confidence will remain of the essence as far as the oil sector of the world economy is concerned. Moreover, the sector will continue to exercise a negative influence on the prospects for recovery and growth in the western economy as a whole. Indeed, it seems likely that the difficulties of the moment will continue to intensify. There are two specific developments in the energy sector which would have a particularly adverse effect on world economic prospect. *First*, the use of increasing amounts of high cost energy (instead of lower cost oil and gas) will absorb an increasing share of scarce resources (of finance, managerial expertise and research and development efforts, etc) and so starve other sectors of the inputs necessary for growth. This phenomenon is already apparent in some rich countries such as France and Belgium where the claims of rapid, high capital cost nuclear power expansion are denying other sectors of an adequate availability of capital. It is even more apparent in many countries of the Third World where, in order to escape from the adverse balance of payments consequences of pricing their oil imports too highly, they are forced to put very scarce capital resources into high cost energy alternatives.

*Second*, it will lead to the regionalization of the international oil industry, as suggested in Figure II-8.1, with the following consequences:

a.      *A continuing decline in international trade in oil and gas, so adding a further element to the retreat from a free-trading world.*

b.      *The attrition of OPEC's membership as various individual member countries evaluate and determine that their interests will be better served by an association with regional (oil) groupings, within the framework of which they hope to be able to secure guaranteed outlets for their production. Examples of this development would be Indonesia's accession to an East/South East regional grouping and Venezuela's incorporation into a Western Hemisphere or Latin American oil region. Even Algeria and Libya could decide that their better interests would be served by an association with a largely self-sufficient Western European oil (energy) region.*

c. *Changes in the operating climate for, and the structure of, the international oil companies. They will be required to regionalize their operations in order to survive the regional structure of the industry. Some of them would no doubt attempt such a sub-optimizing development, but for others an accelerated retreat to 'fortress America' would seem to be more likely.*

d. *An intensification of competition, for the markets which remain open for international trade in energy, between those oil and/or gas exporting countries which will not be in a position to join one or other of the regional entities. This would apply most especially to the oil exporters of the Gulf and to the USSR with its plans for the massive expansion of natural gas exports.*

Clearly the evolution of such a set of largely self-sufficient oil regions would involve high costs for the western world's economic system as a whole. In addition to this potentially large economic burden, however, there would be an equally dangerous political aspect to the development, namely a high probability of enhanced tension in the Middle East as a result of the exposure of the oil exporting countries there to the effects of disappearing oil markets. Alternatively, given the necessity of these countries to compete with the USSR for energy exports there could be a propensity for collusion between them on the basis of their shared economic interests – so leading to further possibilities of political instability. In the final analysis it is thus very unlikely that the regionalization of the international oil industry would produce net benefits for the western economic – or political – system.

## The alternatives

There is no need to stress the practical, let alone the political, difficulties involved in achieving even the beginnings of a global oil/energy strategy based on an agreement between OPEC and OECD – parties which, to date, have generally adopted a confrontational stance towards one another.

There are, however, as has been argued in this article, even greater dangers in not securing an international OPEC-OECD agreement on the longer term oil/energy prospects, given that the 'market' *per se* is unable to cope with as a result of political and behavioural factors involved in the relationships between producer and user countries. It is thus perhaps not inappropriate to suggest that negotiations for an OPEC-OECD agreement represent the North-South equivalent of the East-West negotiations over disarmament. The difficulties which beset the latter have long been presented as approaching the insurmountable, but recent general recognition that negotiations over the issue are a necessary precondition for the survival of both East and West is proving to be a powerful motivation for compromise and agreement. The North-South (OECD-OPEC) oil/energy issue in all its ramifications falls into the same category. The continued failure to resolve it perpetuates and exacerbates the problems of the western politico-economic system and, in the final analysis, threatens its survival – because of the way in which the oil 'problem' now lies behind a whole range of international economic and political issues. The collapse of OPEC, the first successful international economic organization of the South, as a result of increasingly severe disequilibrium in the international oil market, would not help in this respect as it would be interpreted in the South as yet further evidence for continuing economic imperialism. Thus, the uncertainties and difficulties of the present situation, coupled with the impossibility of going back to the *status quo ante* the 1973 revolution in the world of oil power, would seem to provide a strong enough motivation to attempt to create a new institutional framework for the world oil (energy) market. In the context of such negotiations and developments, a long-term solution to the issues, problems and uncertainties discussed in this paper could be found – to the net advantage of both producers and consumers in all parts of the Western economic system.

## B. Back to cheap oil?*

The aim of this paper is to show, first, that the present instability in the international oil market is the result of the mistaken belief that oil would not be able to sustain its domination of world energy markets; and second, that the consequences of policies based on this belief have made the survival of the international oil industry open to considerable doubt.

* **Reprinted from *Lloyds Bank Review*, No.156, April 1985, pp.1–15.**

As a result of the misunderstanding of potential oil supply developments and the structure of the demand for oil, the price of oil has been driven well above its equilibrium level. Thus demand has been severely constrained, while the ability to supply has been greatly enhanced. The inherent instability created has been kept in check by collusion between various interested parties. This has been explicit in respect of OPEC's actions, and implicit in the presentations and actions of many other vested interests. Collectively the parties have sought to maintain the fiction of oil as a high value commodity.

If this fiction is maintained, by means of institutional action on the part of OPEC and others, then the market will remain unstable in the short to medium-term. In so far as this is perceived as unreliability by consumers, it implies the continued necessary decline of the international oil industry. Thus international oil will become progressively less relevant to the western economic system. Eventually, in the long-term, very large remaining volumes of low-cost oil will cease to be resources in anything more than a purely physical sense, given the absence of a demand for them within a time period which makes investment in their recovery a justifiable proposition.

The alternative to this prospect for the long term demise of the industry in the absence of policy changes is a near-future collapse of the oil price. The chance of this is already significant – at, say, a 20–25 per cent probability – and it seems likely to increase year by year (with temporary periods in which the risk is diminished as a result of political problems over oil supply), as the fundamental imbalance between demand and potential supply, with continued institutionalized efforts to maintain present high prices, becomes more acute.

Both the longer term demise and the shorter term collapse of the oil market do, however, have serious implications for the broader western politico-economic system, as well as for the enterprises directly concerned, so that the latters' interests in avoiding both by stimulating policy changes are not without a more general significance. The last part of this paper suggests why and how enterprises within the oil industry have a better opportunity than any other group of entities to reduce the chances of an uncertain future for oil by action which seeks to stimulate the development of the industry – by means of an unequivocal reversal of their misrepresentation of oil as an inherently scarce and high value source of energy.

## The development of instability

This is demonstrated in Figure II-8.2 in which the massive divergence after 1970 between the price of oil in the market and the long run supply price of the commodity is shown. In this illustration the latter curve is drawn on the assumption of an average annual 2.5 per cent increase in the use of oil. Given the continuation of the average annual rate of increase over the last 10 years of zero, this curve could be much flatter. Indeed, it has probably not even turned up at all from its lowest level in 1970, given the failure of demand to increase and the more than adequate availability of low-cost proven oil reserves (see Table II-7.2) to meet all reasonable expectations of the use of oil over at least the next twenty years.[1] I shall return later in the paper to look at the implications of the divergence of the two curves in the Figure but, meanwhile, it is necessary to show how the severe escalation in the price of oil between 1970 and 1981 in the international market led to inherent instability in the system.

Price escalation has had its effects, of course, on both the demand and the supply sides of the equation. On the demand side it has led to both price-generated and policy-encourage attitudes of using as little oil as possible by means of conservation and the substitution of other sources of energy for oil – no matter, in many cases, how expensive such substitution might have been.[2] On the supply side there has been the motivation in many countries to produce as much indigenous oil as required to achieve self-sufficiency and, if possible, also oil exports as a source of foreign exchange.

Both the demand and the supply side developments reflect the perception which was created in policy making and consumers' circles in the 1970s of oil as an inherently scarce commodity – and hence an inevitably rising real-cost commodity, even from the level of a more-than-an-order-of-magnitude higher price by 1980 compared with 1970 – as shown in Figure II-8.2. The strength of this perception is demonstrated in the results of the 1981 survey of international oil price expectations among US-based analysts undertaken by the Energy Modelling Forum (EMF) at Stanford University.[3] This showed year 2000 oil prices (in $1981) ranging from $42–$92 per barrel – with only two forecasts of less than $70 – and expected prices by 2020 in the range of $69 to $152 per barrel. Such views, moreover, were by no means extreme. Indeed, the respondents saw the world oil market remaining fairly slack through the middle of the decade (the 1980s) with nominal prices increasing less rapidly than the general inflation rate.[4] Nevertheless, forecasts for prices in 1985 ranged from over $30 to almost

$50 per barrel – again in 1981 dollars, so that the price of oil in current dollars "ought" now to be between $37 and $65 per barrel!

The EMF's survey work on oil price forecasts was later extended to the international level through the International Institute for Applied Systems Analysis (IIASA) whose major study of the world energy system in the late 1970s/early 1980s had 'warned' of the need to reduce dependence on oil as the dominant source of energy.[5] IIASA's international survey involved 68 projections of the international oil price for 1990, 61 for 2000 and 24 for 2010.[6] The interquartile range of prices forecast for 1990 (in $1981) is $31–43 per barrel: for 2000, $42–55 per barrel; and for 2010, $51–72 per barrel. The median forecasts for the three years are $36, $51 and $60 per barrel, respectively.

The translation of these quantified expectations of future oil prices into required policy action has been summed up by the International Energy Agency in the following way:

> 'The 1980s must be a period of major transition towards a minimum oil economy.... Focussing the necessary adjustments in a timely and smooth way must continue to be a major focus of energy and economic planning.... IEA countries should endeavour to limit net oil imports in 1985 to about 4 million b/d below present projections and by 1990 should be further reduced by about 3 million b/d.... The move towards a balanced world oil market must be a continuous process.... Policies of IEA countries need to be reinforced and extended in order to.... deal with the uncertainties that may continue to arise in world oil supplies.'[7]

The rising real oil price forecasts of the magnitude indicated in the preceding examples and the associated policy evaluation emerged principally from the generally accepted expectation that world oil supply was simply unable to grow for many years beyond the mid-1980s: an expectation which was based on oil company forecasts of the kind such as that made by British Petroleum in 1979, in a widely and heavily promoted presentation, which showed the non-communist world maximum possible level of oil production as having to occur in 1985 – because of supply limitations.[8]

This belief in oil scarcity as a general proposition was strengthened by the more specific belief that OPEC oil was unreliable – because the member countries either could not, or would not, produce enough to meet the rest of the world's demands on their export potential. Dependence on OPEC was thus particularly unacceptable – for a commodity as essential as oil to modern economies and societies. As a result, all importing nations in the industrialized world and many in the Third World strengthened, or initiated, their commitments to indigenous oil production in order to minimize their dependence on 'unreliable' supplies; and if, in this process, a modest excess of production over national demand could be generated, then exports of oil, at the high international prices prevailing and the even higher prices expected, was not a prospect to be ignored.

However, in spite of the perceived scarcity of oil – and an expectation of an inevitable rising oil price – there was still insufficient motivation for these countries to encourage the maximum possible exploration and exploitation of indigenous oil. On the contrary, most governments pursued policies which limited the extent and the intensity of the search for oil (by restrictive licensing policies and/or by restrictions on the sorts of enterprises which were allowed to search). Even when oil was found its exploitation was controlled to a tempo which aimed to ensure an extended life for the fields, so that when the rest of the world's oil was depleted there would still be enough in the ground at home to maintain an essential minimum flow to the economy. Moreover, such physical depletion controls have been matched by equally stringent fiscal restraints on the ability and willingness of enterprises to invest in expanding oil production. Governments have taxed away the bulk of the economic rent emerging from the exploitation of low-cost but high-price oil. It is the 'sophisticated and socialist' countries of Western Europe which have led the way in these respects, with the result that the development of oil production (and natural gas production to an even greater extent) has been held well below the technically possible levels; and, even more important, the prospects for the 'natural' growth of the reserves' potential in such areas has been diminished.[9]

'Fear of scarcity', in other words, begat policies which helped to create scarcity (by limiting both present and future production) so that not even the existence of oil prices far above the long run supply price of the commodity brought forth the level of development of upstream oil industry activities which might reasonably have been expected.

## The present conditions of the industry

The elements I have described above have combined to produce an international oil industry which is quite fundamentally different not only from the one with which the world became familiar in the 1960s and the early 1970s, but also from the OPEC dominated industry which was generally expected from the mid-1970s, following the first oil price shock and the nationalization of the major companies' assets in most exporting countries. The instability to which we are now witness is a consequence of the 'generally unexpected' and is thus very much more, in terms both of its intensity and its likely longevity, than an instability which arises out of a temporary, modest and relatively easily controllable imbalance between oil supply and demand in a period of recession in the western economic system.

I would argue, on the contrary, that the disequilibrium which has developed in the oil industry is much more fundamental and that its intensity can be measured in terms of the three following components – none of which is reversible without deliberate policies and actions on the part of both producing and consuming nations and of oil companies and enterprises.

*First, the degree to which demand for oil has already diverged from earlier expectations.* This can best be summarized by pointing out that only one barrel of oil is likely to be used in 1985 for every two that were expected to be used in that year in the forecasts that were made prior to the first oil price shock in 1973.[10] Stagnation, possibly even a slow decline, in the rate of use of oil has become the new norm so that divergence from what 'should' have been is getting steadily more impressive.

*Second, the fall in demand for OPEC exports.* This has suffered even more dramatic changes compared with the position as it was – and as was expected to be. This is illustrated in Table II-8.1 in which the 1973 role of OPEC oil exports in serving the demand for energy in the rest of the non-communist world is contrasted with the 1984 position. These data make the sharp decline in the importance of OPEC oil over this period crystal clear. From its more than 36 per cent share of total energy supply in 1973 and a position which was almost twice as important as any other component in the total supply situation in that year, its contribution has slumped to under half of what it was. It is now much less important than indigenous oil and less important than both indigenous coal and indigenous natural gas. This dramatic change in so short a period is clearly a function of the strongly negative economic and politico/strategic reactions against OPEC oil and its earlier over-important role in the rest of the non-communist world's energy supply. There appears to be little likelihood that this attitude will change unless there is a sharp fall in the OPEC oil price and a broad degree of acceptance that it is again appropriate to increase the degree of dependence on OPEC oil.

**TABLE II-8.1**
## Sources of energy used in the non-communist world
## (excluding the OPEC countries) in 1973 and 1984

|  | 1973 | | 1984 | |
|---|---|---|---|---|
|  | mtoe* | % of total | Mtoe* | % of total |
| Total Energy Use | 4045 | 100 | 4110 | 100 |
| of which |  |  |  |  |
| a) Imports of OPEC Oil | 1480 | 36.5 | 730 | 17.8 |
| b) Other Energy Imports** | 100 | 2.5 | 230 | 5.6 |
| c) Indigenous Production | 2465 | 60.9 | 3150 | 76.6 |
| of which |  |  |  |  |
| i) Oil | 760 | 18.8 | 1155 | 28.1 |
| ii) Natural Gas | 765 | 18.9 | 735 | 17.9 |
| iii) Coal | 805 | 19.9 | 985 | 24.0 |
| iv) Other | 135 | 3.3 | 275 | 6.7 |

\*          *mtoe = million tons oil equivalent*
\*\*        *Oil, natural gas and coal from the Centrally Planned Economies*


*Third, the proliferation of indigenous oil producers under the combined stimulus of commercial and national interests.* In as far as the price of OPEC oil delivered to a national market provides the baseline against which the commercial validity of investment in indigenous production has to be measured, then, even in the context of relatively small-scale and high-cost developments, some investment has been generally interesting enough for both private and public sector oil enterprises. This phenomenon, moreover, has been strengthened by the increasingly intense and active national motivations to secure the development of indigenous production as a means of reducing the foreign exchange costs and the strategic dangers of dependence on foreign oil. In 1973 there were 16 non-OPEC countries in the non-communist world which produced at least 25 per cent of the oil they used; there are now over 30 and the number is growing year by year. Over the same period, six more non-OPEC countries have joined the three net oil exporting countries in 1973 which were not members of OPEC (*viz.* the USSR, Brunei and Trinidad) as net oil exporters. Overall in 1973, non-communist countries, excluding the USA and Canada, produced about 200 million tons of oil: by 1984 production from this set of countries had increased to about 550 million tons.[11]

The relative importance of these three components in creating the fundamental disequilibrium which currently marks the international oil system is difficult to measure: there is, indeed, a degree of interdependence between them. Taking the three together then their impact can be summarized as follows:

a.  *The emergence of a gross – and growing – overhang of oil production capacity and production potential in the OPEC countries, and in a number of other countries which are finding it increasingly difficult to export oil in an increasingly competitive market.*

b.  *The creation of a strong demand for exploration and exploitation services in non-oil producing and non-oil self-sufficient countries, combined with a reduced demand for the expansion and development of downstream activities. This has altered both the sectoral and the geographical distribution of the activities of the international oil companies.*

c.  *The establishment and continuing development of higher real cost energy supplies than the world need produce; as the world's low-cost (but high-priced) oil resources are backed out of markets in favour of much higher cost, but lower priced (or otherwise preferred) alternatives. This development has been to the detriment of the world's economy and similarly it affects its prospects.*

## Uncertainties in the present situation

The most important of the uncertainties emerging from the inherent disequilibrium of the present situation is the likelihood that the situation will change dramatically. There appears to be a reasonably high level of probability (say a one in four chance) that this will happen within the next five years as a result of the increasing downward pressure on oil prices. Were the price to collapse (to under $20 per barrel in 1984 dollars), then demand restraint would be greatly weakened, and oil use would start to grow strongly again. At the same time there would be a loss of interest in, and/or the ability to sustain, high-cost oil exploration and exploitation and similarly for the viability of high-cost non-oil energy investments.

The return to a much freer – even if not a fully competitive – international market would then seem likely to become the hallmark of the new situation for all or most of the remainder of the century. In this process the main change in economic terms would be the elimination of much economic rent from oil production so that the principal feed-back effect would be on the revenue flows to the governments of oil producing countries – and especially those with upstream oil industries which are relatively high cost.

The prospects for oil enterprises under oil price collapse conditions and a return to a largely competitive international market will necessarily vary, depending on the structure and nature of the activities with which each enterprise is involved. As a generalization all that can be offered in this paper is the suggestion that it is an unwise enterprise that does not have contingency plans already worked out to meet the challenges of the markedly different situation which it would have to face in the aftermath of a price-collapse and a return to the rigours and opportunities of a competitive, but expanding, market for oil over the rest of the century.

The avoidance of a price collapse is now all but beyond the capability of OPEC – unless the members are ready, willing and able to hold their total allowable production significantly below 17.5 million b/d in the short term, and to hold it below that level for as long as it takes for the demand for OPEC oil to strengthen. Happily for the oil exporting countries, however, the consequences of a price collapse – as noted above – affect a large and increasing number of other countries – and entities. Thus, powerful pressure groups in these countries (including many individual oil and gas enterprises) as well as national interests stand to be adversely affected as a result of the consequential undermining of investments in the production of relatively high cost oil and even higher cost alternative sources of energy. It is these forces which seem likely to be able to inhibit the collapse of the oil price – and hence, unless additional policy measures are taken, ensure the maintenance of the present situation of disequilibrium and of uncertainty in the oil market.

The result of a successful attempt to prevent the oil market collapsing will be continuity of the already three-year downward drift in the oil price. There will then be modest increases only in the risk associated with investments in indigenous oil exploitation and alternative energy production. This outlook would not be wholly unacceptable to the oil enterprises involved (given that reduced revenues could be matched, or nearly matched, by reduced costs and lower taxes). However, the maintenance of most of the present high value of oil would ensure that oil demand remains effectively constrained, as there would be little perceived motivation by countries or consumers to change energy use patterns in favour of oil – or to reduce their energy conservation efforts. Indeed, oil in general, and OPEC oil in particular, would continue to become less important in the world energy economy, so that the greatest uncertainty now facing the industry's enterprises would be related to the medium to longer term survival of the oil industry. In the world outside the centrally planned economies, the oil industry would be a declining industry marked by a downward spiral of too high prices and too little demand – and the premature closure (prior to reserves' depletion) of producing areas and the failure of enterprises.

Such a prospect hardly indicates a future for the oil industry which seems likely to inspire confidence within individual enterprises. Even those with indigenous markets big enough to take most of their production could not expect to find much opportunity for continued expansion. Moreover, it does nothing to help the outlook for the western world's economy, as too many resources will have to be devoted to high-cost energy production (while, paradoxically, low cost oil is left in the ground). The prospect also seems likely to lead to an intensification of nationalistic attitudes towards energy supplies – with existing fears for the insecurity of international oil supplies reinforced in their impact on the de-internationalization of the industry by more and more decisions to protect indigenous energy production. There could thus be yet more stimulation of beggar-my-neighbour attitudes and policies, with necessarily adverse effects on the whole of the non-communist world's economic system. In other words, the uncertainties for individual enterprises in the oil industry arising from too high oil prices and too little demand for the actual and potential output of the industry will be generalized into uncertainty about the background politico-economic environment; which, in turn, will have an adverse feed-back effect on the demand for oil and on oil prices – to complete the vicious downward spiral.

## The opportunities for enterprises to reduce uncertainty

The real world background to what, in effect, is a crisis of confidence over the international oil situation is one of serious misunderstandings over basic supply/demand/resources issues on the part of the governments and international organisations (such as the IEA and the EEC), and the impact of such misunderstandings on public opinion and actions about oil. Enterprises in the oil industry are, however, in a position to do something about this – given that they are not entirely unused to co-operation and joint ventures based on a relatively high degree of mutual respect and understanding – even between state and private sector entities. Collectively they are thus in a position to change the perception of the future of oil. In doing so, they would reduce the present instability in the market and indicate an option other than that of a price collapse which, as shown earlier in this paper, carries with it a new set of dangers and uncertainties.

The initial responsibility of the oil industry is to start persuading the world at large that oil is a secure, plentiful, clean and preferable source of much of the world's energy needs for the foreseeable future: instead of emphasising, as over recent years, the resource availability problems and, indeed, even presenting quite pessimistic views on the relationships between reserves development and use.[12]

Table II-8.2 shows the relationship over the last 15 years between annual additions to reserves and oil used. In only four of the years has this relationship been negative (with only two years in which it was anything more than just negative), while over the 15 years as a whole it has been strongly positive. Approximately three barrels of oil have been added to reserves for every two which have been used ($480 \times 10^9$ barrels compared with $314 \times 10^9$ barrels) so that the global reserves to production ratio has improved significantly over this period. Indeed, the ratio now stands at a more favourable level than at any previous time in the history of the oil industry, except for a few years in the mid-1950s when the size of the reserves in a number of the largest oil fields in the Middle East were first revealed. In the non-communist world the position has improved even more dramatically since 1970 with a near 2:1 relationship between oil added to reserves and oil used.

In spite of this highly favourable situation, most of the oil industry has persisted in recent years in overstating its resource problems and in understating the opportunities for further expansion.[13] The facts suggest otherwise. There is a long time horizon over which the world's resources of conventional and non-conventional oil would be depleted even if there were a rate of growth in use as high as 1.5 per cent per annum: the industry would not 'need' to peak under these conditions until almost the middle of the 21st century at a little over three times its present size.

Similarly with statements by the industry on oil costs and prices. Again it is the views expressed by oil companies which have largely been responsible for creating the impression of oil being an expensive, rather than a low cost, commodity. This is basically because of the failure to distinguish between real costs and taxes. Companies have behaved as though high taxes on oil production are immutable rather than, in the

### TABLE II-8.2
### Annual additions to reserves, oil use and net
### growth/decline in reserves 1970–84

|  | Reserves at Beginning of Year | Use of Oil in Year | Gross Additions to Reserves in Year | Net Growth (+) or Decline (–) in Reserves |
|---|---|---|---|---|
|  |  | (in barrels x $10^9$) |  |  |
| 1970 | 533 | 17.4 | 62.4 | +45 |
| 1971 | 578 | 18.3 | 40.3 | +22 |
| 1972 | 600 | 19.3 | –3.7 | –23 |
| 1973 | 577 | 21.2 | 35.2 | +14 |
| 1974 | 591 | 21.2 | 32.2 | +11 |
| 1975 | 602 | 20.2 | 31.2 | +11 |
| 1976 | 613 | 21.9 | 3.9 | –18 |
| 1977 | 595 | 22.6 | 15.7 | –7 |
| 1978 | 588 | 22.9 | 44.9 | +22 |
| 1979 | 610 | 23.7 | 21.7 | –2 |
| 1980 | 608 | 22.8 | 33.8 | +11 |
| 1981 | 619 | 21.3 | 67.3 | +47 |
| 1982 | 665 | 20.1 | 30.1 | +10 |
| 1983 | 675 | 20.0 | 21.1 | +1 |
| 1984 | 676 | 20.6 | 43.6 | +23 |
| Totals, 1970–84 | – | 313.5 | 479.7 | +167 |

final analysis, the residual benefits that flow to governments from the economic rents which can be derived from low cost production and a high value in the market. Cost escalation has also been presented as an axiom – rather than as a function of a retreat from economies of scale and a reduced degree of attention to cost control under inflationary conditions – accentuated by the view that costs did not really matter because the higher the costs, the less the government would take in taxation, so leading to the so-called 'gold-plating' of production facilities. Under the stimulus of falling real oil prices and a more competitive market, it is highly likely that real costs will fall – and perhaps fall quite dramatically – in respect of oil production from most recently exploited environments (notably continental shelf off-shore developments). Oil enterprises would do a great deal to help restore confidence in the oil industry if such reasonable expectations were accorded a greater weight in public pronouncements. In particular, enterprises need to kill the persistent misuse of the so-called 'Hotelling-rule', *viz.* that oil prices must rise *pari passu* with the discount rate because of depletion effects. This is simply not true when the starting point is a price which carries a great weight of economic rent secured as a result of this non-competitive market situation in the 1970s. Present prices have far to fall (to the long run supply price curve in Figure II-8.2) before the rule starts to apply – even if we assume that resources are so constrained as to encourage the deferment of production to earn future benefits.

Finally, enterprises need to stop pretending that the infrastructure of the oil industry is getting relatively more difficult of organize and relatively more costly to develop. This is a pretence which is related to both technical and political factors. It is ironic that whilst these 'difficulties' have been highlighted – perhaps as a means of 'softening-up' both governments and consumers in the context of demands for price rises – the industry has, by and large, continued to do whatever has been required of it in modifying and expanding the infrastructure of international oil. These developments have included massive new delivery systems for offshore oil (and gas) production in several parts of the world; changes in refining/transportation systems to meet the requirements of new capacity in major oil exporting countries; new pipeline systems to avoid having to move oil through areas of conflict and potential conflict; and the modification and rationalization of refining and distribution systems with changing products needs and/or over-capacity as a result of demand limitation.

Enterprises' attitudes and pronouncements in respect of resources, costs and infrastructural difficulties etc. have proved to be virulently negative in terms of the prospects for the continuity of the oil industry. What is now required is the elimination of such negative views and the accentuation of a more positive outlook for oil; first, in respect of its prospects *vis a vis* other energy supplying industries in terms of development opportunities and relative costs over the next quarter of a century; and second, in respect of particular opportunities for expanding the oil industry itself – both generally and in specific geographical areas.

Enterprises could show a more positive response to the possibilities of diffusing the search for oil and its exploitation – in co-operation with national and intergovernmental organizations which see this as a necessary development in much of the Third World.[14]

These countries, and the enterprises – both state and private – that work in them should seek to change the impression created over the last decade that they do not 'need' the money to be earned from oil exports. As indicated by their increasing international debts and imbalances between income and expenditure at home, most OPEC – and non-OPEC – oil exporters are not in this category in any case; but for the few that are, the idea that their resources of undeveloped and/or unproduced oil will somehow produce net benefits in the dim and distant future needs to be scotched. In the absence in the meantime of sufficient production at low enough prices, the oil industry will wither away while the world's energy needs will be met that much sooner by increasing supplies of alternative sources of energy (albeit more expensive ones). Oil which is not recovered in the short to medium-term is in danger of remaining unwanted in the ground.

The world economy stands to benefit all round from a more rapid exploitation of world oil and an enhanced use of oil in preference to more expensive energy to satisfy the higher energy demands of an expanding world economy. Thus, individual oil enterprises – plus others concerned with oil in governmental, academic, and consulting organisations need to get back to 'selling' the oil industry – instead of selling it short. In doing so they will not only help to get the world back to the undoubted advantage of using low-cost oil as the most appropriate source of energy for an expanding economic system, but also reduce the uncertainties which currently face the enterprises within the industry as a result of its decline over the last decade. An apparent 'death wish', in the midst of a plethora of resources and alternatives to oil (and gas) which are more costly to produce and to use, needs to be exorcized for the nonsense that it is. Enterprise-level strategic planning and action to expand production and markets at realistic prices is likely to have the most powerful positive impact on the reduction of today's very substantial oil market uncertainty.

## References

1    See also C W Hope and P H Gaskell, *The Oil Price Without Opec; Back to the $3 Barrel*, a paper presented at the *3rd International Conference of the International Association of Energy Economists*, Cambridge (UK), April 1984, for an econometric analysis which confirms this conclusion.

2    See the International Energy Agency, *World Energy Outlook*, OECD, Paris, 1982, Chapter 3 for a discussion of demand restraint under high price conditions. The IEA's *Energy Policies and Programmes of IEA Countries, 1983 Review*, OECD, Paris 1984, updates the results.

3    EMF *World Oil: Summary Report of the Energy Modelling Forum Working Group*, Stanford University, December 1981.

4    Ibid. p.32.

5    IIASA, *Energy in a Finite World*, Ballinger Publishing Co, Cambridge, Mass., 1981.

6    A S Manne and L Schrattenholzer, 'International Energy Workshop; a Summary of the 1983 Poll Responses', *The Energy Journal*, Vol.5, No.1, January 1984, pp.45–64.

7    International Energy Agency, *Energy Policies and Programmes of IEA Countries, 1980 Review*, OECD, Paris, 1981, pp.13–17.

8    British Petroleum, *Oil Crisis… Again?* BP Ltd, London, September 1979.

9    These negative influences on the exploration and exploitation of significant oil and gas reserves have been at their strongest in respect of the North Sea oil province. See P R Odell, 'Inappropriate Energy Resources Exploitation Policies in Western Europe', *The Times Higher Education Supplement*, London, 1 October 1982 and 'Intervention, Regulation and the Western European Natural Gas Industry', *World Gas Report*, Vol.5, Nos.6 and 7, 1984 for further discussion of the issues and their implications.

10   *Energy Prospects to 1985*, OECD, Paris, 1974, p.41.

11   These data have been derived from series on oil production and use in the *Oil and Gas Journal*, BP's *Statistical Review of World Energy*, the *Petroleum Economist* and the UN *Energy Statistics*, Series J.

12   British Petroleum, 1979, op.cit.

13   We have dealt with these issues at length in P.R. Odell and K.E. Rosing, *The Future of Oil; World Oil Resources and Use*, Kogan Page, London and Nichols Publishing, New York, 2nd Edition 1983. The arguments are stated more briefly in 'The Future of Oil: a Re-evaluation', *OPEC Review*, Summer 1984 pp.203–228; and in 'The Oil Crisis: Its Nature and Implications for Developing Countries', *The Oil Prospect*, Energy Research Group, International Development Research Centre, Ottawa, March 1984.

14   The World Bank's view in this context is already well known. See *The Energy Transition in Developing Countries*, World Bank, Washington, DC, 1983. The United Nations General Assembly has also unanimously approved such initiatives. See the resolution, 'Development of the Energy Resources of Developing Countries', UN General Assembly, 38th Session, 8 December 1983, A/C.2/38/L.106. This was based on a paper prepared by the UN Economic and Social Council, 'Development of the Energy Resources of Developing Countries', E/1983/91, 28 June 1983.

# Chapter II – 9

# The Economic and Political Geography of a Century of Maritime Oil Transport*

## Introduction

Unlike coal, most of which was – and, indeed, still is – used at or near to the points of production, oil has always largely moved to markets. This contrast between coal and oil is partly a function of the generally later involvement of oil in the world's energy economies; at a time, that is, when the earlier use of coal had already led to the establishment of urban/industrial centres of energy demand. It is also partly a function of the inherently greater transportability of oil as a liquid with a high energy value per unit of weight or volume. There are greater difficulties and higher costs involved in moving coal as a solid fuel and with a much lower unit energy value. Finally, it is also a function of the location of most of the world's oil in regions which are remote from centres of energy demand.

## The Oil Industry in the United States

It was this last factor which delayed the large scale use of oil in most parts of the world until rather recently. The most significant exception to this was in North America where the considerable indigenous resources were developed in increasingly extensive and intensive ways after the end of the third quarter of the 19th century to supply a growing share of the United States' rising demand for energy. Appalachian and

* Reprinted from "Met Olie op de Golven", Maritiem Museum 'Prins Hendrik', Rotterdam, 1986, pp.41–54.

mid-Western oil were moved, initially by rail and later by pipelines, to areas of consumption in the early industrialised and urbanised north-eastern United States. With the discovery of the oil resources of Texas and nearby states, the coastwise movement of oil products (and some crude oil) from the Gulf of Mexico to east coast ports became a more attractive proposition economically (compared with overland transport), but its development was later undermined by wartime fears for the safety of the tankers. Instead, 'big inch' pipelines were built across the United States to handle the flows of oil. As a result there was little motivation to use the technically possible larger tankers which were being developed elsewhere in the world for intra-U.S. oil trade. There was thus no incentive in the United States to enlarge terminal and water-depth facilities at loading and unloading ports whereby economies of scale could have been achieved. Overall, the oil industry in the United States (which, in terms of exploration, production, refining and distribution activities was dominant in world oil until after the second World War) thus played only a very small role in the evolution of the industry's maritime sector.

## The Early International Oil Industry

Outside the United States (and Mexico, whose oil industry was run as an adjunct to that of the U.S.), the oil industry remained relatively small and generally very localised until the 1914–1918 war. Oil was produced, on the one hand, in pre-revolutionary Russia, the Dutch East Indies, Burma, Trinidad, Rumania and a number of other countries, mainly for export. There were, on the other hand, generally small specialised markets for oil products in most countries of the world, so generating the need for a diverse international maritime trade in oil products. The development – and acceptance – of the tanker in the last quarter of the 19th century as a safe means of moving oil around the world enabled oil to achieve a high degree of geographical diversity in its pattern of use by dramatically reducing the costs of its international transportation. The earlier 'packaged' movement of oil products (in steel drums) had served to limit oil use to high-value markets only (such as lubrication and lighting functions), unless there was the possibility of overland pipeline transportation – as in the United States. In other words the ocean-going tankers' development was a *sine qua non* for the development of the international oil industry on anything more than a very minor scale.

The first element in changing the potential for oil use as a result of the establishment of tanker transportation came from within the shipping industry itself, *viz.* from the rapid change from coal to oil-fuelled steam turbines for both merchant and naval vessels. This was a consequence of the new ease and relative cheapness with which oil fuel could be made available in the world's ports and naval bases – compared, that is, with the physical and the cost problems that had been associated with coal-bunkering the world's expanding fleets. (In parentheses it is worth noting that the naval use of oil as a fuel greatly enhanced the interests of the European powers – notably the United Kingdom, Germany and France – in securing access to oil supplies. This led to keen attention to, and competition for control of, the potentially oil-rich areas of the Middle East). The provision of a network of oil-bunkering facilities around the world opened up the possibility of oil products being used to replace the use of coal in those countries of the world which, without the development of indigenous energy resources, had to depend on imported coal.

Such markets could now be supplies by tankers plying between the refineries located in the oil producing countries – such as Mexico, Persia, Trinidad and the Dutch East Indies – and the ports of the world where oil import terminals were built. The landed cost of the oil products undercut the prices charged for imported coal, the transport costs of which became much higher than those of oil as consequence of the onset of oil tanker transportation and its steadily developing technology. There was, of course, more to the process of the substitution of coal by oil in the energy-importing countries of the world than simply the impact of falling real-cost maritime oil transport.

First, the low-cost to produce oil in the countries in which oil production was under development contrasted markedly with the rising real cost of coal production in those countries (notably the United Kingdom) which exported coal. This was the result of the increasing cost of labour in their coal-mining industries. Second, unlike the world's coal industry which was well nigh exclusively organised within national frontiers, so that for the companies concerned exports were usually of minor importance (compared with sales in the country itself), the oil industry outside the United States was largely run by a small number of international companies (later to become the 'Seven Sisters' of Shell, Esso, BP, Mobil, Texaco, Chevron and Gulf). These companies deliberately and positively sought to develop their interests in all parts of the world and were thus orientated in their business to the long-distance international maritime movement of oil. In other words their development and their use of tankers for moving oil around the world became an integral part of the *modus operandi* of the oil industry; the norm, that is rather than the exception. By the 1920s, therefore, as oil outcompeted coal in all the then energy-poor countries of the world, the international oil companies established importing terminals and associated facilities to handle the oil products they shipped in by tankers from their areas of production and refining – and where their facilities were deliberately orientated to the export of the oil by sea. Outside North America, the production of oil and its transport by sea were thus in a symbiotic relationship with each other.

## The Expansion of Oil's Markets

Important though oil's take-over of these markets from coal was for the development of maritime oil transport – in that the small oil product tanker became ubiquitous to hundreds of ports in all parts of the world – there was a quantitatively much more dramatic process which got under way after the end of the first World War in respect of the demand for tanker tonnage. This was the process whereby the international oil companies took the opportunity offered by their discovery of large volumes of very low-cost-to-produce oil in a number of countries (initially Venezuela, Persia – now Iran – and the Dutch East Indies, and later the Arab countries of the Gulf plus Algeria, Libya and Nigeria in Africa) to supply oil in volume, to meet general as well as specialised energy needs of the world's industrialised countries, apart from the United States: essentially, that is, to the countries of Western Europe and Japan.

This process expanded relatively slowly in the inter-war period (1919–1939) and then exploded into a massive and prolonged period of development after the end of the second World War. The global-scale geographical separation of the oil producers from what were to become the main oil using countries was of the essence in respect of the growth of maritime oil transportation over the period of about 50 years from the early 1920s to 1973.

These markets for oil were the world's coal producing and using countries *par excellence* whose economies had, for the most part, been industrialised and modernised on the basis of the exploitation of their coal resources. Their dependence on coal did not at first fall in volume terms, let alone collapse under the impact of the growing availability of international oil supplied by the multi-national oil corporations at a generally declining real price. Indeed, at first, oil use was, in most cases, largely restricted to uses for which coal did not compete. This was most notable in respect of the increasing demand for fuel for road transport, the development potential of which had been given a tremendous boost by the mechanisation of the nations' armies in the 1914–1918 war and which in the 1920s and the 1930s enjoyed a period of expansion, under the stimulus of the emerging methods of lower-cost motor-car mass production technology and industrial organisation. Thus, the demand for oil in most of these countries was, in the interwar period, principally a demand for motor fuels – gasoline and diesel oil – so that the maritime transport requirement which was boosted, was the tanker movement of these products from the refineries of the oil producing countries.

It was in this context that the United States' oil industry – which we defined above as one which developed mainly on a national basis – had its principal external relationships. Refineries in the United States, particularly those at or near the Gulf of Mexico coast, close to the oil producing areas of Texas and Louisiana where production expanded rapidly in the 1920s with the discovery of large, highly productive and low-cost-to-produce fields, initially provided the major source for the oil products demanded in Western Europe. Mexico, whose oil industry at that time was in the hands of foreign (mainly United States companies the nationalisation of which and the formation of PEMEX did not take place until 1938), was also a major exporter to Western Europe. The expanding markets for oil products were, however, also served by oil production and refining facilities which had been developed elsewhere in the world largely by the Royal Dutch/Shell group and by a number of smaller European companies such as Anglo-Persian (later BP), Burmah and Compagnie Français des Petroles. The earliest development of this sort had been in Russia, in the Baku region of the Caucasus. This foreign-company produced Russian oil was shipped out of the Black Sea through to the markets in Western Europe and elsewhere.

The Russian revolution and the immediate consequent expropriation of the oil companies' activities in the newly-formed Soviet Union brought this trade to an end for more than a decade, but by the late 1930s, when the Soviet oil industry finally got into its stride, after a period of organisational and technical difficulties, exports of oil products to Western Europe were resumed and, indeed, quickly became second only in importance to products from the United States.

Meanwhile, the maritime movements of oil products from other parts of the world had been developing and these were, by the late 1930s, destined collectively to become the most important source of Western Europe's increasing needs. In the Caribbean, early production and exports from Trinidad and Colombia established the oil trade links, but these were first supplemented, and eventually dominated, by exports from Venezuela whose oil resources were developed rapidly in the 1930s, especially following the nationalisation of Mexican oil in 1938. As an integral part of Venezuelan oil expansion major refineries were established in the nearby islands of the Netherlands Antilles – by Shell in Curaçao and by Esso in Aruba. For these two companies – which were also the two most important oil distribution companies in Western Europe – Venezuelan oil and the Antillen refineries eventually become the main source of supply for their European markets.

For other European oil products' supplies, the Middle East became the more important source region: though in the 1920s and the 1930s the oil industry there was still in its infancy, with only Persia (Iran) and Iraq making a contribution to world oil exports. The British state-owned Anglo-Persian Company (later Anglo-Iranian and eventually B.P.) was the dominant corporate entity in the region between 1919 and 1939, having developed production in Persia in the early years of the century and having taken over former German interests in Iraq in the first World War. Its refinery at Abadan was expanded to become the world's largest (outside the United States). It provided oil products for shipment not only to Europe via the Suez Canal, but also for much of Africa, South East Asia and Australasia. Iraq, where C.F.P. had also secured a share of the Iraq Petroleum Company (created on the basis of previous German exploration and development activities), was a more important source for oil products destined for France and other French controlled or influenced parts of the world.

On the other side of the globe, the growing oil needs of Japan – for expanding its industrial and (in the 1930s) its military sectors – were served in small part only from the Middle East and from the west coast of the United States. In greater part it depended on oil produced and refined in the Far East itself – notably from the Royal Dutch/Shell operations in the Dutch East Indies (from where, incidentally, some oil products were even shipped to Europe), from Burma and, after the Japanese invasion of Manchuria, from the oilfields of China.

## Changes in the Structure of International Oil Transport

Two elements stand out in the developments described so far. First, the growing geographical complexity of an industry which produced oil in a relatively small number of countries and marketed it in many more – with maritime oil tanker transport as the essential link in all parts of the system. Second, the importance in the international oil system – including the ocean transport component – of a small number of international oil companies, the so-called 'Seven Sisters'. Throughout the massive growth and the many vicissitudes of the oil industry over the last 50 years, these two elements have remained constant. On the other hand, all else in international oil trade has been highly dynamic, with many important changes in the geography and structure of maritime oil transport over this period.

As already shown, oil products' movements by tanker were dominant in international oil trade until the post-World War II period. Since then the maritime movement of crude oil has taken over as the most important element of world oil trade. This change in emphasis was related not only to the economics of the oil industry, but also to political and strategic considerations. Until after the second World War refineries in oil consuming countries remote from oil production were, in economic terms, of relatively little interest. This was partly because the demand for oil in an area which could be served from a refinery was not large enough to enable significant economies of scale to be achieved. And it was partly because, as indicated above, the oil products which were most demanded were those used in motor transport (that is, the lighter oil products), so that the other products (the heavier ones) emerging from the distillation process in a refinery could not find sufficient local markets. There was thus imbalance in the demand for the various products from a market orientated refinery, at a time when the technology of oil refining was not sufficiently developed to give scope for much flexibility in terms of the volumes of products which could be produced from a given crude oil.

Thus, little refinery capacity was developed even in the oil-importing industrialised countries, let alone in the oil-importing developing countries in the rest of the world (in spite of their greater relative dependence on oil as the main source of energy). The refineries that were built at this time in oil importing countries – such as Germany, France, Japan and United Kingdom – were built with military/strategic considerations in mind. Such countries saw the need to have their own refineries in order to ensure the supplies of the products they required for their armies, navies and airforces.

Thus, in those days of a relatively limited total use of oil, coupled with the specific lack of demand for heavier products (given the continued general use of coal for electricity generation, industrial steam raising and for home heating), there were few locations outside the oil producing countries themselves (or in nearby locations such as the Netherlands Antilles) where a refinery development was economically justified. One exception was, however, the port of Rotterdam from a refinery at which the oil products made could be transported cheaply by river and canal to extensive parts of Western Europe.

Thus, Pernis was one of the very few pre-1939 refineries in Western Europe which served an economic, rather than a political, function. Even Pernis, however, was, for the Royal Dutch/Shell Group, essentially a supplement only to the much larger and the more sophisticated refineries' facilities of the Group in Curaçao from which, as already mentioned, oil products were shipped to markets throughout Europe. The domination of the maritime oil trade by oil products movements was further intensified during the second World War during which refineries in the Caribbean and the (Persian) Gulf were expanded to cope with the allies' wartime needs in many parts of the world – including, by this time, requirements for oil products (notably fuel oil) in the industrialised areas along the Eastern seaboard of the United States. After 1945, however, the situation changed rapidly and drastically so that within a decade the ocean transportation of oil become dominated by crude oil movements. Oil products' movements by ocean tanker first became of secondary importance, and eventually of minor importance in relation to the total international seaborne trade in oil. This near-demise of maritime trade in oil products has only very recently been reversed – a change to which we shall give attention later in this essay.

## The Causes of the Change

The changed structure of maritime oil transport from its domination by products movements to an overwhelming emphasis on crude oil shipments arose out of a combination of many factors, *viz.*, demand and supply side changes in the economies of the oil market, technological considerations, and political issues. These are described and analysed in the following paragraphs. The rapidly growing demand for oil in the United States after the end of the second World War led, first, to a reduction and, later, to the cessation of the country's traditional oil products' export trade. This meant that the companies which imported and marketed oil products elsewhere in the world – especially in Europe and Japan where expanding post-war energy needs could no longer depend on indigenous coal – faced supply shortages, given that the export refineries in the Caribbean and the Middle East could not be expanded quickly enough to satisfy all demands on them. Moreover, in the changed post-war political climate in the relationship between the international oil companies and the oil exporting countries, the former felt increasingly exposed to threats of nationalisation in the latter so that they were not keen to commit even more of their capital to build refineries in the countries in which they had already heavily invested to achieve additional oil production. They thought it more appropriate to build their refineries elsewhere, so as to diversify their risks. However, safe political locations for refineries' construction, such as Curaçao, Aruba and Trinidad, close to areas of production, were excluded at the time because of Venezuela's objection to its oil being refined elsewhere in the Caribbean, rather than in Venezuela itself. Thus, refining in the oil importing countries themselves became the preferred option for the companies.

This new location preference for refineries was also becoming interesting in technico-economic terms – for two reasons. First, there was increasing evidence of a potential for achieving large economies of scale in oil transportation by the use of much bigger oil tankers than had hitherto been used – provided there were import terminals at which such increased volumes from a single tanker could be effectively handled. This was more feasible, of course, in respect of crude oil, which could be delivered to a large, coastally-located refinery in Western Europe or Japan, than in the case of a single product for which the spatial intensity of demand was seldom great enough to justify the delivery of a large volume to a single terminal. Moreover, the demand for products in many consuming areas in the industrialised world was becoming much more balanced as between lighter and heavier products. In particular, fuel oil markets were building up on a large scale in response to the inability of the indigenous coal industries to supply enough fuel for power generation and industrial steam raising etc. Thus, not only could large refineries be built in importing countries, but the full range of products from them could now be sold locally, rather than some of the products for which there was no local demand having to be re-exported to distant markets – as in the pre-1939 situation.

Finally, for Western European countries and Japan, it was economically important to have oil refineries built at home. First, oil was traded only in U.S. dollars and as the countries concerned were short of dollars in the aftermath of the war, the local availability of a refinery helped their economies considerably as it had the effect of reducing the overall dollar costs of securing the oil products they needed. Second, the establishment of refineries created jobs and stimulated the local production of goods and services, so helping economic growth potential. Indeed, to some degree, the centres of refining activities (to which petro-chemical complexes could be added) became the post-1950 equivalent of the economic growth-poles that the coal producing areas of Western Europe had hitherto been. The development and growth of a number of such complexes continued for more than twenty years.

## The Special Position of Rotterdam

It was in this context that Rotterdam came to pre-eminence as an oil port and a centre for crude oil processing and conversion to basic petrochemicals.

The one (relatively) small Rotterdam refinery of pre-World War II days, *viz*. Pernis, one of Shell's operations, was expanded to be one of the largest and the most complex in the company's international operations. In the 1950s and the 1960s it was, as shown in Fig II-9.1, joined by four other refineries built by four of the remaining six major international oil corporations (*viz*., Esso, Chevron, Gulf and BP). This development in Rotterdam represented the world's most intensive example of the process of market orientated refinery expansion which was stimulated by the combination of factors noted above. It was, however, only a small part of the overall results of the process in which the whole of the coastline of Europe, stretching from the eastern Mediterranean to the North Sea, became 'dotted' with refineries, built to handle the flows of crude oil from the Middle East, North and West Africa, the Caribbean and, to a lesser degree, from the Soviet Union (see Fig. II-9.2). When those flows finally peaked over 20 years later in 1973, the annual tonnage of crude oil being moved into Western Europe by large tankers exceeded 650 million tons: a volume requirement which had necessitated (and, indeed, been made possible by) the evolution of ever-increasing sized tankers in order to secure the considerable economies of scale in the ocean transportation of crude oil.

Rotterdam was able to take advantage of the economic benefits secured by the shift to larger and larger tankers, given the way in which the port's facilities – and the approaches to the port – were continually upgraded over the 25 year period of evolution in crude oil trade. This not only stimulated the size – and the complexity – of the refineries and their associated activities in Rotterdam itself, but also the use of the port of Rotterdam by tankers carrying crude oil for other destinations. This oil was trans-shipped through Rotterdam involving import terminals and tank-storage facilities in the port itself, as well as crude oil pipelines to take the oil on to refineries in Amsterdam, Antwerp and to various places in West Germany (notable the Ruhr and the Frankfurt areas). At the time of the peak of the international long-distance ocean transport of crude oil in the mid-1970s, Rotterdam received more than 200 million tons of crude oil per annum in very large tankers. Of this approximately half went on from Rotterdam to refineries elsewhere (almost all by pipelines). The areas and the facilities involved in these oil related activities in Rotterdam are shown in Fig. II-9.1.

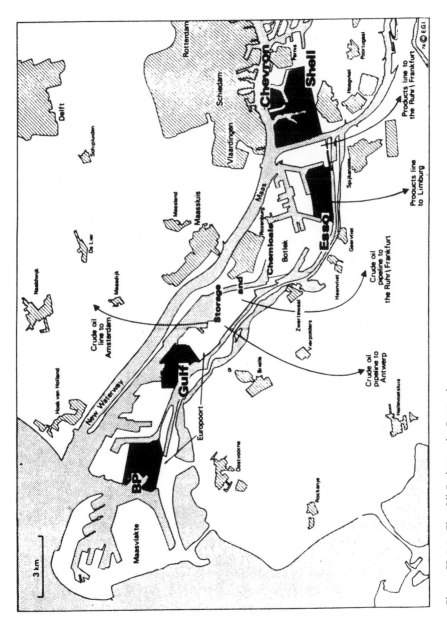

Figure II-9.1: The Oil Industry in Rotterdam

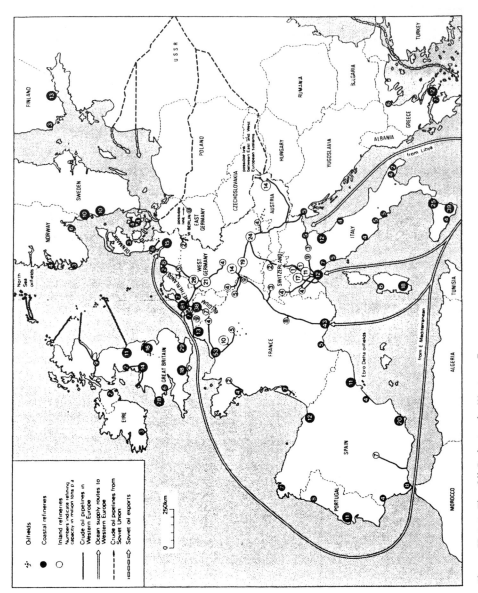

Figure II-9.2: The Oil Industry in Western Europe

## The Post-1973 Situation

The expansion of world trade in oil was brought to a sudden halt by the 'oil crisis' of 1973/1974. By that time flows of oil products over the world's oceans were small by comparison with crude oil movements: moreover, with a small number of exceptions (of which fuel oil shipments from the Caribbean to the east coast of the United States were the most important), international oil products' movements had become largely 'balancing operations' designed to sort out temporary short-falls and surpluses of specific products in different locations around the world. The maritime transport of oil had become dominated – to a degree of more then 95% in terms of ton/kilometres involved – by long distance ocean movements of crude oil in bulk – mainly from the OPEC countries (see Fig.II-9.3) to the major oil-importing industrialised countries, each of which had its own refineries. Since 1974, however, as a consequence of the first oil price shock of 1973/1974 and the second price shock of 1980/1981, the world of international oil – and of the maritime transport activities associated with it – have changed yet again.

First, the high prices charged for oil curbed demand; partly because the oil crises, together with other factors, slowed down the rate of economic growth and hence of energy use. This was partly because higher prices persuaded oil users to use oil more efficiently; and partly because some of the previous users of oil have substituted it by using other sorts of energy. The diminished use of oil has not, however, only been the result of consumers' behaviour. It has also been the official

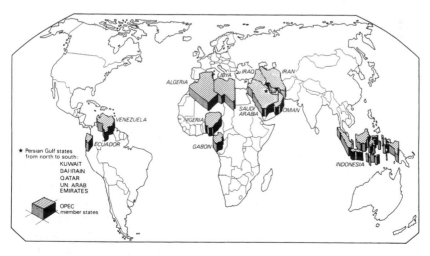

Figure II-9.3: The Members of OPEC

policy of many national governments and of international organisations such as the International Energy Agency (I.E.A.) and the European Community (E.E.C.) to reduce oil use. For example, official policy in Western Europe was to reduce the contribution of oil from over 65% of total energy supplies (as in 1973) to about 45%. By 1985 the share of oil in total energy use had already fallen to only just over 50%, while the amount of oil used was down to less than 550 million tons, compared with about 750 million tons in 1973.

Second, there have been equally significant changes on the supply side. In 1973, outside the United States and the Soviet Union, supply was dominated by oil produced in the member countries of OPEC – almost all of which was shipped as crude oil to the consuming countries of the industrialised world and to the oil importing nations of the developing world. High oil prices and fears for the security and dependability of oil from OPEC greatly stimulated the search for, and the development of, oil production in many other parts of the world. Thus, by 1985 OPEC's oil exports to the rest of the non-communist world had fallen to only about 700 million tons, compared with almost 1500 million tons in 1973. Over the same period, total oil production elsewhere in the non-communist world rose from under 800 to over 1250 million tons.

The dramatic change in the world's geographical pattern of oil supply implied, moreover, a change from the domination of production by countries whose oil had to be transported to markets over long distances by ocean-going tankers. Instead, a much greater degree of importance became attached to oil production which could be used in the countries or regions which produced it.

· The consequence of all these factors has been stagnation and over-capacity in the oil tanker industry. Thus, many tankers have had to be scrapped and there has been no progress towards the even larger tankers (of up to 500,000 or even 750,000 tons) that were prior to 1973 confidently expected to be built. Even so, most companies have had to operate their tanker fleets at a loss over much of the last decade – as prices achievable reflected the over-supply on the market.

The way in which oil markets in Western Europe have developed since 1973 highlights the change. In 1972, the last full year before the first oil price shock and the oil supply problems of 1973/1974, Western Europe's use of oil was about 745 million tons, of which all but 20 million tons was imported. Of the total imports of some 725 million tons, about 50 million tons came over the relatively short sea distances from the producing countries of Libya and Algeria to ports on the north side of the Mediterranean Sea; and rather less from the Baltic by tankers which loaded at the export terminal at the end of the pipeline from the producing areas of the Soviet Union. The rest of Europe's oil – some 650 million tons in the year – was shipped in tankers with loaded journeys of between 4,500 kilometres (e.g. north Africa to Rotterdam) and 22,500 kilometres (e.g.Ras Tanura in Saudi Arabia to Rotterdam via the Cape of Good Hope).

In 1985, as indicated above, Western Europe's total use of oil was less then 550 million tons. Of this much reduced amount, moreover, almost 200 million tons was produced within the continent itself – a tenfold increase in indigenous production since 1973, largely as a result of the rapid and massive exploitation of the rich North Sea oil province (see Fig.II-9.4). A small part of this indigenous oil is exported (mainly to the United States), but the greater part is transported only over distances of a few hundred to a maximum of 1000 kilometres to refineries in nearby locations (such as Rotterdam) in north-west Europe. This is a length of journey which offers little or no economic advantage by using the largest tankers. After allowing for more oil by pipeline from the Soviet Union, trans-Mediterranean crude oil movements from Algeria, Libya and Egypt to southern European ports and an increased import of oil products from refineries in OPEC countries, little more than 200 million tons of crude oil were moved to Western Europe in 1985 over long inter-continental distances by the largest tankers: a dramatic contrast with the 650 million tons of crude oil which flowed to Western Europe in this way in 1973.

Meanwhile, however, there have also been changes in the refining industry which are now influencing maritime transport demands. The decline in the demand for oil has been mainly concentrated on the heavier products (for which there is the greatest competition for oil from alternative sources of energy). Thus, there have had to be a considerable number of refinery closures, particularly of refineries orientated to the production of fuel oils, together with the refurbishing and upgrading of the refineries that remain – to enable them to handle and treat fuel oil so as to convert it into additional supplies of lighter products. These processes have led back to a concentration of refineries in those locations at which it is easiest to receive a range of crude oils (for blending in the most expeditious manner possible, given variable demand conditions), and from which the distribution of the products can be most effectively achieved. These sophisticated and specialised refineries are intended to serve large regions, rather than specific localities (as in the case of the simple refineries of the 1950s and the 1960s). This locational shift has

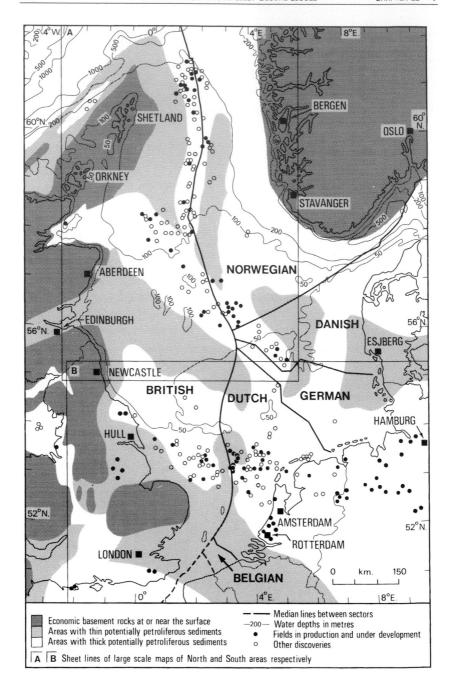

Figure II-9.4: North Sea Oil

created conditions in which there is an increased need for oil products' transport. Transport patterns have also been made much more complex, so enhancing the demand for more specialised oil products' tankers for such duties. Such recent developments in the refining sector and its associated maritime transport needs are serving to enhance the relative importance of high-quality locations such as Rotterdam in the Western European downstream oil industry.

Concurrent with these developments in Europe itself, there have also been changes at the international level in the oil industry which have enhanced the need for the long-distance maritime movements of oil products. First, with the greater flexibility of products' out-turn patterns from advanced refineries (as in Western Europe and North America), there is increasing scope for producing and shipping specific products which are in demand on the other side of an ocean – or even on the other side of the world. With modern communications systems, as well as with the new highly flexible refineries, the international oil companies are able to achieve a very effective utilisation of their refining and ocean transportation facilities.

Second, over the period from 1974 to 1983 when OPEC was able to call the tune over the supply and price of oil, its member countries demanded – and secured – the expansion of their own refinery facilities; sometimes, as in the case of Saudi Arabia, with the financial and technical support and involvement of one or more of the international oil companies. These new and expanded refineries are now coming into production and are capable of delivering large volumes of oil products to Western Europe. These volumes will be increased much more unless Europe introduces tariff or quota restraints on the development. The increase in such traffic is, however, in part the result of arrangements which are being made both between governments (of the exporting and oil importing countries) and by some of the international oil companies to market the products involved.

As a result of these developments an increasing share of maritime oil trade over the world's main supply routes is now – once again – being taken by oil products at the expense of crude oil movements. Indeed, given the opportunities and the challenges created for the long-distance maritime transportation of increasing volumes of oil products, a high motivation has emerged for the development and use of much larger tankers for this trade. This is especially important for those oil exporting countries, such as Kuwait and Venezuela, which are deliberately pursuing policies of expanding their oil marketing and distribution systems in various European countries and in North America.

## Conclusion

The impact and consequences of the changes in the world oil market and in world oil geo-politics since 1973 are for the maritime sector of the industry still emerging. In this respect, the past thirteen years have not been very different from any previous period in the history of international maritime oil transport. Changes, rather than stability, have always been the order of the day as the movement of oil around the world has had to respond to the commodity's changing economic and political geography. The factors involved have been – and remain – highly dynamic, and have thus produced the continuing radical changes in the international transport of oil as described in this paper. Currently, the international oil industry is in a state of serious disequilibrium, with relatively stagnant demand and significant over-supply potential. This has been reflected in the dramatic fall in the price of oil since January 1986 – and this fall in price has, in turn, introduced yet another element of uncertainty into the system.

In this context it is now difficult to forecast how the international oil industry will evolve over the rest of the century as several different pathways of development could be taken. The international market for oil could expand significantly with the much lower prices and, in particular, stimulate the demand for oil from the main exporting countries. This would lead to a resurgence in the volume of long-distance international movements of crude oil. On the other hand, policy makers may try to minimise uncertainty by reaching regional agreements in respect of oil supply demand opportunities as shown in Fig. II-8.1. In this case there could be a continuing decline in the total ton/kilometres requirements for maritime oil transport – and an even stronger propensity for the world tanker fleet to continue to decline in average size. The outlook for maritime oil transport is thus uncertain. Nevertheless, whatever happens, the industry seems likely to remain as dynamic in the future as it has been in its 100 year history to date.

# Chapter II – 10

# The Aftermath of the 1986 Oil Price Collapse

## A. Oil Prices, OPEC and the West: Longer-term Prospects*

### Introduction

It would be a gross misjudgement of the impact of OPEC since the early 1970s to suggest that its inability to keep up the price of oil means a return to the *status quo ante* – in either world economic outlook terms or in respect to international geopolitical questions.

The changes in the Western world's economic situation and prospects arising from the oil price shocks of the 1970s have run deep; the greatest impact has been on growth rates, on employment in the industrialized world and on the aspirations of many countries in the developing world to achieve higher standards of welfare for their inhabitants. The significant changes in the supply and price of a universally required commodity can be interpreted as the single most important factor that turned the Western economic system sour after 25 years of expansion. Even viewed in the narrow terms of price alone, the impact of OPEC on the world economy has been traumatic.

### OPEC's Early Years

OPEC's significance is, however, much deeper and more fundamental than this. For the first decade of its existence, OPEC

* Reprinted from *Geopolitics of Energy*, Vol.8, No.5, May 1986, pp.7–10.

was essentially ignored both by the international oil companies and by the industrialized world's governments; there was no expectation that OPEC could or would have any relevance to the supply and price of oil. In the 1960s the oil companies dealt with OPEC only in respect to technical questions and did not visualize OPEC as a contributor to the strategy of international oil – even when, after 1968, they decided to 'use' OPEC as the scapegoat for the increased price of oil which was needed to finance expansion in the industry. The oil companies thought themselves powerful enough to maintain control over the system.

## Traumatic Changes

The oil companies were thus shocked by OPEC's imposed price increases of 1973–74 and 1979–80. More importantly, developments in the relations between the oil exporting countries and the companies were fundamental in changing the structure of power in the international oil system. The companies were simply dispossessed of the bulk of their assets as, one by one, OPEC's members nationalized the concessions built up by the companies over many decades The compensation received by the oil companies was, in general, derisory as it was generally related to the "written-down" value of the companies' assets – a value related to tax minimisation objectives rather than to the worth of the oil discovered.

Thus dispossession was not only traumatic for the oil companies (which hitherto have been the most powerful set of interests in the world oil system), but it was also irreversible, for it is impossible to conceive of circumstances in which ownership could be regained – except by force. This wholesale nationalisation represented a breakthrough in the relationships between developing countries and multinational corporations as the former achieved a completely new degree of control over the use of their national resources. Given what has happened in the world of oil between the international oil companies and the major oil exporting countries, a firmer base may well have been established for the late 20th century developments which could – eventually – change the shape of the world economic system.

## North-South Relations

The changes induced by OPEC have also been significant for North-South relationships at the intergovernmental level. The failure by the OECD countries to recognise OPEC in its first decade of existence was replaced after 1973 by apprehension concerning the threat posed by

OPEC's actions to the Western economic – and political – system. There was much talk of the need to destroy OPEC before it destroyed the system. In other words there was still little or no industrialized world recognition of the 'right' group of countries elsewhere in the world to join together to secure what they saw to be their own best interests – as, in fact, the industrial countries had done over many decades, and especially since 1974, with organizations such as the OECD and the EEC. Instead, OPEC had to fight to secure recognition and, in a very short period, through increases in prices and the nationalisation of the oil companies' assets (including the oil reserves the companies had discovered), it achieved significant gains for its members.

## Taking OPEC Seriously

The seriousness with which OPEC had then to be taken by the industrial countries was demonstrated in the way energy questions in general, and oil issues in particular, were elevated to a position of prime importance in international discussions – even to the level of the summit meetings of Western leaders. Reaction against OPEC became so frenetic after the second price shock of 1979–80 that the International Energy Agency (IEA) – the industrialized world's countervailing power – centrally directed its efforts to the implementation of policies to minimise the use of OPEC oil no matter what the cost. This objective was spelled out clearly in the IEA's 1980 Annual Review:

> "The 1980s must be a period of major transition toward a minimum oil economy… Necessary adjustments… must continue to be a focus of energy and economic planning. Policies of IEA countries need to be reinforced and extended to… achieve these ends."

The result of this near obsession with the dangers of 'reliance on OPEC oil' was action which, in essence, was contrary to the economic philosophy of the OECD countries – given that the policies involved deliberately sought to restrict international trade (in oil) and advocated decisions to encourage, and even require, the use of alternative, inherently more expensive, sources of energy. The fact that the industrialized West acted in such a manner is a powerful expression of a retreat from the previously unknown concern for the influence on the system of a group of nations in the South. As such, it represented a new departure in the Western system of international economic and geopolitical relationships.

## Protecting OPEC?

Whether the OECD countries are now willing to accept the validity of OPEC's existence and the changes which the oil exporters have brought to the Western economic and political system may never be tested. OPEC is now all but incapable of saving itself, given the rising pressure from non-OPEC supplies in the context of a near-stagnant demand for oil in the non-Communist world outside the OPEC countries. However, a number of OECD countries themselves, either oil exporters or those anxious to maintain revenues from oil consumption, as well as many powerful institutions in the financial and commercial sectors of the industrialized world's economy, are also being adversely affected by the sharp fall in oil prices. Hence, it may now be argued that OPEC's continued existence, as well as the maintenance of the higher economic standards to which the populations of the oil exporting countries have become accustomed, is now a requirement for the overall well-being of the Western system. OPEC, in other words, is now an organization to be protected and cosseted so, in effect, incorporating into the international system a group of countries which hitherto had little claim on its wealth or as participants in its decision-making. Again, suggesting that there is a new departure in geopolitical relationships.

## New Policies Required

Two other aspects of the enhanced role of the oil exporting countries since 1973 are significant. The first relates to the geopolitical, strategic, and cultural importance of the Middle East where most of the OPEC member countries – and most of the world's proven reserves of oil – are located. The predominantly Judæo\Christian-based set of rich industrial countries, with its political and strategic interests in the Middle East generally, and in the Gulf in particular, partly as a result of its proximity to the Soviet Union, has been confronted with an additional important dimension in evaluating its relationships with the region. In essence, the oil companies' loss of ownership of the oil reserves of the region and their loss of control over decisions on their exploitation has necessitated a much more active, responsive, and more carefully considered policy toward the Arab oil exporting countries. In the first instance the needs arising from the oil crises produced a break in the earlier high level of Western solidarity with Israel. Policies have had to become more even-handed. Even more fundamentally, the question of the relations of the Islamic world of the Middle East (and elsewhere, given that two of the other five member countries of OPEC, *viz*. Nigeria and Indonesia, are also partially Islamic) with the West has become an issue of long-term importance, which would not have happened without the transfer of wealth which took place to OPEC's members as a result of the oil crises.

## A West European Problem

This is perhaps most specifically a West European problem – as a result of both history and geography – and, in economic as well as political terms, it points to the validity of policies based on the concept of 'the Middle East as an extension of Europe'. Preliminary discussions with Turkey over its accession to the EEC have already taken place and the issue is likely to become one of real-world politics before the end of the decade. If and when that step is taken, then a geographical barrier will have been broken (and earlier geographical ties renewed). Thereafter, further integration between Western Europe and Turkey's eastern neighbours – the oil exporting countries of the Middle East – would seem to be an unstoppable process, with the recognition of a mutuality of interest in respect to long-term energy markets, on the one hand, and low-cost oil and gas availability, on the other.

## OPEC's Continued Success

Finally, the continued existence and success of OPEC is important for North-South relations. Though the OPEC-generated oil price increases were an economic disaster for the rest of the Third World, there has been OPEC concern about their effect on the rest of the countries of the South. This has been evident by the fact that many of the oil exporters have extended economic assistance to a number of poor countries. Many of these programs have, on a basis related to population or per capita income, exceeded the direct help given to such countries by most of the industrial nations of the North. OPEC's success in partially undermining the power of the rich industrial countries created a great deal of admiration in the rest of the South. Thus, the demise of OPEC would, very generally, be regarded by the South as a retrograde development in international political and economic terms. And should such a demise occur, it would certainly be presented (and not without justification) as a result of a deliberate counter-attack by the North – a counter-attack, moreover, aimed not only at OPEC as such, but also at maintaining the *status quo* between the industrialized and the developing countries in terms of the division of the power in the system. In other words, the continuity of OPEC, as an organization involving a number of the world's developing countries, is required if there is not to be a further exacerbation of the already less-than-satisfactory North-South relationships within the Western system.

## The Future of Oil

OPEC and the oil crises of the 1970s have thus changed much more than the price of oil. Indeed, the latter – measured in real terms – is now falling quickly and the process may not stop before the price returns close to the long-run supply price of the commodity. The longer lasting economic and geopolitical components of the formation and work of OPEC and of the oil crises of the 1970s are more fundamental and may yet mark a turning point in the history of the Western system. Much more so, indeed, than would have been the case if the oil crises had simply been the harbingers of economies obliged to run on sources of energy other than oil. The era of oil as the world main energy source dates back for less than two generations. The future of oil seems likely to be longer than its past, but its future role could well be in the context of a Western system significantly changed by the political-economic forces which resulted from OPEC's initial successes and from the temporary energy supply and pricing crises which it caused.

## B. The Prospects for Oil Prices and the Energy Market*

## Introduction

The price of oil has returned closer to its competitive long-run supply price (see Figure II-10.1) after a decade and a half of super-normal profits from oil taken mainly as government revenues. These profits resulted not only from the temporary successes of OPEC in restricting supply, but also from the general perception of oil as an inherently scarce commodity throughout much of the western world, so creating consumer willingness to accept high prices as inevitable and a propensity by policy makers to invest in alternative high-cost sources of energy.

In spite of the continued success of the world oil industry in adding to reserves at a rate 50 per cent greater than use (as shown in Table II-10.1 some $520 \times 10^9$ barrels have been added to reserves since 1970, while only $355 \times 10^9$ bbls have been used), the prospects for world oil supplies are

* Reprinted from *Lloyds Bank Review*, No.165, July 1987, pp.1–14.

still perceived as being severely limited. Thus, it is still argued, alternatives to oil need to be deliberately encouraged and their development can, moreover, be evaluated in the context of a widely-held expectation of an inevitable renewed serious upward pressure on oil prices in the 1990s. This generally pessimistic appreciation of the future of oil is made more

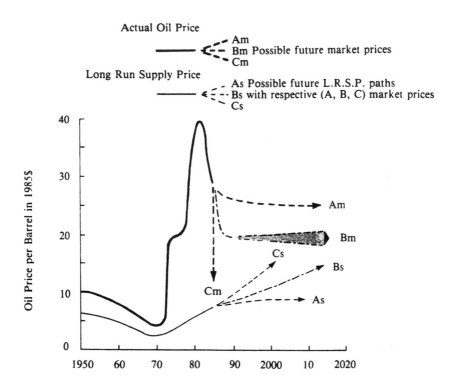

Figure II-10.1: The Long-run Supply Price (L.R.S.P.) and the Market Prices(s) of Oil 1950–2020

Am    *Slow downward drift of oil price under conditions of institutionalized intervention in the market, leading to continued decline in the demand for oil; and continued instability in the market.*

Bm    *Near future (1986) price cut of 25–30 per cent so stimulating demand, but with continued propensity for investment in oil exploration and exploitation. Thereafter, relative narrow long-term price range under conditions of sustainable supply/demand relationships.*

Cm    *Under price collapse conditions the actual oil price returns to a level close to the L.R.S.P. Demand is stimulated with a consequential steeper slope to the L.R.S.P. curve, so creating a potential for relatively near-future renewed upward price pressures and price fluctuations.*

specific – and intensified – by the further perception of the inevitability of a soon-to-be-renewed domination of the international oil market by the members of OPEC – in spite of the post-1973 collapse of their contribution to non-communist world energy supplies, from 36.5% in 1973 to 15.3% in 1985, as shown in Table II-10.2.

The essence of this view (epitomised by the analysis and pronouncements of the International Energy Agency) is its basically confrontational implications, *viz.*, that as OPEC dominates potential oil supply, and as OPEC cannot be trusted to deliver, then negotiations for a *modus vivendi* over the supply and price of oil are simply not

### TABLE II-10.1
### Proven reserves, oil use and net growth/decline in reserves 1970–86

| | Proven Reserves at Beginning of Year | Use of Oil in Year | Gross Additions to Reserves in Year | Net Growth (+) or Decline (–) in Reserves in Year |
|---|---|---|---|---|
| | | (in barrels x $10^9$) | | |
| 1970 | 533 | 17.4 | 62.4 | 45 |
| 1971 | 578 | 18.3 | 40.3 | 22 |
| 1972 | 600 | 19.3 | –3.7 | –23 |
| 1973 | 577 | 21.2 | 35.2 | 14 |
| 1974 | 591 | 21.2 | 32.2 | 11 |
| 1975 | 602 | 20.2 | 31.2 | 11 |
| 1976 | 613 | 21.9 | 3.9 | –18 |
| 1977 | 595 | 22.6 | 15.7 | –7 |
| 1978 | 588 | 22.9 | 44.9 | 22 |
| 1979 | 610 | 23.7 | 21.7 | –2 |
| 1980 | 608 | 22.8 | 33.8 | 11 |
| 1981 | 619 | 21.3 | 67.3 | 47 |
| 1982 | 665 | 20.1 | 30.1 | 10 |
| 1983 | 675 | 20 | 21.1 | 1 |
| 1984 | 676 | 21.1 | 43.8 | 23 |
| 1985 | 699 | 20.5 | 29.5 | 9 |
| 1986 | 708 | 21.4 | 10.8 | –10 |
| 1970–86 | | 355.4 | 520.6 | 166 |

*Sources: Reserves' developments based on data from the annual survey of world oil reserves in the* Oil and Gas Journal, *1970–87; from* World Oil, *1970–83 and from De Golyer and MacNaughton's* Annual Survey of the Oil Industry, *1975–83*

worthwhile. OPEC's members, it is implicitly argued, do not have any inherent justification, arising from their hitherto low level of economic development, to seek to achieve the highest possible flow of revenues from their only marketable resource. Even if they are willing to sell at the price determined by the market, the use of their oil must still be minimised, given that dependence on OPEC oil is politically unacceptable. Most of OPEC's members, it is perceived, are not only ideologically opposed to the Judæo\Christian based western capitalist system, but are also committed to the establishment of a more acceptable situation in the Middle East in respect of the position of Israel *vis a vis* the Arab world.

## The Western Response to the Economic Reality of Lower Oil Prices

In this context the reaction of western policy makers to the oil price collapse is replete with paradoxes. There has been a welcome for the way in which the 'market' has demonstrated that it works and has thus undermined the OPEC cartel, but there are fears about the effect that the

### TABLE II-10.2
### Sources of energy used in the non-communist world (excluding the OPEC countries) in 1973 and 1985

|  | 1973 m.t.o.e.[a] | 1973 % of Total | 1985 m.t.o.e.[a] | 1985 % of Total |
|---|---|---|---|---|
| Total Energy Use | 4045 | 100 | 4115 | 100 |
| of which |  |  |  |  |
| a) Imports of OPEC oil | 1480 | 36.5 | 630 | 15.3 |
| b) Other energy imports[b] | 100 | 2.5 | 220 | 5.3 |
| c) Indigenous Production | 2465 | 60.9 | 3265 | 79.4 |
| of which |  |  |  |  |
| i) Oil | 760 | 18.8 | 1200 | 29.0 |
| ii) Natural Gas | 765 | 18.9 | 750 | 18.7 |
| iii) Coal | 805 | 19.9 | 1045 | 25.2 |
| iv) Other[c] | 135 | 3.3 | 270 | 6.4 |

a.    *m.t.o.e. = million tonnes oil equivalent*
b.    *Oil, natural gas and coal imported from the centrally planned economies*
c.    *Primary electricity (mainly hydro-electricity) and nuclear electricity converted to m.t.o.e. on the basis of the heat value of the electricity produced*

lower prices will have on oil demand. There has also been a welcome for the boost likely to be given to levels of economic activity, but fears for the possibly higher levels of energy use. Similarly, the likely expansion of non-oil sectors of the economy has been welcomed, but this has been tempered by fears for the greater losses which will be sustained in the development of inherently higher cost energies. And finally, the attraction of the reduced debt burden of oil importing countries has been offset – in part at least – by the fears for the impact on western institutions of the intensification of the oil exporters' debts.

Overall, 'faith' in the market is not strong enough to enable western policy makers to accept that all, or indeed, any of these paradoxes will be resolved without intervention. In effect, 'lip service' is paid to market forces, while potential controls are explored for curbing the development of energy supply/demand relationships which appear to threaten pre-conceived notions of what is appropriate for the western industrialized world. Thus, the IEA persists in its plea for as low a use as possible of oil, in general, and of OPEC oil, in particular: even though such pleas run counter to some of the basic tenets of the western system – such as free competition for markets and free commodity trade. This makes the introduction – or the sustenance – of essentially protectionist energy policies that much easier. This is seen already in a number of decisions relating to energy in Western Europe. For example, Chernobyl notwithstanding, there have been declarations of support by governments for their state-owned electricity authorities' insistence on the 'desirability of', or the 'need for', further nuclear power expansion, in spite of prices for fossil fuels, *viz.* fuel oil at $100 per tonne and natural gas at equivalent prices and imported coal at under $40 per tonne) which make the non-competitiveness of nuclear electricity crystal clear. And in the explicit, or implicit, intentions to continue to protect British and German coal production against lower cost coal imports, and/or its possible substitution by oil and gas. And in the commitment of the state owned and/or regulated monopolistic European gas distributors to the purchase of large volumes of Troll gas from Norway even though the economics of the development appears to depend on the gas being marketed at above the present oil equivalent price.

The element in common to all such counter-economic actions is, of course, the perception that it is impossible (and/or undesirable) to have serious discussions with OPEC in order to try to create conditions which ensure an increased supply of the world's lowest cost energy (both oil and gas) from its member countries. This is in spite of the obvious long-term best interests of the oil exporting countries in such a development. The unwillingness even to try to reach an understanding with OPEC lies in the continuing widely-held belief in oil as an inherently scarce commodity so that, it is argued, the inevitable early depletion of the non-communist world's oil, makes it only a matter of time before OPEC will once more be back in control of the international supply and price of oil – and energy. Thus, in the meantime, it is more appropriate to concentrate on building up the supply of alternative energies: a view most powerfully put in the UK Secretary of Energy's recent assertion that there is an overriding need for the western industrialized world to expand its nuclear power capacity – in spite of the fears which its dangers create.

Neither the UK Secretary of State – nor anyone else for that matter – has produced much by way of evidence to support this hypothesis. Instead, we find in such presentations unsubstantiated assumptions of high rates of future growth in energy demand, coupled with expectations of non-OPEC oil – and gas – supply constraints. Before looking in more detail at such views it is worth recalling that they are little different from those in the now effectively discredited massive IIASA study, *Energy in a Finite World* published in 1982. Indeed, the assertion of oil – and other fossil fuels – 'scarcity' and the consequent 'need' for nuclear power, appears to be something of a Pavlovian response by those long committed to the all-electric, nuclear-power based energy future. Such analysts, it will be remembered, forecast that oil (and other fossil fuels') prices would keep going up – uninterruptedly – at three per cent, five per cent or even more per annum, even from the dizzy heights of the early 1980s (see Figure II-10.2)! With oil prices now back to little more than one-half of that level one can only view the 'inherent validity' of their work with a high degree of scepticism!

Ironically, with oil prices now lower (in real terms) than at any time since 1973 (Figure II.10-3), both supply-side constraints on new capacity development, and enhanced demand growth potential, give a little more justification to the analysis. Nevertheless, even as far as oil is concerned, let alone relatively more plentiful gas and much more plentiful coal, the portrayal of inevitable problems in the fossil fuels markets is the result of continuing too low supply-side expectations and too-high demand-side assumptions. It would take a further collapse of the oil price to less then $10 per barrel and no US response to protect its own oil (and other energy) industries, for medium term oil supply/demand imbalance to occur: and the joint probability of such events is low. (In Figure II-10.4 the emergence of such a low oil price is shown as having a ten per cent probability.)

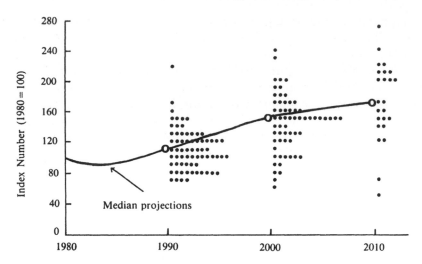

Figure II-10.2: Oil Price Projections for 1990, 2000 and 2010 by Respondents to the Survey Organised for the E.M.F./I.I.A.S.A. International Energy Workshop, 1983

*(Note that 2 of the 4 lowest price forecasts for 1990 and the two lowest forecasts for 2000 and 2010 were made by the author of this article.)*

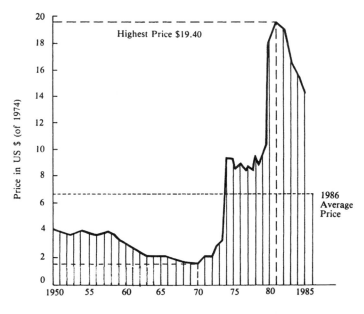

Figure II-10.3: The Per Barrel Price of Saudi Light Crude Oil 1950–86 (in 1974 $)

## Realistic Oil Supply/Demand Relationships
## with Lower Oil Prices

Given a much greater chance of the oil price 'settling down' in the $14–18 per barrel range (See Figure II-10.4), neither rapid demand growth nor a cessation of expansion in total non-OPEC production seem likely to occur. On the demand side, end-user price reductions, after tax changes have been taken into account, will be too modest to change consumers' behaviour dramatically; and neither do the lower prices seem likely significantly to slow down the continuing technology-related improvements in the efficiency of energy use. The impact of these forces from 1973–85 in the energy-intensive industrialized countries is shown in Figure II-10.5. Neither does the substitution away from oil and into other fuels seem likely to slow down very much, let alone to go into reverse. Natural gas and imported coal will in general, and in Europe in particular, be able to continue to compete effectively on the basis of price and/or convenience.

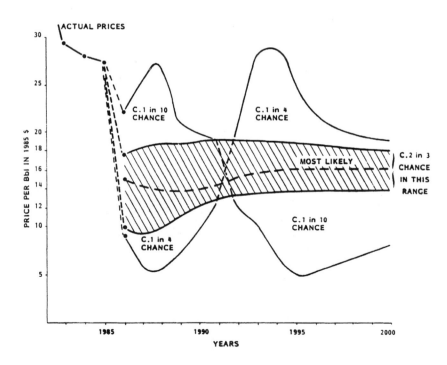

Figure II-10.4: Internationally Traded Crude Oil Probabilities of Alternative Price Paths to 2000. (Prices in 1986 $)

This failure to revert to using oil will, in part, result from the perception of it as a scarce and 'politically-suspect' source of energy. Almost no matter what its price, there will be a high propensity not to use oil (if alternatives are possible), or, if there is no alternative, then to use it as efficiently as possible, especially in the countless cases in which oil has to be imported. Thus, unless oil again goes below $14 per barrel – and stays there for years rather than for months – oil demand seems unlikely to increase by very much more than one-half the rate of economic growth and by no more than two-thirds the rate of growth in energy use. Under these conditions world oil production would finally get back, by the year 2000, to its historic peak (in 1979) of just over 24 x $10^9$ bbls – to only 15 per cent more, that is, than production in 1986.

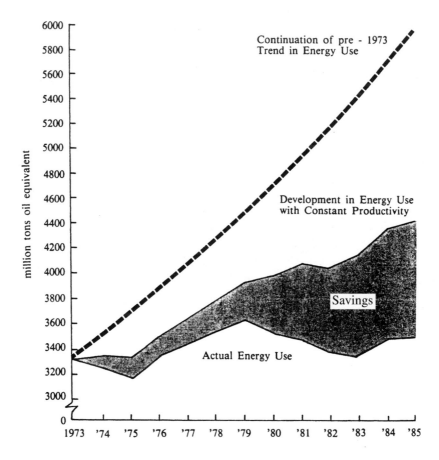

Figure II-10.5: Potential For and Actual Energy Use in the Member Countries of the International Energy Agency, 1973–85

Given these modest prospects for oil demand over the rest of the century, and the more-generous-than-ever reserves position which now exists (as shown in Table II-10.1), there is little likelihood of any pressure on oil supply potential for the remainder of the century. On the contrary, oil supply availability will remain easy in global terms. Thus, a resurgence of prices is possible only with tightening supply/demand relationships for internationally traded oil; that is, for OPEC oil, and oil from what is generally assumed to be a limited number of other net exporting countries. In order to show that this is 'bound' to happen most analyses of oil supply prospects assume little, if any, increase in non-OPEC non-communist world production, together with stagnation, or even a decrease, in communist world exports to the West. Thus by 1995, it is argued, OPEC's production would have to be more than $8 \times 10^9$ barrels compared with its 1985 level of $5.85 \times 10^9$; but note that even $8 \times 10^9$ barrels is well below OPEC's all-time high production of $11.4 \times 10^9$ bbls in 1977). The argument, of course, assumes OPEC's ability to survive in the short-term (largely in the expectation of a strong short-term expansion in the demand for its oil), and that it will be able to continue with the supply controls which it has been applying since the beginning of September. Even these specific assumptions are by no means self-evident – given continuing doubts over the short-term cohesion of OPEC, and the strong medium term motivation for its members to increase production in order to compensate for the present and near-future lean years of low oil prices and low demand for OPEC oil.

There are, however, two more fundamental objections to the hypothesis of an inevitably tightening international oil market. First, it is highly unlikely that communist world net exports will stagnate, let alone decline. After a temporary period of difficulty (due largely to mismanagement), Soviet oil output is once again increasing at the same time as the use of oil in the Soviet Union – as well as in Eastern Europe – is rapidly being substituted by natural gas as a matter of highest priority. The Soviet Union is thus pursuing a policy which, if successful, will make increasing volumes of oil available for export to the non-communist world. Its ability to increase its gas exports to Western Europe is axiomatic. Such export prospects for oil and gas are of high macro-economic significance to the USSR, given the degree to which its external trade balance – and its ability to import essential goods and services – depends on the value of its oil (and gas) exports. Lower oil prices will, in effect, serve to stimulate Soviet efforts to export oil (and gas). Meanwhile, on the other side of the communist world, China's oil exports have increased rapidly in recent years. They have increased by 25 per cent since

1985 and they will, at least, keep on expanding slowly – again under the motivation of the country's need for hard currency earnings to sustain ambitious development plans over the rest of the century. Overall, a minimum increase of 25 per cent, and a more likely 33 per cent increase, in the non-communist world's use of communist oil by 1995 seems likely. This means that the share of communist world oil in the slowly growing non-communist world's total demand for oil will increase.

Second, non-OPEC, non-communist world oil supply will continue to expand. Between 1973 and 1985 it expanded, as shown in Table II-10.2, by over 50 per cent to more than $9 \times 10^9$ barrels. Part of this entered international trade, but little went further than neighbouring or proximate countries (i.e. Canada and Mexico to the US; Norway and the UK to other West European countries). Such oil can be considered as pseudo-indigenous production as it flows essentially within the context of special trading regimes. This is important because indigenous oil will, in general, continue to be accorded priority over other oil – for both political and economic reasons. In this context about 90 per cent of current, non-OPEC non-communist world oil production will be protected – either absolutely, or to a large degree. It will thus not be subjected to much competition from OPEC or communist world exports. The further exploitation of the already proven reserves of oil in the non-OPEC countries will, very generally, also be secured by means of national or regional policy decisions. This will include the reserves of the United States (except for a very limited volume of oil with very high variable production costs), and the reserves of newly emerging producers in the third World – such as India and Brazil. The exposure of North Sea oil potential to the winds of competition from OPEC oil, and to the purely commercial decisions by international oil companies is exceptional. Yet even in the case of the North Sea there is scope for change in the fiscal and other concessionary arrangements between governments and companies so as to stimulate continuing investments in new production. And there is scope for organising preferential arrangements for markets in Western Europe for indigenous oil and gas – as in the case of the Troll deal. Such arrangements could eliminate much, if not all, of the current reluctance of the companies to continue with their hitherto successful efforts to find and develop new reserves as a result of the oil price collapse. The governments of most of the non-OPEC producing (exporting) countries have too much at stake to risk the decimation of their oil industries, as a consequence of a low oil price and too high government tax takes on the companies involved. The 1985 level of about $1.5 \times 10^9$ barrels of Western European oil production seems likely to be a minimum prospect for future production levels, while gas production will increase – on the assumption that oil prices do not go below $14 per barrel (on other than a short-term basis). Elsewhere in the non-communist, non-OPEC world (except the US), one can still look to prospects for increase levels of annual production.

To summarize; the non-OPEC non-communist world which cut back its dependence on OPEC oil by over 50 per cent in the decade to 1985 (see Table II-10.2) is unlikely to pursue policies which permit the earlier level of dependence to be re-established. On the contrary, a high degree of oil independence – either nationally or within regions of inter-related countries – will be a positive objective of policy makers with the support and encouragement, where allowed, of the oil companies, both large and small. The companies do, after all, have to have locations in which to continue to invest in oil activities. There are no shortages of interesting areas for exploration – either in the industrialized world or, even more so, in the Third World, where recent re-evaluations of the potential in all three continents have been positive. Thus, the current 9.3 x $10^9$ bbls of annual oil production of the non-OPEC non-communist world seems unlikely to represent the peak potential, even if the US fails to maintain output at its 1986 level of 3.6 x $10^9$ bbls. Indeed, the faster and further US output falls, the greater the opportunity for Canada and Mexico to make up the short-fall. The political problems associated with such a North American regionalization of the oil industry development pale into insignificance compared with the difficulties the US would expect in the event of other than a minor increase in imports from the Arab world. Avoiding such increases in imports from the Middle East is, indeed, already, the prime aim of US oil (and energy) policy.

## Implications and Responsibilities

Expanding non-OPEC oil production in the non-communist world, plus increased communist world exports means that OPEC's members are unlikely, over the medium term, to achieve more than an average one per cent per annum increase in their oil exports. They could, by 2000, after taking their own now more slowly increasing oil use into account, be back up to a production level of 7.5 x $10^9$ – but this is a lower rate of production than in any post-1972 year, except for the lean years since 1982. It is, moreover, a modest level of production compared both with their potential to produce and with their output preferences. Moreover, as oil's percentage contribution to the non-communist world's energy needs will by the year 2000 have declined yet further, and as the OPEC countries' share of the market seems likely to be less than it is now, OPEC's ability to exercise much pressure on the international economic system will remain limited. Overall, its prospects for reasserting its domination of the international market this side of the year 2000 is negligible.

Indeed, as suggested earlier, concern for OPEC's influence on the western system arising from its renewed control over the energy sector is misplaced. The issues which should be under active consideration are different. Because of the oil/energy supply/demand relationships set out above, OPEC now appears to be all but incapable of reasserting its control over the market. It is thus the prospects for the position and status of OPEC as an international economic organization which need to be reappraised – in terms of its position in the western politico-economic system overall. In this context one can argue that OPEC's continuation as an organization, as well as the maintenance of the higher living standards to which the populations of the OPEC countries have now become used, are objectives to be sought for the sake of the overall well-being of the western system. OPEC is, indeed, an organization to be protected and cosseted, so, in effect, firmly incorporating into the western system a group of countries which have hitherto enjoyed scant respect. This is particularly important from two standpoints; *first*, because of the relationship of OPEC with the Middle East question. This is most specifically a West European issue, given the long historic ties and contemporary economic and potential relationships. These imply a mutuality of interests which could be well served by linking long-term energy markets in West Europe with the long-term availability of low-cost oil and gas in the Middle East. *Second*, the continued existence of OPEC is critical for north/south relations. In spite of the hardships caused to much of the Third World by the high price of oil, the demise of OPEC would, very generally, be regarded as a backward step in international political and economic terms by the nations of the South. The event would certainly be presented as a result of a deliberate counter-attack by the North – aimed not only at OPEC as such, but also as a means of maintaining the *status quo* between rich and poor nations. The destruction of OPEC would, in other words, lead to an exacerbation of the already less-than- satisfactory state of north/south relations.

## Conclusions

A negotiated use of increasing volumes of OPEC oil is thus a requirement for a western economic system in order to ensure that the energy sector does not form an inherently de-stabilizing element. Simply using more OPEC oil seems, however, unlikely to be enough for ensuring stability. In addition, the principle of OPEC's 'right' to exist and to work for its members interests has to be accepted by the industrialized countries. It could, indeed, be seen as the first step in moves towards negotiations at the international level which seek to find an acceptable long-term solution to oil supply/demand and pricing issues in which the world's lowest cost sources of energy (the oil and gas of the OPEC countries) are accorded a more important role. In other words, the present much lower oil price should not be seen as a development which simply takes the world back to the *status quo ante* the first oil price shock; but as a development which provides an opportunity for relating energy questions – and the interests of the energy exporting countries – to the broader long-term prospects for the future evolution of the western economic and political system.

## References

Adelman, M.A., 1972, '*The World Petroleum Oil Market*': Baltimore, the John Hopkins University Press.

— 1986, 'The Economics of the International Oil Industry', in P.R. Odell and J. Rees, eds., *The International Oil Industry – An Inter-Disciplinary Perspective*: London, MacMillan Press.

British Petroleum, 1979, *Oil Crisis... Again?*: London, B.P. Ltd.

Energy Modelling Forum, 1981 *World Oil – Summary of the Energy Modelling Forum*, Stanford University Dept. Of Operational Research.

International Energy Agency, 1981, *Energy Policies and Programmes of I.E.A. Countries, 1980 Review*: Paris, O.E.C.D.

— 1982, *Energy Policies and Programmes of I.E.A. Countries, 1981 Review*: Paris O.E.C.D.

— 1982, *World Energy Outlook*: Paris, O.E.C.D.

— 1984, *Energy Policies and Programmes of I.E.A. Countries, 1983 Review*: Paris, O.E.C.D.

International Institute of Applied Systems Analysis, 1981, *Energy in a Finite World*: Cambridge, (Mass.), Ballinger Publishing.

Kemp, A.G., and D. Rose, 1984, 'Investment in Oil Exploration and Production – The Comparative Influence of Taxation', in D.W. Pearce, H. Siebert, and I. Walter, eds., *Risk and the Political Economy of Resource Development*: London, MacMillan Press.

Kouris, G., 1984, 'Oil Trends and Prices in the Next Decade – An Aggregate Analysis': *Energy Policy*, vol. 12, no.3, p.321–328.

MacKillop, A., 1984, 'Adjustment and Energy – Dangers of a Demand Side Obsession'; *Energy Policy*, vol. 12 no.4, p.380–394.

Manne A.S., and L. Schrattenholzer, 1984, 'International Energy Workshop – a Summary of the 1983 Poll Responses': *Energy Journal*, vol.5, no.1, p.45–64.

Mikdashi, Z., 1986, 'Oil and International Financial markets', in P.R.Odell and J. Rees (eds.), op. cit.

Molle, W., and E. Wever, 1984, *Oil Refineries and Petrochemicals in Western Europe – Buoyant Past, Uncertain Future*: Aldershot, Gower Publishing.

Noreng, Ø., 1980, *The Oil Industry and Government Strategy in the North Sea*: London, Croom Helm.

Odell, P.R., 1982, 'Too Exposed to Political Pressures – Oil and Gas in Western Europe': *Times Higher Education Supplement*, p.13–15.

—     1986, *Oil and World Power*, 8th ed: Harmondsworth, Penguin Books

—     and K.E. Rosing, 1976, *Optimal Development of the North Sea's Oil Fields*: London, Kogan Page.

—     and K.E. Rosing, 1983, *The Future of Oil – World Resources and Use*, 2nd ed.: London, Kogan Page/New York, Nichols Publishing.

—     and K.E. Rosing, 1984, 'The Future of Oil – a Re-evaluation': *OPEC Review*, Summer, p.203–228.

—     and L. Vallenilla, 1978, *The Pressure of Oil – A Strategy for Revival*: London, Harper and Rowe.

Robinson, C and Morgan, J, 1988, *North Sea Oil in the Future – Economic Analysis and Government Policy*, London, MacMillan Press for the Trade Policy Research Centre.

Saunders, H.D. 1984, 'On The Inevitable Return of Higher Oil Prices': *Energy Policy*, vol.12, No.3, pp.310–320.

United Nations, 1983, *Development of the Energy Resources of Developing Countries – Economic and Social Document*, E/1983/91: New York.

World Bank, 1983, *The Energy Transition in Developing Countries*: Washington.

# Chapter II – 11

# The Prospects for non-OPEC Oil Supply*

## Introduction

In an expanding and continually modernizing global economy both
non-fuel and transportation uses of oil will, for the foreseeable future,
constitute growth areas for the demand for relevant oil products. Even
so, the intensification and the geographical diffusion of increasing
efficiencies in the use of transportation fuels and in the oil to
petrochemical conversion processes may even moderate the rates of
growth of these lighter products compared with historical precedents. In
some countries, moreover, particularly in those with the most private
transport orientated systems, there will be attempts to curb oil use in the
transport sector in the context of policies which seek to limit $CO_2$
emissions. Thus, even in these oil-specific markets an overall relatively
low global growth rate in oil products demand seems likely. It is also very
likely that a low rate of growth for those oil products will be entirely or
largely offset by declines in other uses of oil: for uses, that is, in which
oil has to compete with other energy sources. In these latter markets,
moreover, the overall growth rate in energy use will be kept to modest
proportions as a result, first, of changing economic structures (in a shift
to less energy intensive activities) and, second, of improving efficiencies
in the use of energy.[1]

The mid-point values which emerge from our most recent
evaluations of the prospects for overall energy and individual energy

* Reprinted from *Energy Policy*, Vol.20, No.10, October 1992, pp.931–941

sources' growth over the three decades 1990–2020 are set out in Table II-11.1.[2] Primary energy use is forecast to continue to increase at about the historic post-1973 rate of just over 2% per annum for another decade. Thereafter, over the following two decades annual rates of increase fall to

## TABLE II-11.1
### Global economic/energy prospects, 1990–2020[a]

| | Actuals for 1990 | Forecasts for 2000 | 2010 | 2020 |
|---|---|---|---|---|
| Average price of internationally traded crude oils (in 1990 US$) | 21.3 | 17.75 | 21.5 | 26.5 |
| World economic product (1990=100) | 100 | 130 | 175 | 235 |
| Average annual growth rate over previous decade (%) | | 2.7 | 3 | 2.9 |
| Total primary energy use (mtoe) | 7790 | 9770 | 11455 | 12945 |
| Average annual growth rate over previous decade | | 2.2 | 1.6 | 1.3 |
| of which | | | | |
| Oil (mtoe) | 3100 | 3270 | 3520 | 3705 |
| (%) | | 0.5 | 0.8 | 0.7 |
| Natural gas (mtoe) | 1740 | 2760 | 3725 | 4270 |
| (%) | | 4.7 | 2.9 | 1.4 |
| Coal (mtoe) | 2195 | 2660 | 2810 | 2990 |
| (%) | | 2.3 | 0.6 | 0.6 |
| Nuclear power (mtoe)[b] | 185 | 225 | 280 | 400 |
| (%) | | 2 | 2.2 | 3.6 |
| Renewable energy (mtoe)[b] | 570 | 855 | 1110 | 1580 |
| (%) | | 4.1 | 2.7 | 3.6 |

Notes:   a    The mid-point of a range of estimates based on a range of oil price forecasts.

       b    Nuclear power electricity and primary electricity from renewables (eg hydroelectricity) converted to million tonnes of oil equivalent on the basis of the heat value equivalent of the electricity produced.

Source: Derived from the most recent revision of the December 1990 contribution by EURICES to the IIASA/Stanford University International Energy Workshop, Survey of Global Energy Prospects.

1.6% and 1.3% respectively – under the impact of the effective energy policy steps which will be progressively introduced to reduce the emissions of greenhouse gases.[3]

The forecast breakdown of demand by energy source is also shown in Table II-11.1, while in Figure II-11.1 the emerging percentage contribution of each of the five sources is shown in graph form. The growth rate in oil demand is shown at under 1% per annum throughout the period. This is well below the growth rate in energy use so that the share of oil use in total energy use falls from just under 40% in 1990 to only 28.5% by 2020. Towards the end of the first decade of the 21st century oil finally loses its role (held since 1965) as the premier source of global energy. Natural gas then becomes the single most important energy source. (Note, in passing, that natural gas is forecast to become

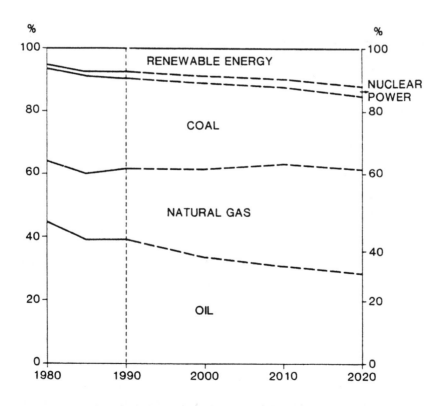

Figure II-11.1: The percentage contribution of the main sources of energy to world energy use, 1980–2020.

*Based on mid-point values for forecasts of world energy use: see Table II-11.1*

more important than coal before the end of the 1990s.) Meanwhile, non-fossil fuels (nuclear power and renewable energy) make relatively slow progress in the overall energy market. They expand modestly from their present share of under 10% of demand in 1990 to just over 15% by 2020.[4] In other words, the global energy market will remain dominated by fossil fuels throughout the next 30 years – in spite of the efforts which will be made to restrain $CO_2$ emissions.

The more important contribution to those efforts will, indeed, be achieved by the faster growth of relatively carbon-poor natural gas at the expense mainly of coal and, to a lesser degree, of oil. Thus, while the indicated 66% increase in energy use from 1990 to 2020 is met to the extent of over three quarters by fossil fuels (almost 4000 million tons of oil equivalent additional use in 2020 compared with 1990), this is at a cost of only an approximate 43% increase in $CO_2$ emissions.[5] The difference between the percentage increase in the fuel burn, on the one hand, and that of emissions, on the other, is a measure of the $CO_2$ emissions' savings generated by the switch to natural gas. This large scale switch to natural gas is, we believe, a hitherto largely ignored component in evaluating the long-term demand prospects for oil.

Given the now highly probable continuation of low rates of increase in energy use, together with the preferred expansion of natural gas, oil supply expansion requirements will be eminently modest. In round figures the required expansion is less than 200 million tons of additional oil production per decade – from the 1990 base of approximately 3100 million tons of global oil production. This implies a less than 20% increase in the total oil supply over the 30-year period. The oil industry thus remains an expanding one for the next generation – but only just. Its designation over this period as a sunset industry would thus be inappropriate but it will, nevertheless, come close to the zenith of its achievements – but as a consequence of demand-side prospects, rather than supply-side limitations.[6] There are, nevertheless, significant supply-side implications which arise from this analysis. It is these which form the subject of the remainder of this paper.

## Supply-Side Prospects

### Global aspects

Given what now appears to be a severe demand-side constraint on the growth of the oil industry, it is ironic that conventional wisdom through the 1970s and well into the 1980s was fearful for the future of oil on the grounds of the near future exhaustion of reserves. Thus, it was then widely argued that, no matter what the price, the rate of production would soon necessarily be constrained by scarcity and shortly thereafter would inevitably have to fall as annual additions to reserves failed to match the increasing use of oil. Proponents of this view of oil included many of the major oil companies – including BP, whose 1979 forecast in this respect is shown in Figure II-11.2. Subsequent events have confirmed the views that we expressed at the time that such prognostications were based on a fundamental misjudgement of world's oil potential.[7]

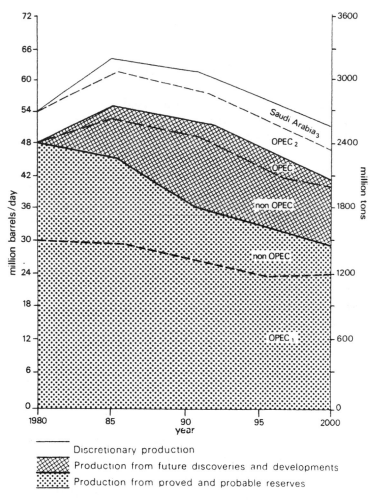

Discretionary production

Production from future discoveries and developments

Production from proved and probable reserves

Figure II-11.2: B.P.'s forecast in 1979 of non-communist world oil production prospects, 1980–2000

*In British Petroleum, Oil Crisis... Again?, London, 1979.*

In Table II-11.2, for example, the annual data on remaining proven reserves, production and gross and net additions to reserves since 1970 are set out. This shows that there have since then been only four years in which there were net declines in reserves; and that over the 20 year period as a whole gross additions to reserves exceeded the cumulative production of oil by a factor of more than two.[8] On graphing this data to cover a longer period of recent oil history and after smoothing the gross additions to reserves curve by means of a calculated five-year running mean (in order to avoid the inevitably large swings in volumes of reserves additions which emerge on a year-to-year basis), one can see, from Figure II-11.3, how the long-term upward trend in annual additions to reserves has persisted even into the post-1980 period when demand (= production) ceased to grow (thus eliminating the need for additional reserves), so leading to highest ever surpluses of proven reserves.[9] This is, of course, reflected in the recent evolution of the highest ever reserves to production ratios based on annual data, *viz*. over 41 years in 1990 – compared with a reserve:production ratio of 25.5 years in 1980 and just over 30 years in 1970.

Thus, the earlier fears for an oil industry which by now was forecast to have been running out of oil have been shown to be entirely groundless. Instead, we have a global oil industry which is still running into oil – at a rate, indeed, which had it not been for demand constraints, would have allowed for higher rates of production than have been achieved.

Given that proven reserves are defined by reference to current costs and prices[10] and that there is now little likelihood that the costs of extracting presently known reserves will rise significantly in real terms, then the continuing producibility of the reserves is in no economic doubt, providing prices remain at or close to present levels. In this case, the world can be certain of a continuing long-term supply of oil which remains adequate to meet the anticipated demand – political issues apart. Moreover, except for the latter considerations, there is no need for higher real prices for the indefinite future.[11] One cannot even put a date on the possible termination of this favourable supply prospect. It can, indeed, only be so dated once there is evidence of a failure by the industry to sustain the running mean of annual reserves additions at or about the annual rate of production/use for a continuous period of at least 10 years. Paradoxically it is, of course, the current high prices of oil (relative to a long-run supply price which is no more than $10–12 per barrel – and perhaps much lower)[12] which sustain the intensive exploration rates for oil around the world. These economic processes are, however, a function of issues and actions which are essentially political, ie. the oil production limitations imposed by OPEC and the reactions to those limitations in various other parts of the global oil industry.

## TABLE II-11.2
## Proven reserves, reserves/production ratio, oil production and net growth/decline in reserves over the 20 year period from 1970

| Year | Proven reserves at beginning of year (R/P ratio in brackets – in years) | Production of oil in year (in barrels x $10^9$) | Gross additions to reserves in year | Net growth (+) or decline (–) in reserves in year |
|------|------|------|------|------|
| 1970 | 533 | 17.4 | 62 | +45 |
| 1971 | 578 (33.2) | 18.3 | 40 | +22 |
| 1972 | 600 (32.8) | 19.3 | –4 | –23 |
| 1973 | 577 (29.9) | 21.2 | 35 | +14 |
| 1974 | 591 (27.9) | 21.2 | 32 | +11 |
| 1975 | 602 (28.4) | 20.2 | 31 | +11 |
| 1976 | 613 (30.3) | 21.9 | 4 | –18 |
| 1977 | 595 (27.2) | 22.6 | 16 | –7 |
| 1978 | 588 (26.0) | 22.9 | 45 | +22 |
| 1979 | 610 (26.6) | 23.7 | 22 | –2 |
| 1980 | 608 (25.7) | 12.8 | 34 | +11 |
| 1981 | 619 (27.1) | 21.3 | 67 | +46 |
| 1982 | 665 (31.2) | 20.1 | 30 | +10 |
| 1983 | 675 (33.6) | 20.0 | 21 | +1 |
| 1984 | 676 (33.8) | 21.1 | 44 | +23 |
| 1985 | 699 (33.1) | 20.5 | 30 | +9 |
| 1986 | 708 (34.5) | 21.4 | 67 | +45 |
| 1987 | 753 (35.2) | 21.9 | 113 | +91 |
| 1988 | 844 (38.5) | 22.8 | 99 | +76 |
| 1989 | 920 (40.3) | 23.5 | 72 | +48 |
| 1990 | 968 (41.2) | 23.8 | 40 | +16 |
| 1970–90 | | 488 | 900 | +452 |

*Sources: Reserves' developments based on data from the annual survey of world oil reserves in the* Oil and Gas Journal. *1970–90; from* World Oil, *1970–91 and from De Golyer and MacNaughton's* Annual Survey of the Oil Industry, *1975–83. Annual production data from the* Petroleum Economist, *1970–91.*

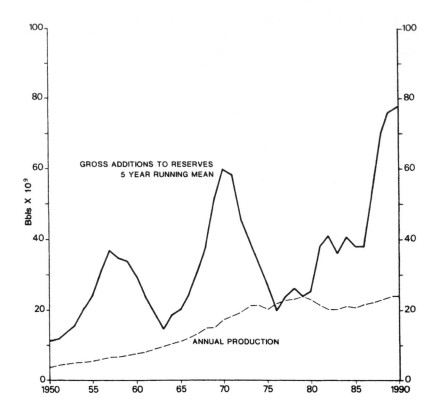

Figure II-11.3: Annual world production and the five-year running mean of gross additions to reserves, 1950–2000

## Supply decisions to the early 1970s

Over the period of low and declining real prices for oil – between 1950 and 1970 – decisions on supply were taken largely by the multinational oil corporations. They were, of course, motivated well nigh exclusively by commercial considerations, little constrained by governments' intervention. The principal exception to this was in the case of the USA where protection of the indigenous industry, through import quotas, sustained national production at a level between 30% and 50% higher than would otherwise have been possible.[13] Thus, the expansion of non-communist world oil supply between 1950 and 1970 was concentrated in the lowest cost producing areas. This involved only a small group of countries, essentially those which were to become the founder members of OPEC in 1960. This group was, indeed, responsible in 1970 for over 1000 million tons of additional production compared with 1950. Its additional output, moreover, represented over 80% of the total increase in the world outside the non-communist countries and the USA, showing clearly how they dominated the growing market for oil. Even countries – such as Brazil and India – which had, during the period, created or intensified state oil monopolies to inhibit the influence of the multinational oil corporations at home, preferred to import increasing volumes of oil produced in the low cost exporting countries by the multinational oil corporations, rather than to find and develop their own reserves and production. Indeed, with oil imports readily available to them at under US$2 per barrel, it would have been grossly uneconomic for such countries to have used scarce financial and managerial resources to stimulate indigenous production. Thus, in 1973, at the time of the first oil price shock, but by when internationally traded crude oil prices had already risen by 50% over 1970 levels, as a result of higher government tax takes and of tightening markets, 82% of non-US/non-communist world oil production originated in the member countries of OPEC.[14]

## Post-1973 changes

There has since been a radical change in that domination of non-US/non-communist world oil supply by OPEC members' production. By 1985 their share of the total had fallen to under 55% and, in spite of some recovery since then, their 1990 share was still only 60% of the total.[15] In spite of this shift away from OPEC oil there are oft expressed fears that renewed OPEC domination of supply is destined to return. Such fears are usually formulated in terms of an expected relatively near future re-dependence on OPEC oil as a function of a rising oil demand which can be met only from the faster depletion of the large volume (105,000 million tons) and the high percentage (76.6%) of the world's oil reserves which are located in the OPEC countries.

Yet, in the context of current oil policy objectives throughout much of the world most of those reserves of OPEC are irrelevant to production prospects for some decades to come. In essence, because of OPEC's high oil price policy and because of fears for the reliability of production from most of the member countries of OPEC, oil imports from OPEC have become very widely viewed throughout most of the rest of the world as the supplies of last resort. There was undoubtedly some degree of mitigation of this strongly held view in the aftermath of the 1986 oil price collapse; and in the context of efforts by most of the oil exporting countries to try to establish themselves as reliable traders who accepted, albeit, in some cases, reluctantly, the unreality of their earlier claims for much higher oil prices than the market would bear. Unhappily for OPEC's improving prospects as a result of these post-1986 developments, Iraq's annexation of Kuwait and the subsequent political and military conflict in the Gulf in 1990–91 once again served to undermine part of the 1986–90 relaxation of concern over too high a degree of dependence on Middle East oil among oil importing nations. This has led to the reintroduction and/or the reintensification of 'away from oil' – and, especially, away from OPEC oil – policies.[16]

## Prospects for supply developments

M iddle East politics and OPEC's declared strategic aims still seem, however, to be less important as a curb on OPEC's export potential than the supply side economic implications of an international oil price which has stuck at around US$20 per barrel. At this price there is a great deal of oil potential elsewhere in the world in which investment is worthwhile – both for countries and for oil companies. For the former, any oil which can be found at home and produced either to substitute imports or to generate an export surplus is likely easily to justify the investment in its development. Not only can full costs be recovered, but revenues for governments will arise from the economic rents (surplus profits) which can be generated by the investments against an oil price of US$20 per barrel.[17] For the international oil companies, for which upstream investment in most OPEC countries remains foreclosed by those countries' oil industry nationalization measures of the 1970s and 1980s, their commercial interests are best served by investments which are made in inherently less productive oil reserves developments elsewhere. By so doing they achieve a larger availability of guaranteed profit equity crude. They thus become less dependent on crude oil purchases from the OPEC countries from the sale of which the companies' level of profits depend on the volatile and uncertain products/crude oil price differential and/or on high risk trading activities.

     Thus, from both the public and the private stand-points, a continuation of the increasingly significant post-1973 efforts to enhance oil production from non-OPEC sources remains relevant and appropriate – unless and until OPEC policy changes, or the organization falls apart. The continuing build up of dependence on non-OPEC crude production depends, for the short to medium term, on there being adequate oil reserves to be produced and, for the longer term, on the existence of additional resources to be discovered and developed.

The availability of adequate non-OPEC oil reserves for the short to medium term is not in doubt. A two and a half fold increase in their production between 1973 and 1990 has been accompanied by an approximate threefold increase in proven reserves so that the overall current reserves: production ratio (outside the USA) is of the order of 20 years – allowing modest scope for enhanced production, even in the unlikely event of there being no further net additions to reserves. Even in Western Europe, where production has increased tenfold since 1970, proven and probable reserves are adequate to keep production moving slowly ahead at least through the 1990s – ultimately to some 220–225 million tons per year.[18] This ensures that there will be little need for any proportional increase in the region's oil imports before 2000 at the earliest. Indeed, in the context of a highly likely near stagnant oil demand, there could even be a small decline in the absolute quantities of oil imports from OPEC sources.

There is a similar situation for the developing countries outside OPEC. The rapid expansion of their production from 146 million tons in 1970 to almost 425 million tons in 1990 is shown in Table II-11.3. With a total production over the 20 year period which has exceeded 5500 million tons, they have, nevertheless, since 1970 achieved a gross addition to their proven reserves of oil of more than 14,000 million tons. Their remaining proven reserves are currently about 12,600 million tons of oil so that they now enjoy a close to 29 year reserves:production ratio. They thus have a short to medium-term prospect for significantly enhancing their collective annual production – if they so need, or wish.

The pre-2000 possibilities of increased oil output from Western Europe, the non-OPEC developing countries and from other OECD countries apart from the USA, thus seem likely to be enough to offset some short-term declines in other non-OPEC oil availability – notably from the continuation of the slow decline in US output and the now likely diminution of total flows from China and the former USSR to the market economies. Given, as shown above, a high probability that oil demand will grow only modestly over this period, the additional call on OPEC supplies will thus be relatively small – and will probably be of no consequence as far as price determination is concerned. International oil price changes over this period are more likely to be random – as a result of accidents, either natural or political, which temporarily upset supplies

– or the consequence of shifts in OPEC policies, either towards or away from the controls it exercises over total production. In a situation of relatively stable demand and an absence of down-side elasticity of supply among non-OPEC suppliers – in that most of these will continue with near capacity production policies – the impact of unpredictable accidents affecting supply, and of equally unpredictable OPEC policy changes, will be the main destabilizing elements on world oil prices over the coming decade.

For the longer-term (post-2000), assuming for the moment neither major changes in OPEC policies nor the break up of the organization, and thus a continuation of present prices in real terms, the slow increase in oil demand will be met by the combination of two main supply-side developments: first, the renewed expansion of the oil industry of Russia and other former Soviet republics – and thus of their oil exports to the rest of the world; and, second, the continued expansion of production in the non-OPEC developing countries.

The former (the re-expansion of Russian production and exports) will emerge from the initial rehabilitation of the industry through western investment, management and technology in the aftermath of the break-up of the USSR and under the aegis of the European Bank and the

## TABLE II-11.3
### Non-OPEC developing countries: oil production, 1970–90
(million tons)

| Region | 1970 | 1980 | 1990 | Current R/P ratio (years) |
|---|---|---|---|---|
| Africa | 31.5 | 50.0 | 93.1 | 15.0 |
| of which Egypt | 16.3 | 29.8 | 45.5 | 13.5 |
| Asia (excluding Middle East) | 40.3 | 43.5 | 84.2 | 20.6 |
| of which India | 6.8 | 9.4 | 34.8 | 29.5 |
| Latin America | 74.6 | 172.6 | 246.1 | 37.1 |
| of which Mexico | 24.3 | 107.3 | 147.1 | 52.5 |
| Total non-OPEC LDC's | 146.1 | 286.1 | 423.4 | 28.9 |
| OPEC production | 1156.6 | 1357.6 | 1205.4 | 87.0 |
| Non-OPEC developing countries production as % of that of OPEC | 12.7% | 24.0% | 35.1% | |

European Energy Charter organization (the former already in place and the latter agreed in principle at the time of writing). Politics apart, there is a great deal of optimism concerning the long-term hydrocarbons potential of Russia and a number of the Central Asian republics.[19] By the beginning of the next century there will have been time enough to enable that optimism for higher levels of production to have been converted into reality. Oil (and gas) exports should then be on a strongly rising trend as indigenous demand is curbed through enhanced efficiencies in use,[20] and as the foreign producing companies preferentially exploit their equity crude potential in Russia and other republics.

Within the non-OPEC developing world the political risks associated with continuing rapid hydrocarbons exploitation are much more diverse and diffuse. While there will undoubtedly be failures arising from adverse political circumstances – either within individual countries or between governments and foreign oil investors – these seem likely, in the changed international political circumstances in the 1990s, to be the exception rather than the rule. The rule will be one in which the ability of private foreign companies both to invest in oil exploration and development and to incorporate the oil produced into the countries' development efforts, either by substituting energy imports or by exporting all or some of their locally produced oil and gas, is enhanced. And even if foreign companies are not so involved in some developing countries – either because they continue not to be welcome, or because they decline to accept the risks they perceive – then national companies will have a higher chance of success in the future than they have had in the past. This is basically because in so many cases there has already been initial success in the hydrocarbons' exploration and development phase, so that revenues are now flowing and are thus available, at least in part after governments have used some for other purposes, to sustain continuing investment for the further expansion of the industry.[21] Likewise, international institutional money – notably from the World Bank – has become much more readily available for investment in the developing countries' hydrocarbon sectors, so also helping to overcome the hitherto often intractable problem of inadequate investment availability.[22]

Meanwhile, of course, the oil demand motivated interest for securing the expansion of indigenous production will continue to operate. Additionally, continuing technological advances in both the exploration and the field development phases serve not only to reduce the risks associated with upstream oil investments, but also to reduce the costs involved.[23] Thus, hitherto poorly rated oil and gas prospects in Latin America, Africa and Asia are now securing more favourable evaluations so increasing, over the longer term, the numbers of countries and regions which are oil and gas prospective. Relative to the rest of the world's potentially prospective hydrocarbons basins, those in the developing countries remain underexplored and/or underdeveloped. As the expansion of oil demand will be increasingly concentrated in these countries over the next 30 years – accounting for up to two-thirds of annual additional demand by 2020[24] – one can estimate that this will by then lead to a 75% increase in oil supplies from the developing countries outside OPEC – from the current level of about 425 million tons per annum.

Table II-11.4 summarizes our views on the long-term evolution of global oil supplies. This shows a relatively stable contribution by the present members of OPEC at present levels. For each year, this represents the residual amount required from them to clear the market in the context of the global demand (separately estimated, as shown above, on the basis of energy market considerations). It does, of course, imply that OPEC will continue in existence with its present members and that it will persist in its present policies of constraining supplies so as to maintain prices.

## An Alternative Prospect

### OPEC seeks to maximize its market share

Faced with the prospect of little better than a stagnant demand for its oil at a price which rises little in real terms for the next 20 years, OPEC could decide that it must take action to expand the demand for oil, in general, and for its own oil, in particular. This would, however, involve both deep price cutting and investment in capacity expansion so that, even if this were politically acceptable to all OPEC's members, it could still be an economic impossibility for many – or perhaps most – of the countries concerned. In essence, it would separate the Middle East Gulf members from the rest (with the possible exception of Libya and Venezuela), and thus, on the one hand, undermine the solidarity of the organization while, on the other, raising the spectre among the world's oil importing countries of specific dependence, for an economically and societally vital commodity, on the oil reserves of the Middle East.

In many respects it would represent a return to the situation in the 1960s – and could even incorporate the re-involvement of the major multinational oil corporations in respect of the investment, expertise and managerial input which the Gulf producers would require for the rapid and large-scale expansion of their upstream oil industries.[25] Such large-scale input requirements to production expansion seem unlikely to be available to the countries concerned in any other way. Certainly, the combination of the companies' know-how and financial strength, with the – in practice – limitless oil reserves of the Gulf region (over 90,000 million tons), would provide a powerful combination in economic terms for the re-establishment of international oil as a growth industry, and thus effectively undermine the constraints on long-term prospects for global energy developments along the lines forecast in the earlier part of this paper. Its evolution would, however, pose questions as to its

### TABLE II-11.4
### The evolution of global oil supplies, 1990–2020
(in million barrels per day)

|  | Actuals for | Forecasts for | | |
|---|---|---|---|---|
|  | 1990 | 2000 | 2010 | 2020 |
| World total | 64.9 | 68.7 | 73.4 | 77.3 |
| of which: |  |  |  |  |
| FSU, China and other CPEs | 14.9 | 13.8 | 16.2 | 18.8 |
| of which: Used internally | 12.6 | 12.4 | 13.2 | 14.6 |
| Exported | 2.3 | 1.4 | 3.0 | 4.2 |
| Present members of OPEC (assuming the continuation of OPEC) | 24.8 | 25.7 | 26.7 | 25.7 |
| Rest of the world | 25.2 | 29.2 | 30.5 | 32.8 |
| of which: USA/Canada | 10.8 | 10.6 | 10.8 | 10.7 |
| Rest of OECD | 4.8 | 5.4 | 5.4 | 5.3 |
| Developing countries outside OPEC | 9.6 | 13.2 | 14.3 | 16.8 |
| Total supplies to market economies including imports from CPE | 52.3 | 56.3 | 60.2 | 62.7 |

Source: 1990 data from BP Statistical Review of World Energy, London, June 1991. For forecasts see Table II-11.1.

acceptability elsewhere in the world and lead to reactions which could quickly serve to undermine its viability.

## Reactions to a resurgence of Middle East oil market domination

Apart from powerful and general political concerns for this prospective development of the oil market – which, in themselves, would probably generate protective steps against the threatened exposure – there would also be reactions specifically related to economic considerations.

First, there would be a plethora of actions to safeguard the future of energy supplying industries in many countries which have supported, or even developed, such industries in order to achieve lower energy import dependence. Second, the non-Gulf member countries of OPEC, finding the viability of their oil industries in danger (compare the decline of the oil industry of Venezuela in the 1960s as it was out-competed by Middle East oil)[26] and facing prospective severely adverse effects on their economies, would be obliged to seek alternative alliances and relationships which reflected their concerns.

The stage would thus be set for a potential regionalization of the global oil industry based on the perceived mutual economic and political interests of contiguous or proximate exporters and importers. The most important elements of the prospective regionalization would be related to the likely adherence of non-Gulf major oil exporting countries – both former OPEC members and those which never became members of the organization in spite of the importance of their oil exports (eg Mexico and Egypt) – to trading blocs built around the three geographical groupings of OECD countries ie North America, Western Europe and Japan/Australasia.

Precursors of such a regionalization already exist in more general politico-economic policies which seek to sustain and develop such regional linkages. These include the North American free trade area proposal, to which Mexico is already seeking accession and membership of which the oil exporting countries of northern South America would not find an unacceptable proposition.[27] In the Far East, both ASEAN and, more specifically, Japanese policy makers already visualize an intensification of regional economic relationships.[28] Within the context of such relationships, intra-regional trade in oil (plus gas and coal) would fit very nicely. There seem to be no overwhelming reasons why appropriate terms and conditions for such trade would not be readily negotiable. For Europe, the links with the successor states of the USSR, with their potential for larger-scale oil (and, even more so, gas) exports, are already being established, partly under the influence of a self-evident mutuality of interests and partly in the context of the recently signed European Energy Charter.[29] North and even West Africa's association with his regional groping is already an important component in the oil and gas exports of the OPEC members lying within the region, *viz*. Algeria, Libya, Nigeria and Gabon. These relationships could be much further developed, given the prospects for good additional resource developments and the possibility of new infrastructure developments such as the construction of oil and gas pipelines in North and West Africa and the expansion of pipeline capacity for gas across the Mediterranean.[30]

Clearly this prospect for a set of regionalized oil – and energy – markets offers an alternative to a globally organized oil – and energy – industry, dominated by a renewed combination of Middle East oil resources and their development with the cooperation of the international oil corporations. A strong motivation to proceed to such regionalization would thus arise from the go-it-alone policies of the oil-rich Gulf exporters, whose possible intentions along that pathway would thus be constrained. Such a constraint can thus be interpreted as a stabilizing influence on the potential reaction of the Middle East oil states to the limitations they will have to face on the expansion prospects for their oil industries. In this respect the apparently widely held belief in the ability of these countries to redominate the international oil market is not soundly based. They cannot go it alone, without running the risk of countervailing developments based on the ability of most of the rest of the world to secure the volumes of energy it needs, at prices which it can afford, without recourse to more oil from the Middle East.

The more open question is whether the rest of the world would seek actively to regionalize its energy supply/demand relationships without such positive provocation from the Middle East. It could happen simply because the Gulf producers become too 'greedy' in respect of the share of the slowly growing international oil market which they demand – to the disadvantage of OPEC and non-OPEC producers elsewhere, so that the latter seek protection in regional arrangements. It could also emerge if new destabilizing events in the Middle East yet again disrupt – or threaten to disrupt – oil supplies and prices, as a result of which major importing and consuming countries would be positively encouraged to seek to minimize their dependence on the region by maximizing their energy trading relationships with their nearer neighbours – so offering deals and long-term prospects which the latter would find difficult to refuse.

There are, in other words, alternative ways, in the course of the coming decades, in which the Middle East core of OPEC could be undermined as a leading or even a significant supplier of oil to the very slowly expanding world market. The expansion and diffusion of production in oil rich – and potentially oil rich – regions in other parts of the world, in formal or informal relationships with neighbours or near neighbours whose markets could be quite adequately and as cheaply served at real prices little or no higher than today's US$20 per barrel, is beginning to emerge as a viable alternative for the provision of the world's oil requirement by 2000 – and beyond.

## References

1   For a recent overview of energy demand developments, see Lee Schipper, 'Improved energy efficiency in the industrialized countries: past achievements, $CO_2$ emission prospects', *Energy Policy*, Vol.19, No.2, March 1991, pp.127–137.

2   The mid-point values of future energy use are from a range of values which reflect varying future energy prices. The energy prices used reflect the 2 in 3 probability for future international oil prices. (As reported to the annual Energy Modelling Forum (Stanford University) survey of future energy use and prices by the Centre for International Energy Studies. Erasmus University Rotterdam (EURICES), 1991.)

3   P. Criqui, 'Trends in world energy demand in the face of possible global climate change'. *Energy Studies Review*. Vol.1, No.3, 1989, pp.258–267.

4   It is important to note that the conversion of primary electricity (from renewables) and nuclear power to oil equivalents is made according to the heat-value equivalent convention as developed and used by the United Nations for publication in its annual statistical series (Series J) on energy. The alternative fossil fuels input equivalent convention for conversion (as used by the International Energy Agency) gives an apparent contribution by primary electricity and nuclear power to total energy supply which is approximately twice as great (eg in 1990, 19.4% in place of 9.7%; and in 2020, 28.5% instead of 15.3%).

5   This is calculated on the basis of an assumption that there will be no significant technological breakthrough in reducing the amount of $CO_2$ released to the atmosphere per unit of fossil fuels burned.

6   See below for earlier concern over the adequacy of oil resources and the prospects for supply limitations within a period of 10 to 15 years.

7   P.R. Odell and K.E. Rosing, *The Future of Oil: A Simulation Study of the Inter-Relationships of Resources, Reserves and Use, 1980–2080*, Kogan Page, London, 2nd Edition, 1983.

8     Gross additions to reserves comprise two elements, *viz.* first, estimates of reserves in new field discoveries made during the year and second, the appreciation of proven reserves in fields discovered in earlier years. The latter is a process which continues over the full period of a field's development and depletion period. This can be in excess of 20 years. Strictly speaking the second element is a net figure after allowing for the occasional downgrading of reserves estimates for previously discovered fields.

9     The concept of a surplus of proven reserves relates to the commercial requirement for approximately 10 years of reserves at current levels of consumption. If use is growing and that growth is expected to continue, then the reserves to production ratio must be higher: for example, with a 5% expected growth rate in demand, the reserves:production ratio must be about 12.6 years. Reserves over and above those which are required commercially (as defined above) can be designated as surplus reserves. With the current near zero growth rate in global oil demand almost three-quarters of the world's proven reserves may be defined as surplus.

10     The American Petroleum Institute's definition of proven reserves is those quantities of reserves which geological and engineering information indicate with reasonable certainty can be recovered in the future from known reservoirs under existing economic and operating conditions. The definition thus has economic as well as physical components.

11     A slowly rising long-run supply price curve from today's price level of no more than US$12 per barrel and with demand increasing at about 1% per annum suggests a post-2020 date before the market price for oil 'needs' to rise above present levels.

12     M.A. Adelman has estimated the long-run competitive equilibrium price at 'about $5 per barrel': see M.A. Adelman, 'Mid East governments and the oil price prospect', *The Energy Journal*, Vol.10, No.2, April 1989, p.24.

13     P.R. Odell, *Oil and World Power*, 8th edn, Penguin Books, Harmondsworth, 1986, p.40.

14     British Petroleum Company Ltd, *BP Statistical Review of the World Oil Industry 1973*, London, 1974, p.6.

15     British Petroleum Company Ltd, *BP Statistical Review of the World oil Industry 1990*, London, 1991.

16     For the European Economic Community this is reflected in the renewed attention to issues related to the security of oil imports (see EEC, 'Energy for a new century', *Energy in Europe*, Special Issue, July 1990) and in the enhanced attention given to the cooperation with Russia (and other republics of the former USSR) through the aegis of the European Energy Charger as a means of 'helping to improve the EC's security of supply by reducing... dependence on the Gulf' (*Energy in Europe*, No.17, July 1991, p.9).

17     With oil at US$20 per barrel new North Sea oil and gas developments, for example, can move profitably ahead. See A.G.Kemp, 'Development and production prospects for UK oil and gas post-Gulf crisis: a financial simulation', *Energy Policy*, Vol.20, No.1, January 1992, pp.20–29.

18     *Ibid.*

19     With respect to oil, this optimism is based on Western companies' anticipation of the much higher productivities achievable following the application of the latest technology from fields already in production; and for natural gas on the massive volumes of proven, but undeveloped, reserves.

20     See R. Caron Cooper and Lee Schipper, 'The Soviet energy conservation dilemma', *Energy Policy*, Vol.19, No.4, May 1991, pp.344–363.

21     See Kameel I.F. Khan, ed, *Petroleum Resources and Development: Economic, Legal and Policy Issues for Developing Countries*, Belhaven Press, London, 1987. A number of contributions to this book are concerned with investment in the oil sectors of developing countries' economies.

22     P. Bourcier, 'The World Bank programme to accelerate exploration in developing countries', in United Nations, *Petroleum Exploration Strategies in Developing Countries*, Graham and Trotman, London, 1982. Also H.S. Zakariya, 'The public international sectors and petroleum development in the Third World', in *op cit*, Ref.21.

23     See papers by P. Kassler, D.H.N. Alleyne and UNCTAD in United Nations, *op cit*, Ref.22.

24     As indicated in the surveys of global and regional energy demand undertaken by the International Energy Workshop (IEW) based at the Energy Modelling Forum, Stanford University, Stanford, CA.

25     This has been very clearly indicated by Sheikh Ahmed Zaki Yamini, Chairman of the Centre for Global Energy Studies. See his paper. 'A general agreement on oil' presented at the 2nd Annual Congress of the CGES London, September 1991.

26     *Op cit*, Ref.13, pp.77–83.

27     Latin American countries have long sought special economic relationships with the USA whereby they would secure preference for their exports of primary commodities in the USA. (Under the Mandatory Oil Import Program from 1959 to 1973, special preferences were accorded to Western Hemisphere crude.)

28     'World status: a grid for East Asia'. *Energy Economist*, Financial Times Business Information Ltd, London, No.124, February 1992, pp.15–22.

29     The Charter was signed by 45 countries in The Hague on 17 December 1991. The agreement to the plan in principle remains (in April 1992) to be worked up into a legally binding basic agreement (for

ratification by member countries): beyond that, implementation of various elements in the agreement depends on the acceptance of a number of protocols, one of which will be concerned with hydrocarbons.

30    There is already an operational natural gas pipeline from Algeria through Tunisia to Italy. Expansion of its capacity is already under way. Further west, work has recently started on a gas line from Algeria through Morocco to Spain, while to the east Libya and Italy have begun discussions on another trans-Mediterranean gas line. It is suggested that this latter line could continue on in to east and south-east Europe. All the lines would connect into the EC Eurogas transmission system, the extension and intensification of which forms part of the gas sector proposals for the EC single energy market. Meanwhile, the North African countries from Morocco to Libya are in the initial stages of discussions of an east-west gas line system across the whole of North Africa.

# Chapter II – 12

# The Global Oil Industry: the Location of Production. Middle East Domination or Regionalization?*

## Introduction

Even under the still widely accepted theory of the biogenic origin of petroleum (North, 1985), the occurrence of oil is, as shown in Figure II-12.1, a widely dispersed phenomenon. This map of the actual and potential availability of oil in the world's sedimentary basins does not, however, take us very far towards a description, let alone an explanation, for the historical or contemporary geographical patterns of global oil production (Odell, 1963). As successive publications by Michael Chisholm, starting with his *Rural Settlement and Land Use* (Chisholm, 1962), have shown, the choice of location for all sorts of economic activities at all scales from the global to the local emerge out of fundamental questions of relative costs of production in different locations and the incidence of transport costs in bringing the production to the centres of demand: with the theoretical optimal patterns tempered by forces of a non-economic character. These latter non-economic influences are especially significant in influencing global-level geographical patterns of production in a world of nation states and of international disputes. In such a world, protectionism, security of markets and of supply and issues of geopolitics become major inputs to locational decisions on development. (Gibb and Michalak, 1994)

This is the case even with commodities and products which are "optional" in overall consumption patterns (for example, bananas and

* Reprinted from *Regional Studies*, Vol.31, No.3, 1999, pp.311–322

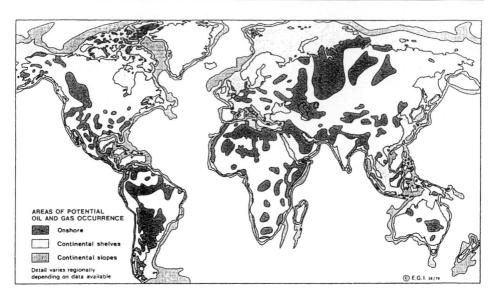

Figure II-12.1: The world's petroliferous areas

kiwi fruit). When an input – such as oil – is of fundamental importance to virtually all the world's economies (Gillespie and Henry, 1995), non-economic considerations are necessarily given relatively much more importance in production development decisions. This attention to such factors is not only extended by the world's sovereign states, but it is also an inherent part of decision taking by the multi-national oil corporations in which such a high percentage of the world's expertise in oil supply is concentrated (Odell, 1986; Hartshorn, 1993).

The purpose of this short essay is an attempt to put world oil supply issues of the past 40 years or so in a conceptual framework which implicitly recognises those organising forces of location. In particular, the phenomenon of the very large, extremely low cost to produce and strategically vulnerable oil reserves of the Middle East will be examined in respect of their changing importance over time *vis a vis* the exploitation of the world's other reserves.

## Middle East Oil: a Unique Oil Supply Phenomenon

Outside the United States and the former Soviet Union (FSU), oil from the Middle East was, since the early 1950s, the source of about 45% of total supply (BP, Annually). Its enormous contribution to global oil supply has, even so, been overshadowed by the region's share of proven reserves. This has ranged from a low of about 61% (in 1981) to a high of 78% (in 1960), suggesting that the prolific nature of oil occurrence in the Middle East has never been matched by an equivalent dependence on its production. This is in spite of the extremely low cost of oil discovery and development in the region (Adelman, 1993) and the continuing indication for most of the past 40 years of a flat – or nearly flat – long run supply price curve for Middle East oil up to potential levels of production which, if they had been achieved, could have served most of the world's demand outside the former Soviet Union and the United States. One also notes that the role of Middle East oil was critical to the industry's ability to meet the burgeoning demand for oil during the 25 year period at a near 8% per annum rate of increase in the use of oil from the early 1950s. Aspects of the changes are shown in Table II-12.1. In 1955, total world production of oil was 786 million tons of which about 37% was traded internationally. The Middle East produced just over 150 million tons – under 20% of the total – but it was already responsible for almost half of the oil entering world trade. By 1975 world output was almost three times higher (at 2734 million tons) of which the Middle East produced almost 36%. The latter's output of almost 1000 million tons was, however, six times up on 1955 – and by 1975 it accounted for over 60% of internationally traded oil, *viz*. over 900 million tons, compared with only 145 million tons twenty years previously. Of the 1950 million tons of *additional* oil demanded in the world economy by 1975 (compared with 1955), the Middle East was responsible for 42%, while 63% of the additional 1200 million tons which was traded internationally originated in the Middle East.

These data show that the international oil companies, which dominated all sectors of the oil industry over the period from 1955 to 1975 (Odell, 1986) – including the reserves' development process and the expansion of production capacity – clearly recognised the ability of the Middle East to produce the rapidly increasing amounts of oil demanded and, given the low costs of finding and developing Middle East oil, that it could most profitably supply the levels of production required (Adelman, 1995). The companies' success is also demonstrated by the way in which they built up the level of the remaining discovered reserves in the Middle East. These were increased from 24.5 billion tons in 1995 to 55 billion tons in 1975: in spite their having been of a cumulative production during the twenty years of over 10 billion tons. It is, moreover, worthy of note that the companies' attraction to, and development of, the prolific and low-cost oil resources of the Middle East continued in spite of the region's political instability, most notably, of course, the Israel/Arab conflicts of 1956/7, 1963/4 and 1967, but also including radical changes of government in the producing countries (Stocking, 1971; Adelman, 1972). It was also the period of intensive Arab nationalism and the overthrow of British and French colonial rule in the region, yet all this led to no more than temporary supply problems, as a consequence of the intermittent interruption of oil flows from the countries involved (Penrose, 1983).

A general and persistent situation of developed over-capacity within the Middle East meant that lost production from any one country could quickly be made up from other countries in the region. As a result there was no year in the 1955 to 1975 period in which Middle East production did not increase above that of the previous year. It was the attractive economics of Middle East oil production, relative to oil production everywhere else (except for limited low-cost output from Venezuela and Libya), which was the primary cause of the increasing geographical concentration of the world's upstream oil industry. This supply price function's favourable influence was, however, enhanced by a relative decline from about 1965 in the cost of moving crude oil to markets over long distances. These changes, as seen in Table II-12.2, reduced the impact of transport costs on the total delivered costs of Middle East oil in remote markets (such as Rotterdam) and thus made the oil relatively more competitive. Through to 1971 the f.o.b. price of Saudi Arabian light crude oil remained at under $2 per barrel (see Figure II-12.2). In the context of such low f.o.b. values, transport costs savings of a few cents per barrel on Middle East oil were not unimportant in terms of enhancing the region's international competitiveness (Hartshorn 1993).

At the beginning of the 1970s the process of the geographical concentration of supply on the Middle East seemed set to continue. Additional companies had moved in, as and when conditions allowed, to exploit the oil of the old producing countries (as in Iran in the 1950s and in Iraq in the 1960s), while a number of new low-cost Middle Eastern producers – viz. Abu Dhabi, Dubai, Qatar and Oman – had also granted concessions to companies which were new to the international oil industry and were thus anxious to secure access to production rights for some of the world's lowest cost oil. Company/government negotiations over their relationships in the Middle East – as well as in North Africa – in the early 1970s had in essence confirmed the perceived mutually advantageous relationships, albeit in the context of higher excise taxes imposed by governments on the concessionary companies (Turner, 1983; Adelman, 1995).

Such changes in favour of the host governments were, however,

**TABLE II-12.1**

**Global, middle eastern and regional oil production and trade, 1955–95 (in millions of tons)**

|  | 1955 | 1960 | 1965 | 1970 | 1975 | 1980 | 1985 | 1990 | 1995 |
|---|---|---|---|---|---|---|---|---|---|
| Global oil production: | 786 | 1079 | 1450 | 2352 | 2734 | 3082 | 2796 | 3187 | 3260 |
| of which Middle East | 164 | 264 | 385 | 689 | 975 | 927 | 517 | 867 | 972 |
| as % of Global production | 20.8% | 24.2% | 26.5% | 29.3% | 35.7% | 30.1% | 18.5% | 27.2% | 29.8% |
| Oil traded internationally: | 291 | 456 | 677 | 1263 | 1508 | 1588 | 1264 | 1612 | 1780 |
| as % of total production | 37.0% | 42.3% | 46.7% | 53.7% | 55.2% | 51.4% | 45.3% | 51.2% | 54.8% |
| Middle East exports: | 145 | 229 | 340 | 631 | 918 | 864 | 447 | 704 | 815 |
| as % of internationally traded oil | 49.8% | 50.2% | 50.2% | 50.0% | 60.9% | 54.7% | 35.4% | 43.7% | 45.8% |
| Intra regional trade: |  |  |  |  |  |  |  |  |  |
| Total | 106 | 190 | 238 | 527 | 402 | 433 | 539 | 570 | 615 |
| of which: |  |  |  |  |  |  |  |  |  |
| a. Western Hemisphere | 87 | 129 | 124 | 180 | 177 | 168 | 182 | 199 | 255 |
| b. Europe/W Africa | 11 | 42 | 158 | 310 | 175 | 193 | 308 | 313 | 265 |
| c. East Asia/Australia | 8 | 9 | 7 | 37 | 50 | 72 | 49 | 58 | 95 |
| as % of international trade | 36.4% | 41.7% | 35.2% | 41.7% | 26.0% | 27.3% | 42.6% | 35.4% | 34.6% |
| as % of M.E. exports | 73.1% | 83.0% | 70.0% | 83.5% | 43.7% | 50.1% | 120.6% | 81.0% | 75.5% |

*Sources: Data from sequential issues of B.P.'s Annual Statistical Review of the World Oil Industry (1955–1980) and its Statistical Review of World Energy (1985–96).*

matched – or even more than matched – by similar tax changes elsewhere so that companies producing oil in the Middle East maintained their cost advantages. Thus, they continued to expect further growth in the demand for oil from their Middle East concessions. For example, ARAMCO, the single concessionary company in Saudi Arabia (jointly owned by four of the US oil majors, *viz*. Exxon, Mobil, Texaco and Chevron), foresaw its output rising from 400 to 1000 million tons per year over the rest of the century (Adelman, 1995). The Iranian Oil Consortium and the Iraq Petroleum Company both visualised the doubling of their output over the same period. It thus seemed highly likely at the time that the companies would continue to exploit the Middle East's oil resources to a degree which would further enhance the region's contribution to global oil supplies for the indefinite future, providing only that the oil exporting countries did not fundamentally change the way in which the production and development processes were organised.

### TABLE II-12.2
### Crude oil transport costs (in $ per barrel) Persian Gulf to Rotterdam relative to the price of Saudi Arabian light crude oil
### (also in $ per barrel)

| Year | Transport Cost[1] | Crude Oil Price[2] | Transport Cost as % of Crude Oil Price |
|------|-------------------|--------------------|----------------------------------------|
| 1955 | 0.65 | 1.80 | 36.1% |
| 1960 | 0.61 | 1.50 | 40.7% |
| 1965 | 0.53 | 1.17 | 45.3% |
| 1970 | 0.69 | 1.77 | 39.0% |
| 1975 | 0.86 | 10.87 | 8.0% |
| 1980 | 1.10 | 28.53 | 3.8% |
| 1985 | 0.96 | 27.54 | 3.5% |
| 1990 | 1.29 | 20.48 | 6.3% |
| 1995 | 1.09 | 16.80 | 6.5% |

Source: 1.   *1955–70 from Adelman, 1972: 1975–80 Transport costs from Adelman, 1995. 1985–95 transport costs from Worldscale, Persian Gulf/Rotterdam base and calculated from monthly averages of actuals in the* World Tanker Fleet Review *(6 monthly).*

2.   *Crude oil prices 1975–95 from B.P. (Annually).*

## Restraints on Middle East Supply

In the event, however, the organisation of Middle East oil supply was radically changed post-1975 as, in the aftermath of the first oil price shock (1973–4), the countries concerned decided to terminate the concession-type arrangements and re-institute national ownership over the resources and national control over levels of production (Odell, 1986: Hartshorn, 1993; Adelman, 1995). Within a few years, total or controlling ownership of both the oil reserves and the producing facilities in all the main and some of the minor exporting countries in the Middle East had reverted to the nation states and their state oil companies. The previous concessionary companies often retained some influence through management agreements with the state corporations and in some cases secured preferential rights to the supply of agreed minimum/maximum volumes of oil. Such arrangements, however, were generally very short-lived so that, by the end of the 1970s, all important

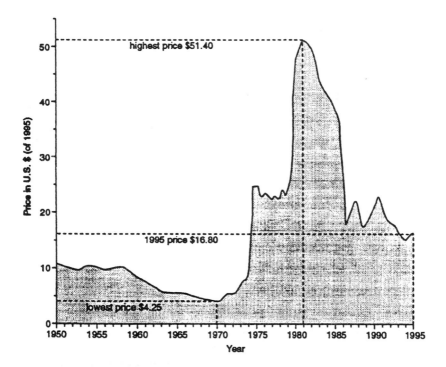

Figure II-12.2: The price of Saudi Arabian light crude, 1950–95
(in 1995 values)

decisions on Middle East oil production and supply were in the hands of national entities: rather than, as previously, mainly taken by the western international oil corporations.

Importing nations' fears for the security of oil supplies from each individual country of the region now became greatly enhanced. There was, moreover, concern for the security of Middle East supplies overall, in the context of the main producers' membership of OPEC and its decision to operate as a cartel, with a quota for each member's output. This development eliminated the flexibility of supply which the companies had previously enjoyed by virtue of their ability to play one nation off against another. These supply uncertainties were further intensified by the concomitant decline in the west's political influence in the region. This emerged partly because of fears for possible expansionist policies into the region by the Soviet Union; partly because of a heightened concern for the installation of radical regimes in individual countries; and partly because of the rise of inter-state rivalries within the region for both political and religious reasons. The Soviet Union's incursion into Afghanistan in 1979; the overthrow of the Shah in Iran in the same year and the declaration of war by Iraq on Iran in 1981 illustrated the validity of the concerns (Peterson, 1983).

The consequential "away from Middle Eastern oil" policies by many countries (in the more general context of "away from oil" policies because of its high price and the widely held perception of oil as a scarce commodity), expressed the unacceptability of too much oil from the Middle East. Moreover, the high price of oil after 1974 (five times up on 1970 prices) and its further doubling by 1980, coupled with the production/export limitations imposed by producers themselves, helped to make Middle East oil unaffordable or even unobtainable. Given that OPEC had determined to price its oil up to the cost of alternative energy supplies, hitherto low-cost oil supplies from the Middle East became available only at prices which included very high excise-taxes imposed by the governments of the producing countries. These taxes eliminated the economic advantage of buying Middle East oil. Additionally, political concerns over the security of its supply made sure that it would not be bought if any alternatives were available. Middle East oil was, moreover, on average, more distant from markets than all alternative sources, so that the geography of international trade in oil would inevitably respond to the transport-costs related unfavourable location of the region's oil (Weiner, 1991).

Thus the Middle East's share of world oil production peaked in 1975 – the year following the first oil price shock – at 35.7% (see Table

II-12.1). It then fell year by year to 1985 when it reached a low of only 18.5%, with a volume of production which was only a little more than half that of 1975. Table II-12.1 also shows exports from the Middle East at a peak in 1975, when they accounted for 61% of all internationally traded oil; by 1985, the volume of its exports had more than halved and its share of world trade had fallen to only 35%. Clearly, the dramatically changed situation which had emerged in the international oil industry had undermined the hitherto dominant position of Middle East oil in world markets. Likewise, the nationalisation of the oil companies' assets brought the long-ongoing expansion of Middle East proven oil reserves (from under 20 billion tons in 1956 to 55 billion tons by 1974) to an end. By 1980, the proven reserves of the region had fallen to under 50 billion tons and they did not recover to their 1975 level until 1987. The changed Middle East industry thus became unable to sustain the reserves' discovery process, in spite of the huge economic rent which the low cost oil producers of the Middle East earned against the very high unit price of oil over this period (Adelman, 1995). Against a full cost of production of no more than $1 per barrel for the most expensive barrel required, the unit price (measured in dollars of 1973) rose from about $3 per barrel in 1973 to a high of over $20 in 1981 and still as much as $13 in early 1986, immediately prior to the price collapse later that year.

## Post-1986 Market Developments

In the preceding sections we have shown how the previously economically dominant and highly competitive low-cost oil potential of the Middle East was seriously undermined by the changes in the international oil system from the mid-1970s. It was the recognition of the increasingly severe implications of these changes for the Middle East producers that eventually led to Saudi Arabia's decision in 1986 (supported by Kuwait and the UAE) to sell its oil for whatever price it would fetch in the market place (Hartshorn, 1993: Adelman, 1995). As a result the average price of internationally traded crude oil fell from $27.10 per barrel in January 1986, to a low of $8.96 per barrel in July. This price collapse was followed, thereafter, by a modest recovery, as OPEC succeeded in restraining the production of its members. Since then – except for some months in 1990/91 during the Gulf War – the international oil price has 'stabilised' at about two-thirds of its level in 1985, measured in current dollars. In real terms, however, it is now priced at only about 40% of its 1985 value (see Figure II-12.2). Steep though the price fall has been, it was, nevertheless, contained by two factors in the period to 1991; first, by the interruptions to supplies caused by the Iraq/Iran war from 1981 to 1988 and, second, by the events of the Gulf War in 1990/1. Since then other factors have sustained the price, *viz*. the continuing absence of Iraqi oil from the market as a consequence of the UN embargo on that country's exports; and the OPEC-agreed and generally observed limitations on production by its members.

In spite of these constraints, Middle East oil has made a comeback in the market. This is, in part, because of the re-expansion of the global demand for oil in the context of lower prices. From 1985 to 1995 world output grew from 2800 to 3250 million tons. In part, however, it arises from the specific ability of the Middle East to take advantage of the location of much of the growth in oil demand in the Far East, where alternative sources are less readily available than in the Atlantic Basin. Thus, as shown in Table II-12.1, over the 10 year period to 1995, Middle East output has grown by 85% and its share of world oil production from 18.5 to 30%. Likewise, its share of world trade grew from 35% in 1985 to over 45% by 1995. Just as emphatically, the Middle East's declared proven reserves have increased from 54.8 billion tons in 1986 to 89.2 billion tons in 1995: an increase of 63%. Proven reserves in the rest of the world over the same period increased by only 33% from 28.3 to 37.6 billion tons. The Middle East's reserves as a percentage of the total moved up from under 66% to 71% in this period.

On examining this data and in the context of an expected continuing expansion of oil demand, many analysts now predicate the near-future re-domination of the international oil market by oil from the Middle East (International Energy Agency, 1995). This cannot be excluded as the most likely potential development, when viewed from the standpoint of the economics of oil supply (Adelman, 1995), but there are alternative prospects when a combination of economic and political factors are taken into account.

## Alternative Prospects for World Oil Markets

A return to the pre-1975 situation of the oil importing countries' heavy dependence on the Middle East for their oil needs would again necessitate the heavy involvement of the major multi-national oil corporations in the development of the supply potential of most Middle Eastern countries (Odell, 1994). The companies' managerial input and, above all, their ability to supply the high level of investments needed to make much higher production in the Middle East possible would, however, again raise the spectre of a collusive arrangement between producing countries and companies – as in the early 1970s (Odell, 1971) – and so rekindle the importing countries' fears of such dependence.

Apart from the purely political concern in many importing countries over a prospective international oil market situation in which Middle East supplies once again come to dominate the international oil market, there could also be opposition based on national energy sector considerations in higher-cost energy producing countries elsewhere in the world. In these countries, actions to safeguard the future of their indigenous energy industries can be expected. Additionally, the non-Middle East oil exporters, finding the viability of their own oil industries in danger, could also decide to seek alternative alliances and relationships which reflect those concerns.

Under these circumstances, the stage would be set for a potential regionalization of the global oil industry based on a perception of mutual economic and political interests between contiguous or proximate exporters and importers. The essential elements in such a regionalization process seem likely to be the adherence of the non-Middle East major oil exporting countries – both former OPEC members and those exporters which never became members of the organisation, in spite of the importance of their oil exports – to the three oil trading blocs built around the main geographical groupings of the OECD countries, *viz.* North America, Western Europe and Japan/Australasia. A possible "shape" for these blocs and the allocation within them of large oil exporters/potential exporters is shown in Figure II-12.3. Precursors to such a regionalization of what has hitherto been viewed as a global oil

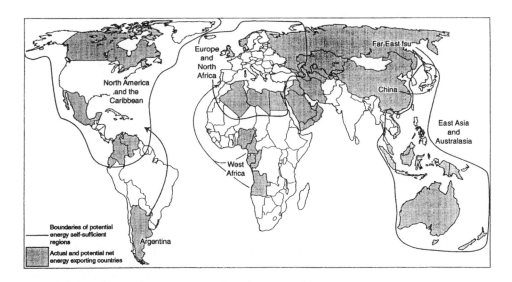

Figure II-12.3: The prospective regionalization of the global oil system

market already exist in the more general politico-economic agreements which are already in place for the development of such regional geographical linkages (Gibb and Michalak, 1994).

*First*, there is the North American Free Trade Area (NAFTA) of which both Canada and Mexico, important non-OPEC oil producers and exporters, are already members (McConnell and MacPherson, 1994; Gwynne, 1994). The oil exporting countries of northern South America, *viz*. Venezuela, the third most important OPEC producer and exporter (and a founder member of the Organisation), Colombia and Ecuador, seem unlikely to consider the membership of NAFTA an unacceptable proposition, providing they too secured unfettered and preferential access to the US oil market. But NAFTA even seems likely to evolve into a broader regional grouping with the accession of other western hemisphere countries including Argentina, Chile and Brazil. Argentina would add to the export potential of the group, while Chile and Brazil will continue to be net oil importing countries.

Intra-regional trade within the Western Hemisphere has already built up significantly over the past 15 years: from 168 million tons in 1980 to 255 million tons in 1995. Over the same period the region's oil exports to the rest of the world have declined and in 1995 were only 60 million tons. Western hemisphere – mainly US – imports of oil from the rest of the world do still remain more important, at just over 300 million tons, but, nevertheless, constitute only about one-quarter of the region's total use of oil. The degree of integration into the global oil system – as opposed to the integration within the Western hemisphere system – is thus already at a very modest level. The regional production level will, moreover, continue to increase in spite of a stagnating, or even a slowly declining, output in the United States itself. This is because all other currently important producers will continue to expand their output, while both Colombia and Argentina are at the beginning of a period of highly-likely rapid growth in output. Venezuela, in particular, aims to increase its production capacity by about one-third by 2000, but, in order to make use of this, its continued membership of OPEC looks increasingly uncertain. Venezuela's continued acceptance of the Organisation's production quota system would inhibit such an expansion of its output. In such circumstances it could be highly motivated to choose for involvement in an hemispheric oil market. Such participation would enable it to achieve the oil industry expansion which it seeks and, indeed, requires to sustain its economic development.

*Second*, in the Far East, there are a number of regional organisations which are already involved in generating intensified economic relationships within a region stretching from Japan in the north to Australia and New Zealand in the south; and usually incorporating the rest of Asia, except for China and the Indian sub-continent (Hodder, 1994). The high economic dynamics of this region involve high growth rates in oil demand – of the order of 6% per annum. Thus, since 1986 annual oil use in the region has increased from 320 to 580 million tons. Very little oil has ever been traded out of the region (up to a maximum of about 20 million tons, mostly from Indonesia), but since 1985 intra-regional trade has almost doubled to 100 million tons per year. Even together with indigenous production used nationally (about 160 million tons), the region is, however, still only 45% self-sufficient in oil. Thus, in 1995, about 320 million tons of oil was imported, mostly from the Middle East.

The medium-term aim of the Far East's regional energy policy thus cannot be oil independence. Indeed, unless Vietnam's potential is developed quickly (as the only country within the region with a resource potential capable of sustaining significant indigenous oil production increases before 2000), there will continue to be an increased requirement for imported oil – of upwards of 25 million tons more per year (Fesheraki, 1996). Thus, energy-sector co-operation in the Far East seems unlikely to go any further in the short to medium-term than an attempt to eliminate some of this anticipated additional requirement for oil. This will be through the mechanism of a co-operative effort to enhance the production, transportation and utilisation of alternative energy resources which are potentially available within the region. Coal from Australia and Indonesia is already traded on a large scale within the region and its expansion will continue. Nevertheless, it is unlikely to make anything more than a small dent in the growth in oil demand (for technical and environmental reasons), in spite of its availability at well-below oil equivalent prices. Natural gas expansion, on the other hand, offers considerably better prospects. Australia, Brunei, Indonesia and Malaya already export 80 million tons oil equivalent (m.t.o.e.) to other parts of the region, almost entirely as LNG, and produce as much again for indigenous consumption. The region's already proven reserves of over 6000 Bcm of gas can sustain much higher production rates, but the exploitation of the reserves remains in its early stages throughout most of the region.

The major constraint on expansion lies in the geographical separation of production from areas of consumption and in the costs of liquefaction/regasification and tanker transport for liquefied natural gas and/or of pipelines' construction and use for the movement of gas as gas.

Nevertheless, the intra-regional trade in LNG has grown from zero to almost 100 Bcm over the last decade and another 100 Bcm by 2005 is projected/planned. A pipeline system (proposed by ASEAN – the Association of South East Asian Nations) to interconnect the region (except for Australasia) at an estimated cost of $10 billion is under study. The first elements in this system – involving additional intra-regional trade of up to 50 Bcm – should be in place by 2005. By then, national consumption of indigenous gas production will also have doubled to almost 200 Bcm. The net additional regional gas production/use of about 250 Bcm (= about 220 m.t.o.e.) expected within a decade will have increased regional gas use partly at the expense of imported oil and, in doing so, will have expanded the region's energy self-sufficiency to 50–52%. This is a modest improvement on the current 45% use of indigenous energy, but, thereafter, given the progress in the expansion and further reticulation of the intra-regional pipeline system, more rapid steps towards a higher degree of regional energy self-sufficiency seem likely to be achieved.

Oil/energy regionalization in the Far East cannot, however, be structured in the context of current supply/demand relationships. The relative immaturity of most of the economies concerned requires, on the one hand, continuing high energy use growth rates in a situation of a generalised under-development of the region's hydrocarbons resources. The speed of resources' exploitation requires to be accelerated and must depend largely on external expertise and capital, given the limitations which are imposed by the under-development of upstream oil and gas companies in the region and the greater priority given by most of the governments in the region to investments in the manufacturing and public infrastructure sectors of the economy. On the other hand, it seems unlikely there will be any objections in principle to the concept of preferential regionalization in the energy sector, in the context of the very widely accepted view of the need for government involvement in activities which give direction to developmental potential. The great interest in, and support for, the projected regional network of gas transmission lines is an important indication of a willingness in the region to facilitate developments which seem likely to assist the speed and effectiveness of economic progress (Paik, 1995).

*Third*, the remaining potential regional market is centred on OECD Europe (Wise, 1994). Until quite recently it would have comprised that set of countries, plus the oil exporting countries in North Africa and, less firmly, West Africa. With the fall of the centrally planned regimes of Eastern Europe and the Soviet Union, it must now be extended east to embrace the rest of the continent of Europe as far as the Ural mountains in Russia and to the western shore of the Caspian Sea (including the former Soviet republics of Azerbaijan, Armenia, Georgia, the Ukraine, Belarus, Moldava and the Baltic Republics (Michalak, 1994; Bradshaw, 1994). To Eurocentrics the size of this region may seem overwhelming. It is, however, as shown in Figure II-12.3, of roughly the same geographical scale as the Western Hemisphere and the Far Eastern regions dealt with in the preceding sections.

It is a region with a highly developed upstream oil industry: notably in north and west Africa and the former Soviet Union, but now also including another two of the world's eight largest oil producers, *viz.* Norway and the UK. Thus, in spite of the high intensity of oil use in both Western Europe and most of the formerly centrally planned economies, there is already a close balance between production and use – at about 1000 million tons per year. Note, however, that the region as defined is by no means isolated from the rest of the world oil industry: approximately 250 million tons are exported and some 300 million tons imported. Nevertheless, there is a very high degree (about 95%) of net self-sufficiency. Within the region as defined, there is a considerable volume of intra-regional trade. This is currently also of the order of 250 million tons, flowing between the component groupings within the region (Western Europe, FSU, North Africa etc.). The region's imports of Middle East oil (about 190 million tons) now represent under 20% of total use and, as shown above, these imports are more than offset by exports of indigenous oil to the rest of the world: notably by West European and West African oil which flows mainly across the Atlantic to North America.

The region as defined is not, of course, economically integrated and, unlike the other two regions, the Western Hemisphere and the Far East/Australasia, there are no existing political initiatives to develop such integration across the whole region. On the other hand, there are initiatives between a number of the region's component parts. First, in terms of the very recently opened opportunities between Western Europe and the formerly centrally planned economies. These are particularly significant with respect to energy questions (given Russian and other former Soviet republics' energy resources' potential). They have already been formally approached through the initiative of the European Energy Charter (Doré and De Bauw, 1995; Estrada, 1995). It proposes enhanced physical integration of the transmission facilities within Europe (mainly through additional pipeline capacity) and the development of acceptable conditions for investments by energy companies. In any event, the previously severe constraints on linkages, imposed as a result of the earlier political divide between east and west, have been lifted so that energy inter-relationships based on commercial considerations are already building up. In particular, much additional Russian gas will flow to the rest of Europe (Stern, 1995). This will add to the indigenous gas supply element and it will, in large part, replace oil imports.

Secondly, the North African energy sources already move overwhelmingly into Europe (80% of oil and 95% of natural gas) as the "natural market." Thus, the new political initiative on cohesion in the Mediterranean basin, involving both the European and the African countries, will, with respect to energy, need to do little more than merely formalise a pre-existing integration of supply/demand relationships across the Mediterranean Sea (European Commission, 1995). Given, however, the much enhanced east-west energy flows in Europe which can be expected in the changed political circumstances (see above) and the increasing potential of west Europe's own resources, then North African energy suppliers could well need to seek specific political recognition of their interests in a share of the European energy market as part of a Mediterranean Basin agreement. This would help to cement the cohesion of the region we have suggested for this study, even though for Algeria and Libya such a development would necessarily be at the expense of their relations within OPEC.

The links between West African producers – most notably Nigeria, a member of OPEC, and Europe have become more tenuous in recent years. Significantly more West African oil is now exported to North and South America than to Europe. This is partly because West Africa is disadvantaged in respect of the Euro-market by the availability of similar quality oils from both the North Sea and North Africa; and partly because the US embargo on oil from Libya creates a larger market there for oil from West Africa. Additionally, West African oil is of a quality and type which is in strong demand in the Far East, where neither many indigenous nor Middle East crudes match the markets' requirements. Thus, closer links between west Africa and Europe with respect to oil now seem to be less likely than a decade ago. West Africa's exclusion from the potential regionalization of the oil industry could, however, prove to be disadvantageous, as there would be no safeguarded areas for its exports. Thus, competitive marketing of all the oil produced there would be required, so posing a particular problem for West Africa where, for both geological and infrastructure reasons, costs of production are high, compared with the Middle East. Lower transport costs to Western hemisphere markets offset only part of this competitive disadvantage.

## The Implications of Regionalization

The potential re-structuring of the international oil market, as set out above, is predicated in the context of a continuing widely shared concern for the security of oil supply in both the short and the long-term. The concern does not arise from the fear of scarcity. On the contrary, it is the plenitude of discovered oil and the industry's success over the past 25 years in adding two new barrels to reserves for every one that has been used (leading to a contemporary record-high near-45 year reserves to production ratio) that has, paradoxically, created the interest in the potential security of supply which can be achieved through regional organisation (Odell, 1994). Nevertheless, given the heavy concentration of the discovered reserves in the Middle East where, moreover, an even greater percentage of the lowest cost to produce oil is available, then re-dependence on the Middle East could quickly emerge. This would happen if the Middle East producers decided to sell their oil at prices well below the current price.

In the absence of a strong enough commitment in the main consuming areas to the continued use of indigenous or regional oil, supply from these alternatives to Middle East exports could quickly be brought down to levels which make Middle East dependence inevitable. Each downward movement in the price of oil would further diminish the supply of alternatives to the low-cost Middle East oil. Such cheaper supplies of crude oil would, of course, have favourable short-term macro-economic effects for the importing countries (Adelman, 1995), but these would seem likely to be offset in the medium term by the uncertainties of political repercussions at the international level, or by the economic disruptions caused by the curtailment of the production of inherently higher cost oil and other energy production in many importing countries, or as a result of the potential political instability that the developing situation might engender in the Middle East itself.

The stimulus which events since 1975 in the structure of the international oil industry have given to the geographical diffusion and diversification of oil production have shown the world to be much more generally rich in hydrocarbons than had previously been thought (Odell, 1994). This has produced the changes, shown in Tables II-12.3 and II-12.4, in the ability of both the OECD countries and those of the non-OPEC developing world to enhance their oil production and the discovery/additions of new reserves. More recently, the oil price fall since 1986 has led to productivity increases and to the application of new technologies in the new oil producing areas. These developments have led to significant reductions in production costs. To date, this has been most evident in the exploitation of the North Sea's reserves, where much higher levels of reserves discovery and of production than was generally considered possible at an earlier stage of development have been achieved (Odell, 1996). The continuing diffusion of such developments to many other areas offers the prospects for enhanced regionalised oil and gas production, thus heightening the potential for resistance to Middle East re-domination of global oil supply.

The potential regionalization, as indicated in this paper, is not comprehensive: in part, because some areas – such as much of Africa and the Indian sub-continent – are neither rich enough as potential producers, nor important enough as users to enable them to fit into the three regional groupings suggested above. Nor can they constitute effective additional regional entities in their own right. Their continuing dependence on the Middle East producers may well be disadvantageous in the short term: but in as far as these producers will, under the regionalization of much of the rest of the international market, be anxious to secure "captive" markets, then the excluded areas could, in the longer term, secure their oil import requirements at highly competitive prices: and, in so doing, secure an advantage for their further industrialisation and development.

## TABLE II-12.3
### Oil reserves, production and use in the OECD countries (excluding the US), 1973 to 1997

| Period | Proven Reserves at Beginning of Period (billion bbls) | Production in the 5 year Period (billion bbls) | Use in the 5 year Period (billion bbls) | Production as Percentage of Use (billion bbls) | Gross Additions to Reserves in Period (billion bbls) |
|--------|------|------|------|------|------|
| 1973–77 | 24.9 | 5.4 | 38.9 | 14.0% | 16.6 |
| 1978–82 | 36.2 | 8.3 | 35.7 | 23.1% | 4.3 |
| 1983–87 | 32.2 | 10.9 | 33.1 | 32.9% | 10.8 |
| 1988–92 | 32.1 | 17.4 | 41.4 | 42.0% | 10.8 |
| 1993–97 | 25.5 | 21.7 | 45.3 | 47.8% | 27.2 |

*Source: Sequential issues from 1973–1980 of B.P.'s annual* Statistical Review of the World Oil Industry *and of its successor publication,* Statistical Review of World Energy, 1981–1998. *Forecasts are the author's own. Note that Mexico, which became an OECD member in 1995, has not been included in the analysis.*

China has also been excluded from the analysis. The continuing centrally planned nature of its economic system, coupled with its essentially autarkic policy towards energy production and use to date in the context of a rapidly growing economy, puts it in a category apart. It seems likely to continue effectively to control the rate of growth of oil demand and, on the supply side, to do enough in respect of exploiting its

## TABLE II-12.4
## Oil reserves, production and use and import dependence in the non-OPEC developing countries, 1973–1996

| Year | Reserves at Jan 1 | Production | Reserves/ Production Ratio | Use | Net Oil Imports | Imports as a % of i. Prod. | ii. Use |
|------|------|------|------|------|------|------|------|
| | (bill.bbls) | (mill. b/d) | (years) | (mill. b/d) | (mill. b/d) | % | % |
| 1973 | 35.3 | 3.50 | 27.4 | 6.28 | 2.78 | 79.4 | 44.2 |
| 1974 | 35.0 | 3.44 | 27.9 | 6.46 | 3.02 | 87.8 | 46.7 |
| 1975 | 36.0 | 3.76 | 26.1 | 6.42 | 2.66 | 70.7 | 41.4 |
| 1976 | 35.2 | 3.94 | 26.9 | 6.72 | 2.78 | 70.6 | 41.4 |
| 1977 | 44.9 | 4.18 | 23.0 | 7.00 | 2.82 | 67.5 | 40.3 |
| 1978 | 44.5 | 4.46 | 27.3 | 7.30 | 2.84 | 63.7 | 38.9 |
| 1979 | 47.5 | 5.14 | 25.3 | 7.60 | 2.36 | 47.9 | 32.4 |
| 1980 | 57.1 | 5.70 | 27.4 | 7.98 | 2.28 | 40.0 | 28.5 |
| 1981 | 71.5 | 6.34 | 30.9 | 8.06 | 1.72 | 27.1 | 21.3 |
| 1982 | 83.0 | 7.00 | 32.3 | 8.48 | 1.48 | 21.1 | 17.5 |
| 1983 | 78.1 | 7.36 | 29.0 | 8.52 | 1.16 | 15.8 | 13.6 |
| 1984 | 82.0 | 7.88 | 28.4 | 8.82 | 0.94 | 11.9 | 10.6 |
| 1985 | 82.0 | 8.24 | 27.1 | 9.04 | 0.80 | 9.7 | 8.8 |
| 1986 | 89.0 | 8.25 | 29.5 | 8.94 | 0.69 | 8.2 | 7.6 |
| 1987 | 86.9 | 8.68 | 27.4 | 9.30 | 0.62 | 7.1 | 6.7 |
| 1988 | 96.4 | 9.02 | 29.2 | 9.96 | 0.94 | 10.4 | 9.4 |
| 1989 | 98.6 | 9.54 | 28.3 | 10.52 | 0.98 | 10.3 | 9.3 |
| 1990 | 90.5 | 9.98 | 24.8 | 10.98 | 1.00 | 10.0 | 9.1 |
| 1991 | 91.3 | 10.32 | 24.2 | 11.52 | 1.20 | 11.6 | 10.4 |
| 1992 | 92.2 | 10.54 | 24.9 | 12.26 | 1.72 | 16.3 | 14.0 |
| 1993 | 95.5 | 10.94 | 23.9 | 12.82 | 1.88 | 17.3 | 14.6 |
| 1994 | 96.5 | 11.38 | 23.3 | 13.33 | 1.95 | 17.1 | 14.6 |
| 1995 | 100.2 | 11.87 | 23.1 | 13.80 | 1.94 | 16.3 | 14.1 |
| 1996 | 101.5 | 12.87 | 21.6 | 14.43 | 1.56 | 12.1 | 10.8 |

Note: *Data includes Mexico*

Sources: Worldwide Reserves Report, Oil and Gas Journal Annual *1972–1995 and* Annual Review of the World Oil (Energy) Industry, *BP, 1973–1997.*

own still largely underdeveloped resources to maintain a high degree of independence from the mainstream of international oil supplies. Alternatively, it could deliberately seek to co-operate with the Asia/Australasian region – as defined and described above – to ensure the balanced petroleum sector for which it strives.

## Conclusions

The geography of the exploitation of a set of widely distributed petroleum deposits across the earth's surface would, under competition, reflect the incidence of the most favourable costs of production, modified by contrasting costs of transportation to the markets from the various production locations. Under the increasingly competitive conditions in the oil industry from the late 1950s to the early 1970s, the geographical distribution of oil production gradually adjusted to reflect these economic forces. This led to an emphasis on the exploitation of the low cost reserves of the Middle East and the gradual marginalizing of production from elsewhere (e.g. the United States, Venezuela and Indonesia). The availability of growing volumes of low-cost Middle East output also inhibited the search for, and the development of, reserves in other parts of the world. By 1975 Middle East oil accounted for over 60% of all international traded oil and for over 35% of all oil production, compared with only 18% twenty years previously. The increasing strength of Middle East oil in the world oil system (the world outside the US and the USSR) from 1955 to 1975 can also be quantified in the more than 40 billion barrels which were added to reserves in that small region around the Gulf over the twenty year period, compared with under 20 billion added everywhere else in the world. By 1975, in spite of an eight-fold increase in annual production since 1956, the Middle East's oil reserves still represented over 50 years of production at the 1975 rate so that, in the context of an accepted requirement for no more than a 15 year reserves to production ratio, there was plenty of scope remaining for a further significant increase in production, even if little more oil were found in the area.

Elsewhere in the world, the reserves to production ratio was barely 25 years, so that even near-future production increases necessitated near-instant and ongoing additions to reserves to ensure the continuity of the industry. Adelman's earlier estimate of the long-run price in the Middle East, as lying in the range of 10 to 20 cents per barrel, compared with $1.20 for the United States and 64 cents for Venezuela (Adelman, 1972), suggested that relative costs too would continue to favour the Middle East very strongly, so that there would be a further geographical concentration of production for the foreseeable future. The economics of international oil thus clearly indicated that its economic geography would develop with a Middle East core of production servicing most parts of the global market; in some countries completely, but everywhere to a high and increasing degree. Supplementary supplies came only from those geographically dispersed production facilities where production costs were low enough to give protection against Middle East competition after transportation costs were taken into account; for example, in markets in the interior of the US or in markets in close proximity to the only somewhat more expensive oil of Libya or Venezuela.

Wars, threats of war and other serious political problems in the Middle East in the 1950s and 1960s did lead to some expressions of concern for the security of nations and of economic systems increasingly dependent on oil, in general, and Middle East oil, in particular: culminating, for example, in Japan's 80% dependence on imported oil and 61% dependence or Middle East oil in 1973; or, for Western Europe as a whole, 59% and 41% dependence, respectively (Schurr and Homan, 1971) It was not, however, until the Arab/Israeli war of 1973, leading to the oil embargo, price shock and expropriations of the oil companies' assets, that policies were ultimately produced, at both national and corporate levels, to reverse the trend to oil dependence, in general, and to dependence on Middle East oil, in particular. From these policy changes and from the Middle East's subsequent unwillingness to sell its oil except at very high prices, oil from elsewhere became almost equally – or, at least, adequately – profitable to exploit. As a result the geography of global oil production has, as shown, undergone a quite radical change. The new emphasis is on supply diversity, security of supply arrangements and, where possible, the development of indigenous or regional availabilities. The diffusion of production and of exploration efforts has since spread through most of those parts of the (former non-communist) world. Companies have been attracted, by reasonable-enough concession systems or by contractual and taxation arrangements, to make investments in oil exploitation in many alternative locations. As a result, the world is now a patchwork of oil producing areas across almost 100 countries of which as many as 45 produced over 5 million tons in 1995. Some 20 countries outside OPEC are net oil exporters (Odell, 1994).

Even after the price collapse in 1986, many importing nations' continuing concern for security of supply, the maintenance of prices by OPEC at a level which still made much "high cost" oil profitable to produce and the recognition that there was much scope for cost reductions in the non-Middle Eastern upstream industry as a result of technological developments, has collectively maintained much of the enthusiasm for finding and developing geographically diffuse supplies. As a result, non-OPEC oil production has continued to expand, and has thus limited the markets for OPEC, including most Middle Eastern, oil. In this context reserves' bases have been established for energy policies which encourage oil (and other energy) production within the regions that would otherwise become re-dependent on the Middle East. These, as described in this paper, conveniently fall into place within the context of regional politico-economic entities for which there is already a degree of inter-governmental and/or popular support (Gibb and Michalak, 1994). Moreover, within the regions the encouragement of oil and gas production and the utilisation of that production is generally mutually advantageous for most member countries: especially when contrasted with the sometimes intense national rivalries that are generated between such countries in respect of the national division of investments and jobs in activities which are free of physical location constraints. We would thus hypothesise the continuation of the regionalization of the international oil (energy) market. Though this will not be to the extent of the deliberate exclusion of all oil from the Middle East, it will, in effect, constrain the opportunity for Middle East oil to come to re-dominate the global market for the foreseeable future – except in the context of an unlikely return to an oil price level no higher than that (in 1995 $ terms) of 1970, *viz.* about $4.45 per barrel, compared with a weighted average price for internationally traded oil in 1995 of $16.75 per barrel.

## References

Adelman, M.A. (1972), *The World Petroleum Market*, The Johns Hopkins University Press, Baltimore.

Adelman, M.A. (1993), *Economics of Petroleum Supply*, M.I.T. Press, Cambridge, Mass.

Adelman, M.A. (1995), *The Genie out of the Bottle: World Oil Since 1970*, M.I.T. Press, Cambridge, Mass.

Bradshaw, M. (1994), 'The Commonwealth of Independent States' in Gibb, R. and Michalak, W. (Eds), *Continental Trading Blocs*, J. Wiley and Sons, Chichester.

British Petroleum, (Annually), *Statistical Review of World Energy*, London.

Chisholm, M. (1962), *Rural Settlement and Land Use*, Hutchinson, London.

Doré, J. and De Bauw, R. (1995), *The Energy Charter Treaty*, R.I.I.A., London.

Estrada, J., Moe, A. and Martinsen, K.D. (1995), *The Development of European Gas Markets*, J. Wiley and Sons, Chichester.

European Commission (1995), *Energy in Europe*, No.26, pp.88–91.

Fesharaki, F. (1996), 'The Outlook for Oil Demand, Supply and Trade in the Asia-Pacific Region to 2005' in CGES Report on its 6th Annual Conference, *Can OPEC Survive in the Face of mounting Competition*, London.

Gibb, R. And Michalak, W. (Eds.) (1994), *Continental Trading Blocs: The Growth of Regionalism in the World Economy*, J. Wiley and Sons, Chichester.

Gillespie, K. and Henry, C.M. (Eds) (1995), *Oil in the New World Order*, University of Florida Press, Gainesville.

Gwynne, R. (1994), 'Regional Integration in Latin America: the Revival of a Concept' in Gibb R., and Michalak, W. (Eds), *Continental Trading Blocs*, J. Wiley and Sons, Chichester.

Hartshorn, J.E. (1993), *Oil Trade: Politics and Prospects*, Cambridge University Press, Cambridge.

Hodder, R. (1994), 'The West Pacific Rim' in Gibb, R. And Michalak, W. (Eds), *Continental Trading Blocs*, J. Wiley and Sons, Chichester.

International Energy Agency (1995), *World Oil Outlook*, O.E.C.D., Paris.

McConnell, J. and MacPherson, A. (1994), 'The North American Free Trade Area: an Overview of Issues and Prospects' in Gibb, R. And Michalak, W. (Eds), *Continental Trading Blocs*, J. Wiley and Sons, Chichester.

Michalak, W. (1994), 'Regional Integration in Eastern Europe' in Gibb, R. and Michalak, W. (Eds), *Continental Trading Blocs*, J. Wiley and Sons, Chichester

North, F.K. (1985), *Petroleum Geology*, Allen and Unwin, Boston.

Odell, P.R.(1963), *An Economic Geography of Oil*, Bell, London.

Odell, P.R., (1971), 'Against an Oil Cartel', *New Society*, No.437, pp. 230–232.

Odell, P.R.(1986), *Oil and World Power*, Penguin Books, 8th Edition, Harmondsworth.

Odell, P.R.(1994), 'World Oil Resources, Reserves and Production,' *The Energy Journal*, Vol.15, Special Issue, pp.89–114

Odell, P.R. (1996), 'The Exploitation of the Oil and Gas Resources of the North Sea: Retrospect and Prospect' in MacKerron, G. And Pearson, P. (Eds), *The UK Energy Experience: a Model or a Warning?* Imperial College Press, London, pp.123–133.

Paik, K-W (1995), *Gas and Oil in Northeast Asia: Policies, Projects and Prospects*, R.I.I.A., London.

Penrose, E., (1983), 'International Oil Companies and Governments in the Middle East' in Peterson, J.E. (Eds), *The Politics of Middle East Oil*, Middle East Institute, Washington.

Peterson, J.E. (1983), *The Politics of Middle East Oil*, Middle East Institute, Washington.

Schurr, S.H and Homan, P.T. (1971), *Middle Eastern Oil and the Western World: Prospects and Problems*, American Elsevier Publishing, New York.

Stern, J. (1995), *The Russian Natural Gas Bubble*, R.I.I.A., London.

Stocking, G.W. (1971), *Middle East Oil: a Study in Political and Economic Controversy*, Allen Lane, London.

Turner, L. (1983), *Oil Companies in the International System*, Allen and Unwin, London.

Weiner, R.J. (1991), 'Is the World Oil Market "One Great Pool?', *The Energy Journal*, Vol.12, No.3, pp.95–108.

Wise, M. (1994), 'The European Community' in Gibb, R and Michalak, W. (Eds), *Continental Trading Blocs*, J. Wiley and Sons, Chichester.

# Chapter II – 13

# The Final Years of the Century of Oil – Changing Structures

## A. International Oil: a Return to American Hegemony*

### Oil: the American Industry

For most of its history the world oil industry was an essentially American phenomenon. It was, in part, an industry consisting largely of powerful American corporate entities. U.S. dominance was, however, also in part a consequence of highly active international oil diplomacy by the State Department. This was not only supportive of the companies in their search for concessions, but also active against the international oil interests of Britain, France and The Netherlands whose large empires and other areas of effective political control (as in the Middle East) inhibited the degree to which the U.S. oil majors were able to extend their neo-colonialist activities.

In the 1930s and during the Second World War and its immediate aftermath, the attrition of the strength of the European colonial powers enabled American hegemony in the organisation and operation of the world oil industry (outside the Soviet Union) to be established. Though Royal Dutch/Shell, Anglo-Persian and Compagnie Française des Pétroles (CFP) survived, they did so only by the direct involvement of the British and French governments in the cases of the latter two; and, in the case of Royal Dutch/Shell, partly through the help of the British and Dutch governments and partly because Shell, given its important

* Reprinted from *The World Today*, Vol.50. No.11, 1994, pp.208–210.

investments in the oil industry in the United States itself, knew exactly when and where to adapt to American policies. This was demonstrated, for example, in its acquiescence over severe limitations on oil trade with communist China, in Venezuela in 1948 when the oil industry helped to topple the newly elected left-wing government of Acción Democrática, in the boycott required by United States policy to Castro's Cuba in 1958 and in its acceptance in the late 1960s of the strong American opposition to the expansion of Soviet oil exports to Western Europe.

Ultimately, Britain's military withdrawal from East of Suez, following the debacle of the Anglo/French invasion of Egypt (undertaken, ironically, to try to ensure the continued availability of the Suez Canal for Middle East crude oil moving to Europe), left the United States in *de facto* political control over most of the non-communist world's reserves of oil; and thus over the supply potential to meet the burgeoning demand for oil imports throughout the non-communist world.

American oil companies – including not just the 'majors', but also a score of 'independents' – now went international with such a vengeance, in order to exploit the world's low-cost oil reserves, that they outdid Coca-Cola in their universality and General Motors plus Ford in their influence.

## The Decline and near Demise of U.S. Power

The American dominated oil industry, as described above, eventually changed the hitherto coal-based economies of Japan and Western Europe into oil dependent countries and established internationally traded oil as the principal source of energy throughout most of the rest of the non-communist world. This success marked the zenith of the companies' influence and it ensured profitable outlets for the American companies' low-cost oil production. Inevitably, however, the American oil companies' success also initiated powerful reactions which subsequently led to a period of significant decline in their fortunes as they became subjected to a double squeeze.

On the demand side, oil import dependent countries began to question the validity of their exposure to oil imported from politically unreliable countries and brought in mainly by American companies. Responses by way of regulation, state oil trading and/or state oil companies became the norm in Japan, Europe and in many developing countries. On the supply side, all the oil exporting countries toughened the terms under which their oil could be exploited and/or initiated moves to re-establish national ownership of the reserves and control over their development. The already almost ten-year-old Organisation of the Petroleum Exporting Countries (OPEC) provided a means for coordinating these actions. As a result, average tax-takes per barrel were enhanced at the expense of the companies' profitability and led ultimately in 1970, in the context of a tightening international market, to a reversal of the 20-year-long period of falling real oil prices: to the consternation of captive customers.

The subsequent first oil price shock in 1973/4 temporarily took the economic pressure off the companies as the much higher prices demanded by the exporting counties enabled the companies significantly to increase their profit margins. Simultaneously, moreover, importing countries quickly recognised and accepted the central role which was played by the companies in ensuring their oil supplies so that the companies were accorded an important collective role in the creation of emergency supply systems under the auspices of the International Energy Agency (IEA). This rebound in the companies' fortunes proved, however, to be short-lived, particularly when the producing countries determined to accelerate the previously agreed slow process of nationalising the companies' upstream assets. This dramatically reduced the availability of equity crude on which their profitability so largely depended. Thus forced into the open market for high priced supplies, they were then further hurt by the turn-down in the demand for oil, leaving them with expensive under-utilised downstream facilities Additionally, more intensified regulation and price control (most notably in the United States itself, where international prices remained unacceptable for more than half-a-decade), further undermined the companies' confidence.

Finally, under the impact of worsening circumstances in which they had to operate, the companies persuaded themselves – and others – that the days of oil were numbered: because the world was running out of oil. This, on the one hand, led to ill-considered forays into diversification – with consequential largely unrewarding outcomes. On the other hand, it persuaded governments and the international energy organisations (notably the I.E.A.) to pursue "away-from-oil" policies, so exacerbating the rate of decline in oil use after 1979. By the beginning of the 1980s international oil began to look like a "sunset" industry in which the very survival of some of the companies became an issue. Which, if any of them, would survive; and for how long? American control over international oil was disappearing fast.

## Resurgence Unlimited

Yet, now, only a little over a decade later, there has been a return to the *status quo ante*. The post-1982 fall in the oil price back down to levels of the mid-1970s in real terms, followed by some renewal in the growth in oil demand, has put a premium on the cost-effectiveness of oil developments in all sectors of the industry, from exploration to marketing. This has been accompanied by a resurgent recognition that "old-fashioned" vertical integration is critical; not only for the operational economies it delivers, but also for its risk-minimisation role in an industry in which too much financial exposure to any one sort of oil activity is inherently dangerous. The traditional, vertically-integrated international (for which read "mainly American") oil companies are now back at centre stage, ready to respond to a wide range of perceived needs for the expansion of the industry. Their technical expertise and their capabilities for generating and arranging investment finance are thus now in demand throughout the world. Such demands emerge from the near-limitless opportunities offered, first, in the geographical and still politically and legally unfriendly realms of the former Soviet republics; second, through the still centrally-planned, but, nevertheless, the strongly reformist, economies of China, Vietnam and even Cuba; third, in many of the non-OPEC developing countries whose pressing needs for additional oil which can be produced, refined and marketed in an efficient way is critical for their economic prospects; and last, but certainly not least, in the members of OPEC, both in the Middle East and elsewhere, whose state oil enterprises have been unable to keep up with the industry's burgeoning high-tech exploration and production developments and consequently find themselves losing out in the required dash for capacity expansion.

Thus, around the world, in a way unseen since the halcyon days of international oil in the 1950s, the companies are now beginning to enjoy a plethora of options for investment on terms which can be profitable, as the negotiations are taking place in the context of competition between countries for the limited resource-availabilities. The opportunities exist, moreover, not only in the oil sector, but also in natural gas. The expansion prospects for gas are, indeed, brighter than those for oil, for both resource availabilities and environmental reasons, and it is the major international oil corporations which are almost always the best placed, from both the technical and commercial standpoints, to undertake that expansion.

## The Jewel in the Crown

The high prospectivity for the international expansion of the American oil corporations (and, of course, for their few counterparts in Europe – and even fewer elsewhere) is exciting enough in themselves, but there is, as in the pre-1960s period, a particular opportunity which stands out above all others as a joint function of American foreign policy and the collaborative role of the major American oil companies.

This extraordinary prospect is the emerging geopolitical reality of a long-term United States-Saudi Arabian compact in which the oil component is a central element. For such a compact to exist does not require it to be formally set out in a secret accord. That political issue is hardly of the essence. What really matters is the clear existence of a fundamental mutuality of interests between the two countries in the aftermath not only of Iraq's temporary occupation of Kuwait and in the context of the demise of the Soviet Union, but also out of the position of the United States and Saudi Arabia as the world's largest importer and exporter of oil, respectively; and as the two countries with the greatest potentials for enhancing their respective importing and exporting roles.

Prior to these developments, but after the nationalisation of ARAMCO, Saudi Arabia had pursued an oil policy which sought independence from the United States. Likewise, the United States sought security against Soviet and Iranian expansionism in the Gulf through its support for Iraq. Neither country got very far in establishing its alternative policy, but, in any case, what little had been achieved in both respects by 1990 was wholly undermined by the consequences of the Gulf War and the disintegration of the Soviet Union. These events, in effect, threw the United States and Saudi Arabia back into each other's arms so that over the past two years the conditions have been established for the re-cementing of the special relationship between the two countries.

In this the following main elements are of significance:

*First*, the mutual acceptance by the parties that the military power of the United States will continue to be required to defend the national sovereignty of Saudi Arabia and the authority of the present regime.

*Second*, Saudi Arabia's willingness to ensure that the United States has access to whatever volume of oil it needs to import at approximately the 1993 level of prices (in real terms). The United States in turn agrees to generate a minimum volume off-take of Saudi oil in order to protect Saudi Arabia from the consequences – as in the mid-1980s – of falling revenues emerging from the country's role as the international swing producer of crude oil.

*Third*, the elimination of any near-future serious downward pressure on oil prices by the continuing unwillingness of the United States to accept a resumption of Iraqi oil exports. This serves to keep Iraqi oil off the market until mid-1995 at the earliest, when agreement to some exports could become possible in order to relieve upward pressures on prices; in the context of an expanding market for OPEC oil resulting from a strong recovery in the economies of the industrialised countries.

*Fourth*, Saudi Arabia, for its part, will control longer-term upward pressures on the international oil price by its willingness of to produce sufficient additional oil to stabilise the market. This will be achieved by investments to secure the more intensive exploitation of the country's existing fields and, as and when necessary, by bringing new fields into production.

## The Impact on International Oil

Success in the implementation of this set of joint policy aims will be mutually advantageous for the two countries concerned, but will also have an impact on the international oil industry. In this respect the following elements stand out:

> *The American-Saudi* entente *will effectively up-stage OPEC, whose other members will become virtually powerless to influence prices as long as the United States and Saudi Arabia remain committed to the maintenance of current price levels. The most OPEC will be able to do is make the best of the bad situation by adjusting its policies to the consequences of the agreement.*

> *In the absence of other traumatic events, the world oil market will stabilize in respect of price and of organisation. The power required to overcome the stabilising impact of the United States-Saudi agreement seems to be too great for any other set of actors to match.*

> *The old ARAMCO partners (Exxon, Mobil, Chevron and Texaco), will secure the first bite of the global oil investment cherry. By virtue of the low costs of producing additional barrels of Saudi oil from present capacity and the relatively modest capital costs of developing additional capacity, they will secure a significant competitive up-stream advantage against*

*other companies. They could also exercise pressures, if they so wish, on other sectors of the industry: in cooperation with Saudi ARAMCO, which has declared interests in downstream investments and in the establishment of a major role in international markets.*

*Much potential oil industry expansion elsewhere will be put "on hold". When and where developments do go ahead, it will have to be in the context of the parameters established by the United States and Saudi Arabia for the evolution of the international oil market. These parameters will also set the baseline against which the validity of oil and energy sector policy objectives in most other parts of the world will have to be tested.*

## Conclusion

In brief, after three decades during which American companies' domination of international oil was undermined and, indeed, almost eliminated, the wheel of fortune now seems set to come full circle. The hegemony within the industry is being re-established. The essential involvement of Saudi Arabia in an alliance of strong political and commercial interests with respect to international oil could well prove as powerful a set of forces as those which previously established effective and long-lived systems in the world oil industry (the Rockefeller Empire, the Achnacarry Agreement and OPEC). If so, then it could be well into the 21st Century before the newly emerging structure comes under serious threat.

## B. Higher Oil Prices: the Result of a Political Agenda*

### US Intervention in the Market

The much higher oil prices in 1996 have generally been attributed to short-term market considerations, *viz.* an upward 'blip' on the demand curve coupled with reduced stocks held by the companies attempting to work with 'just-in-time' deliveries. Even the now imminent return of

---

\* Reprinted from *Geopolitics of Energy*, Vol.20, No.10, 1996, pp.2–5.

some Iraqi oil exports has not brought prices down by more than 50 cents per barrel.

After more than 4 years of price weakness the return to higher enhanced energy costs will have to come as a shock for consumers, particularly as there have been no traumatic events affecting oil supplies. The conventional explanations are, moreover, inadequate: short-term factors cannot explain an eight month phenomenon – with no respite in view.

In reality, the high prices seem much more likely to emerge from a highly significant development of which I warned over two years ago (see my article in *The World Today*, November 1994, "International Oil: a Return to American Hegemony."). This was a strategic decision by the United States to intervene in the international oil market in the aftermath of the Gulf War and following the disappearance of countervailing Soviet power (which would otherwise have inhibited such intervention) after the break-up of the former Soviet Union.

The aim of the new US policy became orientated to securing an international oil price which was sufficiently high to provide the revenue needs of its allies in the Gulf, most notably Saudi Arabia, so that they would have the ability to pay for the military power required to defend themselves against any future attempt to take-over their oilfields. The 1993 and 1994 oil price of only $15–16 per barrel was woefully inadequate for this objective to be achieved.

Thus, the US took an unbending stance against the resumption of oil exports from Iraq, even in respect of exports under UN supervision to serve humanitarian needs in the country. Until recently such limited Iraqi exports would, it was feared, have quickly undermined oil prices. Meanwhile, the US has also imposed barriers on oil exports from Libya and Iran and has introduced sanctions against US and other international investment in their oil industries. Even though such steps may not achieve large reductions in oil flows in the short term, they do lead, nevertheless, to marginal falls in supply volumes and can thus make a significant difference to traded oil markets. And in case these actions were insufficient, the US even threatened sanctions against the Nigerian oil industry.

There are also two other elements to the US' manipulation of the oil market. First, as US oil companies remain large producers of non-OPEC oil, then decisions by those companies modestly to shave output could also serve US policy interests, by helping to produce higher oil prices. Such action would not, of course, be adverse for the companies' upstream earnings. Second, can it be merely a coincidence that US oil geologists related to Federal institutions have chosen 1996 to resurrect the ancient – and long discredited – notion that the world is about to run out of the ability to sustain current levels of oil production because of reserves' limitations? Such misinformation was previously used 20 years ago to create fears of oil scarcity amongst governments and consumers, so that price increases came to be viewed as an acceptable proposition.

In brief, by early 1996 the various elements of a highly pro-active new US strategy towards international oil were in place; and so created the necessary conditions for the oil price to increase to over $20 per barrel; compared with the $16 or less at which it had previously been sold, in the context of an excess of supply potential over prospective demand.

## The Implications

The impact of the United States' actions on the international oil price will, of course, have been unwelcome to oil users whose prospects depend on the continuation of low prices for fuel and feedstocks. Their strategies now seem likely to require adjustment. In part, renewed efforts will be required to use energy more efficiently; and, in part, the companies concerned will need to remind governments at home that higher producers' prices for oil (and, of course, for gas, the price of which is closely linked to those for oil products) will mean much more limited opportunities for the enhancement of taxes on energy use, whether these are intended simply for government revenue reasons or are the result of governmental concerns for environmental issues. As a result, the financial gains which have been anticipated by large energy users from the downward pressures on prices arising from market liberalisation could also disappear. Such gains could, indeed, be replaced by additional burdens, as market competition (rather than the continuation of regulation) could, under circumstances of restrained supplies, produce a rising marginal cost curve, rather than the falling one hitherto expected.

There is, nevertheless, a possible important offset against the higher prices that US policy has currently produced; namely that the US not only intends to use its power to eliminate weakness in the international oil market. It is also ready, willing and able to prevent oil prices from increasing to too high levels. If such a danger were to become apparent then the United States will be able to exercise its influence over its allies in the Gulf. All of them (except Kuwait) have unused supply capacity in the short-term and, even more important, much greater additional supply potential for the medium and longer-term. The United States' intends to ensure that the increased production of which they are – or can be made – capable will be used to keep international oil prices under control.

The emerging return of American hegemony in the international oil system is thus a force for a much higher degree of stability in international oil prices than would otherwise have been possible. Such power would appear to be readily exercisable except in the event of traumatic developments, such as a new Middle East war, or the collapse of the Russian Federation. In essence, what is being established is a commodity price control arrangement, emanating from the United States' benevolent exercise of its power in the system. And as it is being implemented by the world's only super-power, rather than by cooperation between a group of sovereign states, it seems likely to offer a much greater assurance of a stable price than OPEC could ever have hoped to achieve. The preferred price will, of course, remain subject to short-term trading fluctuations, but it seems more likely that fluctuations can eventually be held within a narrow range. A central value of about $20 per barrel seems acceptable enough for the long-term, not only to both producing and consuming countries, but also to the international oil companies. Viewed in this context, then 1996 may well prove to have marked the termination of the extended period of a quarter of a century (since 1971) during which uncertainty over the supply and price of oil was the name of the game.

## C. Changing the face of the international oil market*

During the past 18 years, world oil demand has been severely constrained as a result of competition from alternative energy sources and a near-global emphasis on the more efficient use of oil resources. Global demand peaked in 1979 and then fell for five of the following six years. By 1985 oil use was under 90% of its 1979 level – a level to which it finally recovered only as recently as 1994.

Supply side developments over the same period have been more dynamic. Most notably, production from the Former Soviet Union (FSU)/eastern Europe has fallen by over 40%, from 610 million tons to 360 million tons, while the Middle East's output is still only 90% of its peak 1979 level. Despite these supply constraints, the international oil

* Reprinted from *Petroleum Review*, Vol.52, No.613, 1998, pp.22–24.

price remains weak at under 50% of its level in 1980/81 in current dollars and at less than 35% in real terms.

Middle East producers were fortunate in that they shared the 'misery' with the FSU producers, particularly Russia. Had the net availability of oil for export from the former communist countries, including China, to the non-communist world not fallen from about 170 million tonnes to under 80 million tons since the mid-1980s, Middle East output would have been almost 200 million tons less than 1979, rather than only 100 million tons. Under these circumstances, the price of oil would possibly not have recovered from its collapsed level of 1986. At worst, it would have continued to fall post-1986 to the long run supply price of the commodity of no more than $10 to $12 per barrel in 1997 dollar terms.

Thus, the objective of oil price maintenance has been saved by events in the former Soviet Union, aided in more recent years by the constraints of UN/US embargoes on oil supplies from Iran and Iraq. It is, as usual, politics rather than economics which have continued to generate an oil price that is both well above its competitive level and more or less high enough to satisfy the exporters.

## Short-term prospects

The UN/US embargoes remain critical for stable oil price prospects in the short term. However, two further conditions also need to be satisfied. First, even the non-embargoed exporters may still need to restrain their levels of production below those achieved in 1997, rather than increasing them as decided at OPEC's December 1997 meeting. Second, the now re-emerging excess of Russian supply over internal demand will need to remain limited to no more than China's modestly rising net imports so that they cancel each other out.

Elsewhere, the non-OPEC, non-Russian and non-Chinese supply/demand relationships will in the aggregate remain 'as is' for the rest of the 1990s as total production from the countries involved keeps pace with the group's increasing demand for oil. Under such circumstances the prospect for a significant increase in the demand for Middle East oil – and the maintenance of the oil price at current levels – is no better than 50/50. That is unless the US determines that its further positive intervention in the market is required to keep international supply/demand in balance in order to serve its national interest in oil price stability. It is highly ironic that it is only the revitalized US interest in exercising its hegemony over international oil that can now secure some degree of order in world oil markets through to the turn of the century.

## Looking longer term

The short-term prospect for increased supplies from the Middle East and the prolongation of close-to-present price levels over this period thus depends on the exercise of US hegemonic power in favour of such an outcome. Even so, the acceptability to most Middle East countries, for more than another few years, of a system which effectively limits their revenues and their contributions to the total required supply, seem unlikely.

The Middle East's share of world proven reserves, at 65%, is almost 2.2 times its share of world oil production at 29.7%. This is perceived to be inequitable and the region's producers may seek instead to re-establish their earlier domination of international trade. As shown in Table II-13.1, this was 61% in 1975 compared with only 45% in 1996. The Middle East's theoretical ability to achieve this objective over a five to 10-year period is not in doubt, given the much lower production and development costs compared with most alternatives for expanded oil supply. In a resurgent competitive market in which the Middle East producers/exporters secure the necessary and profitable reinvolvement of the major international oil corporations in the expansion of their output and exports, a return to the pre-1975 situation of the oil importing countries' heavy dependence on cheap oil from the Middle East for their oil needs cannot be excluded.

Apart from the purely political concern in many importing countries over the possibility of an international oil market dominated by Middle East supplies and the major international oil corporations, there would also seem likely to be consternation about the future of indigenous energy production in many countries. Actions to safeguard the future of such nationally oriented energy industries could be expected. Equally important, the non-Middle East oil exporters, finding the viability of their own oil industries in jeopardy from much expanded Middle East production, could decide to seek alternative alliances and relationships which reflect their concerns. If this happened, there would then be a powerful motivation for a potential regionalisation of the hitherto global oil industry, based on a perception of mutual economic and political interests between contiguous or proximate oil exporters and importers. The essential elements in such a regionalization process seem likely to emerge from the adherence of the non-Middle East major oil exporting countries – both former OPEC members and those exporters, such as Mexico, Egypt and Malaysia which never became members of the organization – to three oil trading blocks built around the main geographical groupings of the OECD countries of North America,

Western Europe and Japan/Australasia. (See Chapter II-12 for an evaluation of these regional entities.)

Given such a development, then market opportunities for Middle East oil exporters are likely to be modest relative to the area's supply potential. The countries will, in essence, have the choice of 'hanging together' or 'hanging separately' in respect of their relations with the oil importing world. A failure by the Middle East suppliers to cooperate would likely lead to a price war for the limited markets and thus to temporary success in expanding their international market share. This would, however, be followed quickly by the inevitable international political and strategic concern over dependence on the Middle East and the consequent strengthening of the regional arrangements described. To avoid this, the Middle East producers' cooperation could be developed specifically in respect of the growing Asian markets. This would most likely involve the establishment of long-term supply agreements based on pricing formulae of a kind which fit in well with the statist approaches of most of the Asian countries with respect to the economic sector and rapid economic growth.

### TABLE II-13.1
### Global, middle eastern and regional oil production and trade
(in million tonnes)

|                                   | 1955 | 1975 | 1995 | (1996) |
|-----------------------------------|------|------|------|--------|
| Global Oil Production             | 786  | 2734 | 3266 | (3362) |
| of which, Middle East             | 21%  | 36%  | 30%  | (29%)  |
|                                   |      |      |      |        |
| Oil Traded Internationally        | 291  | 1508 | 1815 | (1911) |
| of which, Middle East Exports     | 50%  | 61%  | 46%  | (45%)  |
|                                   |      |      |      |        |
| Intra-Regional Trade              | 106  | 402  | 643  | (707)  |
| as % of Middle East Exports       | 73%  | 44%  | 79%  | (83%)  |
|                                   |      |      |      |        |
| Trade:                            |      |      |      |        |
| a) within Americas                | 87   | 177  | 267  | (298)  |
| b) within Europe and NW Africa    | 11   | 175  | 296  | (306)  |
| c) within Far East                | 8    | 50   | 80   | (103)  |

Source:     P.R. Odell, *The Global Oil Industry: the Location of Production – Middle East Domination or Regionalisation?* Regional Studies, *Vol. 31, No.3, 1997, pp.331–322 (Data for 1996 from* BP Statistical Review of World Energy, *London, 1997)*

## Future prospects

The Middle East exporters' future thus depends on their ability and willingness to generate the necessary collective political clout to constrain output in order to secure the maintenance of real prices. It is likely that the US will covertly support any such attempts as it exercises its hegemonic power in favour of the status quo. Oil trading would continue under such a scenario, although it would be constrained by tight limits as trading ranges became defined by the overriding political wish for stability in the international oil system.

Should the Middle East oil exporters not reach an accord, in spite of support from the US, an alternative structure for the international oil markets will need to be developed. This too would be likely to be based on the precept of ordered markets, as in the halcyon days of the international companies control over the system. A free-for-all oil system with prices emerging from the largely unfettered play of markets is barely conceivable except as a short-term phenomenon. It would not suit the longer-term energy objectives of any of the significant players – be they country or company – in international economic affairs.

In this context, the non-Middle East members of OPEC (Venezuela, Nigeria, Algeria, Libya and Indonesia) would divest themselves of their commitments to OPEC as it becomes an organization which is unable to deliver the requirements for oil volumes and values which satisfy their economic needs. Their more effective links will lie in their commitments to the regional markets, whereby they can be guaranteed some degree of protection against low-cost, marginally priced oil from the Gulf.

Consequently, the Middle East oil exporting countries will need to reorientate their economic and political interests to the markets that remain outside those which are regionally organized, such as the Indian subcontinent. In particular, they will need to use active diplomacy to secure a rapprochement with the expanding economies of the region, including that of China.

Thus out of a structured Gulf States/South Asia/West Pacific Rim nations' dialogue, a new international energy organization could emerge to replace OPEC. It is likely that such a new body would not confine its concerns with oil supply and transportation issues, but also more generally to influencing the evolution of the energy markets of the world east of Suez.

The West was never willing to make such a commitment to the oil exporters, but the time now seems to have come for a 'producer/consumer dialogue' within the increasingly important region of the world lying beyond the Americas and Europe. The aim of such a dialogue would be to create an international organizational structure capable of minimizing energy supply and market uncertainties. The actors involved would comprise, on the one hand, those Middle East countries with limitless supplies of oil, but with economies with little else going for them and, on the other hand, countries with near limitless energy market opportunities.

Any organization developed to ensure the delivery of both supply and demand among the countries concerned will require a structure in which international diplomacy and intergovernmental regulation are the basic elements. Nevertheless, in a system of the size and complexity outlined in this article, fulfilling the markets' requirements down to the last barrel of oil or the final cubic metre of natural gas would seem likely to need some measure at least of traded markets' operations to ensure equilibrium in supply and demand at all times. This, however, is a far cry from the West's present treatment of oil as merely just another commodity.

## D. OPEC: too Soon to Write it off*

The analysis in the August 1999 issue of the *Petroleum Economist* of recent oil price increases simplistically suggests that they result from OPEC having done "an astonishing job". This fails to recognise a much more powerful force which, in effect, used OPEC to secure a price for oil whereby a shock for a world economy, then struggling to emerge from the destabilising effect of the earlier fall in the price of oil back to pre-1974 levels, was avoided. All major oil exporters faced potential disaster from the low prices of 1997/8 and, moreover, needed "saving" to inhibit national/regional unrest in critical areas of the world. Even more important, however, was the threat to the global politico-economic system from the undermining of Russia's already fragile economic viability and even its cohesion as a country, given its dependence on revenues and exchange earnings from oil and gas to almost the same extent as the most exposed OPEC countries.

Thus, from August 1998, the US government reiterated its interventionist policy towards the international oil market (based on the concept of oil as "just another commodity"). It thus initiated concurrent

* **Reprinted from *Petroleum Economist*, Vol.66, No.10, 1999, p.2.**

negotiations with the three key OPEC members whereby their agreement to cut production significantly was sought, in return for offers of important concessions and/or help from the US. It took all of six months for the US negotiators to secure agreements with the three countries to accept production cuts which they had previously considered to be contrary to their national interests. The agreement with Saudi Arabia involved financial and military aid; that with Iran, a partial lifting of sanctions; and that with Venezuela, economic stabilisation help and guarantees concerning Petroleos de Venezuela's downstream activities in the US.

All other OPEC members, and some non-OPEC exporters as well (especially Mexico and Norway), agreed to fall into line. The major international oil corporations, also under pressure from the US government, accepted that they should not take advantage of state oil corporations' cutbacks by marketing more of their equity crude production.

In other words, the US chose to exercise its hegemonic powers in the interests of global economic stability and brought the price of oil back into line with its post-1986 levels. Is it not significant that hardly a voice has been raised in protest at its initiative, even though it puts the US firmly in place as a supporter – or even the saviour – of a cartel, contrary to the country's usually strong views against such barriers to free trade!

Given this background to today's newly ordered oil market, fears for "the quotas to be bust" are not strongly based. Prices will be politically controlled except, of course, for the narrow range of a few dollars per barrel within which the markets will operate. Indeed, simultaneous action by the US to strengthen the International Energy Agency in its market watching functions and its enhanced co-operation with OPEC's own monitors, makes the provision of timely warnings on the need to adjust quotas or governments' stockholdings much more effectively available.

Lastly, perhaps as just a bonus for the deal or, more likely, as an integral part of it, actions by energy producers in the US to secure the re-imposition of US import quotas on oil and/or the introduction of countervailing tariffs to protect indigenous oil, gas and coal producers against low-cost or even "dumped" oil – have been avoided. 1999, it seems, may well go down as yet another one of those years (such as 1957, 1973, 1979 and 1986) when the International oil game "changed its spots".

## A Postscript*

The oil price rise in 2000 to its highest level since 1991 could be perceived as undermining the validity of the views expressed in the article above. On the contrary, it can be argued that the recent speculation affecting the oil price will enhance the chances of a return to an 'ordered' market. The high and volatile prices of 2000 could well represent the 'final fling' of those who view oil 'as just another commodity'.

Misrepresentations and/or misunderstandings were of the essence. Oil demand was said to be 'booming', when it was, in fact, virtually stagnant. Supply was said to be limited by capacity constraints, when it was, on the contrary, increasing rapidly. And, finally, stocks were reported to be at a near all-time low when, in reality, they were close to an all-time high and capable of sustaining a million barrels per day of additional supply to the market for over 10 years!

Overall, in 2000 world oil supply was up by over 2.2 million b/d, while demand increased by only about 1.5 b/d. Reality has been very different from the media's presentation of a pending supply crisis; and also from the situation in 1999, when supply fell by 2.3% while demand increased by 1.6%.

The world economy has regretfully paid a heavy price for the misrepresentations. On the other hand, there was in 2000 a positive and significant offsetting development, viz. OPEC's demonstration that it not only had the wherewithal to be a powerful player in the market, but also that it was ready, willing and able to use that power responsibly.

In particular, it defined a range of prices within which it wished the price of oil to lie, viz. from $22–28 per barrel. At this price level, OPEC argued, oil could be a growth industry for many decades into the future; while, simultaneously generating revenues so far above production costs as to provide economic and social development funds for the countries concerned.

OPEC's proposal provides a foundation on which an ordered, rather than a free-for-all, market can be based. While a return to an ordered market might be an anathema to the "free marketers", to whom liberalisation is of the essence, its development seems likely to have overwhelming support in most oil importing nations – and amongst oil consumers fed up with massive price fluctuations. Indeed, the extreme volatility of prices since 1998 for a commodity as central as oil to the well-being of nations has posed problems ranging from economic development issues, to issues of social disorder, such as that experienced

* Published in **Petrochem**, February 2001, p.62

in many European countries in late-2000 from the disruptive protests against high prices. The elimination of such threats to modern economies and affluent societies would be widely welcomed.

Any such implied international oil agreement – encompassing both producer/exporter countries and consumer/importer nations – has, of course, no chance of emerging without the involvement of the United States; not only because of its unique strategic and political status, but also because of its direct and important interests in oil per se. It accounts for 28% of international trade in oil; for 25% of the world's use; and even for about 12% of global supply.

In this context the inauguration of President Bush's administration is critical, given that an earlier Republican administration – that of President Eisenhower – was responsible in 1959/60 for the radical action of imposing formal quotas on US oil imports, so leading to a fundamental change in the global oil market at that time. Early reports from Washington suggest that international energy policy decisions will be a priority for the new US government – with an emphasis on supply diversity and security. The country's oil diplomacy will likely be in the hands of Vice-President Cheney and Secretary of State Powell, both with significant Middle East experience – rather than being the responsibility of the much lower-ranked Energy Secretary, as in the Clinton administration.

There is thus a clear indication that US policies in the Gulf, in particular, and in the Middle East, more generally, will become more oil orientated. If so, then negotiations for securing long-term supplies in increasing volumes at reasonable prices (say, in a range of $20 to $25 per barrel and thus overlapping OPEC's defined range), will be on the cards. Saudi Arabia and Iran (plus Venezuela) will be the principal negotiating partners, given that these countries not only have abilities to expand production, but are also committed to price stability. They, in turn, recognise the key role that the US has to play in formulating and implementing the kind of ordered markets on which their futures depend. In this context the rest of the world will have little option but to fall into line with the 'rules' of the ordered system.

## E. The International Community's Perception of the Future of Middle East Oil from the 1960s to 2000*

### Introduction

I t is more appropriate to begin this interpretation of the past somewhat earlier than the 1960s. By then, the most important criteria for the international communities' perceptions of Middle East oil had already been established as a consequence of the exhilarating and complex set of developments and events in that newly emerging oil province in the 15 years since the end of World War II.

## The Exposure of the Middle East's Oil Wealth

Between 1945 and 1960 the Gulf region plus contiguous areas of Iran and Iraq were revealed as an area of extraordinary – even unique – oil resources. Hints of such a phenomenon had emerged already in the 1930s, but confirmation of the situation was delayed by wartime conditions. Thus, in 1945 Middle East oil reserves were declared at a mere 3000 million tons, while the year's oil production was only 50 million tons, with this coming largely from Iran and Iraq, and constituting less than 10% of global production.

By 1960 production had increased 10-fold and to 25% of world output. Likewise, declared oil reserves had grown enormously to almost 25,000 million tons, with the Gulf States, rather than Iran and Iraq, accounting for most of the massive increase. The Middle East's reserves were now six time those of both the US and the USSR and they also gave the Middle East a close-to-100 year reserves to production ratio. Thus, the international community at that time had no difficulty in accepting the idea of "oil in the Middle East" as a "strawberries and cream" phenomenon.

## The Structure of the Oil Industry in the Middle East

Over the same period – and in the context of their much greater knowledge of the actual and potential wealth of the Middle East, all of the world's eight major oil companies, *viz*. the "seven sisters plus the Compagnie Française de Pétroles (CFP), participated in the frenetic action to secure concessions: nominally in competition with each other, but, in practice, usually carving up the countries and their oil-rich areas in negotiated "interests". This process not only reflected their relative strengths as companies, but also the contrasting political and diplomatic weights of the United States, the United Kingdom and France.

As demonstrated in her seminal mid-1960s work, *The Large*

* See page 454. Reprinted from the proceedings of the Institut d'Economie et de Politique de l'Energie and the Institut Français de l'Energie seminar on *The Future of Middle East Oil in a Globalized World*, Paris, October 2000.

*International Firm in Developing Countries* (Penrose, 1968), these companies exercised joint control over most Middle East reserves and production. Their complex, inter-weaving linkages in the Middle East's upstream oil industry at that time are shown in Fig.II-13.1. The international community also readily perceived the critical role and influence of these companies in the future prospects for Middle East oil.

## An Unholy Alliance?

The combination of the phenomena described above – from the oil itself, to the countries under which it lay and further to the eight international petroleum corporations and their mother countries – came to be widely perceived as constituting an 'unholy alliance' wielding a concentrated and critical mass of political and economic power. Its potential dangers were already clear and were perceived as getting worse in the global context of a rapidly increasing oil use, largely at the expense of indigenous energy – notably coal. Concern for an undue dependence on Middle East oil was thus already established a long time ago.

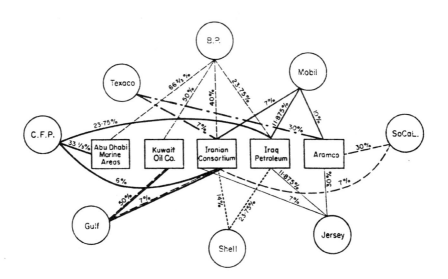

Figure II-13.1: Ownership links between the major international oil companies (including Compagnie Française des Pétroles) and the major crude-oil producing companies in the Middle East 1966

*Source: E.T. Penrose,* The Large-International Firm in Developing Countries, *Allen and Unwin, London 1968*

## Security of the Middle East's Oil Reserves and Production

The concern has, since 1960, thus been more or less continuously exacerbated by issues of security, *viz*.

*First*, through "Cold War" fears of an intervention by the Soviet Union in the region: either, as argued over many years, from a perceived Soviet need for access to Middle East oil in order to sustain its expanding and increasingly energy-intensive economy; or, as a somewhat later point of view, from fears of the exposure of the western world's economy to the Soviet Union's potential ability to deny it access to Middle East oil supplies. Even the Soviet invasion of Afghanistan in 1980 was interpreted as an attempt to achieve that objective.

*Second*, there has been more or less continuing concern for the internal stability – and viability – of individual Middle east oil producing countries: from Iran in the 1950s to Saudi Arabia in recent years. These concerns have been realised from time to time, leading to interruptions in oil flows and to the renunciation of oil concessions with both sorts of events raising the risk profiles of the countries involved.

*Third*, there has been recognition of the prospect for the use of oil as a weapon in the Arab countries' conflict with Israel, given that these countries view the existence of that country as a consequence of Western actions and inaction.

The negative perceptions of the security of oil reserves of, and/or supplies from, the Middle East have proved to be more than conceptual and theoretical. On the contrary, they have been justified by interruptions to exploration and exploitation, in general, and to supplies, in particular, on many occasions since 1950, *viz*. in 1951/4 from a revolution in Iran; in 1956/7, 1963 and 1967 over Israel; in the late 1960s over the exploitation of Iraqi oil; in 1973/4 with the Arab countries' embargo on supplies to importing countries; in 1979/81 following the revolution in Iran; in the early/mid-1980s from the Iraqi/Iran war, in 1990/1 as a result of the Gulf War and since then by the United Nation's embargo on Iraq.

Quite apart from these political/military reasons, there have also been constraints on supplies for purposes of price manipulation: in the 1950s and 60s by the collusion of the operating companies over production levels (as shown by Professor Penrose); and in more recent decades by the influence of Middle East producing countries on OPEC policy making, notably in respect of quota limitations on output.

The demise, first, of the companies' joint control over oil supply and pricing mechanisms and, second, of their "ownership" of most of the region's oil reserves and resources in the 1970s did not lead – except occasionally and for short periods – to an uncontrolled flow of oil from the region's so-prolific available reserves and its well developed production and transportation infrastructure. Not only have individual counties, at different times, inhibited production as matters of policy or internal disturbances; but also various combinations of countries – sometimes as allies, but more often out of enmity between them – have done likewise. And beyond this, collective decisions by all or most of the countries as OPEC members have repeatedly kept supplies off the market: to a very much greater extent, indeed, than they have permitted their oil to flow freely in the context of a competitive market in which prices would have declined towards the long-run low supply price of Middle East oil (Adelman, 1970 and 1993). Overall, the world's 40+ years' experience of restraints on Middle East oil supply and exports has served, more or less continuously, to reinforce the international community's perception of the dangers of dependence on oil from the region.

## The Middle East's Changing Role in Global Oil Supply and Trade

### a. 1955–1985

Table II-13.2 shows the evolution of the Middle East's contribution to global oil supply and trade over the period since 1955. Over the first 20 years to 1975 the oil companies' high degree of – but never absolute – control over Middle East oil ensured a rising exploitation of the reserves of the region so that annual production during that period rose six-fold, from 165 to 975 million tons. Over this period, the region's contribution to international oil trade rose from just under 50% of the 291 million tons traded in 1955, to almost 70% of the 1508 million tons traded in 1975: giving an almost 800% increase in the volume of Middle East oil moving into world markets. In essence, the international oil corporations exploited Middle East oil to the degree that it suited their strategies and enhanced their global after-tax profits, particularly after taxes paid to producer governments were made allowable against taxes at home (Penrose, 1968). Even so, by 1975 the Middle East's reserves to production ratio had been reduced only from its earlier 100 year plus level, to a still much more than adequate 75 years.

After 1975 – by which date most of the oil companies' equity interests in Middle East upstream oil had been eliminated by state actions, the region's production and its contribution to the world supplies started to fall; eventually, by 1985, to little more than half of their 1975 peak values. Over this decade a near-stagnant global oil industry effectively turned its back on the Middle East in the context of the uncertainties concerning the region created by the new national

## TABLE II-13.2
## Global and middle eastern oil production and trade, 1955 to 1999
### (in millions of tons)

| | 1955 | 1960 | 1965 | 1970 | 1975 | 1980 | 1985 | 1990 | 1995 | 1999 |
|---|---|---|---|---|---|---|---|---|---|---|
| Global Oil Production | 786 | 1079 | 1450 | 2352 | 2734 | 3082 | 2790 | 3179 | 3252 | 3452 |
| of which Middle East | 164 | 264 | 385 | 689 | 975 | 927 | 514 | 862 | 967 | 1052 |
| as % of Global Production | 20.8% | 24.2% | 26.5% | 29.3% | 35.7% | 30.1% | 18.4% | 27.1% | 29.7% | 29.7% |
| Oil Traded Internationally | 291 | 456 | 677 | 1263 | 1508 | 1588 | 1264 | 1612 | 1815 | 2025 |
| as% of total production | 37.0% | 42.3% | 46.7% | 53.7% | 55.2% | 51.4% | 45.3% | 51.2% | 55.9% | 58.7% |
| Middle East Exports | 145 | 229 | 340 | 631 | 918 | 864 | 447 | 704 | 825 | 909 |
| as % of internationally traded oil | 49.8% | 50.2% | 50.2% | 50.0% | 60.9% | 54.7% | 35.4% | 43.7% | 45.5% | 44.9% |
| Intra Regional Trade: | | | | | | | | | | |
| Total | 106 | 190 | 238 | 527 | 402 | 433 | 539 | 570 | 643 | 737 |
| as % of international trade | 36.4% | 41.7% | 35.2% | 41.7% | 26.0% | 27.3% | 42.6% | 35.4% | 35.4% | 36.4% |
| as % of M.E. exports | 73.1% | 83.0% | 70.0% | 83.5% | 43.7% | 50.1% | 120.6% | 81.0% | 79.2% | 81.1% |

Sources: Data from sequential issues of B.P.'s Annual Statistical Review of the World Oil Industry from 1955 to 1980 and its Statistical Review of World Energy from 1985 to 2000

ownership and control of oil supply and prices, the Iranian revolution and the political and military disruptions caused by the Iran/Iraq war. The whole region, in more general terms, was also full of uncertainties and adversely affected by the impact of the USSR's intentions in Afghanistan. The international community's adverse perceptions of the future of the Middle East had materialised with a vengeance.

## b. 1985–2000

Initially, with the post-1986 oil price collapse and the re-emergence of growth in global oil demand, Middle East oil supply boomed. By 1990 output was up by two-thirds from its 1985 low and its exports rose by almost 60%. It seemed as though the attractions of the Middle East's available production and transportation capacity and of competitive prices in a low-cost environment were successfully overcoming the negative international perceptions of re-dependence on the region's oil industry.

This was, however, abruptly cut short by the traumas of Iraq's take-over of Kuwait, its threat to the Saudi Arabian oilfields and the consequential Gulf War. Since then there has been almost a decade in which the Middle East has only modestly improved its position in the global oil market: with an increase in production of 12.2% compared with a 10.9% global increase; and a 29% growth in exports compared with 25.6% overall expansion in global oil trade.

These data thus appear to sustain the hypothesis that, post the oil crises/revolutions of the 1970s and the early 1980s and the subsequent Gulf War, the Middle East is struggling to regain its international status. Its share of global production – at 29.7% – is still well behind its 35.7% in 1975 while, even more emphatically, its share of world oil trade remains stuck at about 45% (43.7% in 1990; 44.5% in 1995; and 44.9% in 1999) compared with 69.9% in 1975 and even 49.8% in 1955 when Middle East oil development on a large scale was only just getting under way. Oil consumption within the region is now almost 15% of production, while an even higher percentage – 25% – of the additional output since 1990 has been for local use.

In these circumstances, the near doubling of the Middle East's oil reserves from 1975 (to 91.5 billion tons at the beginning of this year) is an unimportant phenomenon. Those additions to reserves are, quite simply, given the current reserves to production ratio of 87 years, required neither now, nor for the next 20 years (unless there are radical changes in the region's prospects in the short term). Thus, while the 65.4% share of the Middle East in global oil reserves may be an interesting fact, it is, nevertheless, an unimportant one for the future oil development prospects for the region. It may impress the "international community", but only in the sense that it more clearly perceives the continuing need to find and re-evaluate more reserves elsewhere, so as to continue to maintain the rest of the world's approximately 20 years R/P ratio; in the context of a non-Middle East annual supply that is now 2.25 times greater than that of the Middle East.

## Conclusions

Middle East oil developments can thus be perceived by the international community as potentially important in the long term – on the assumptions that oil demand will continue to grow over the long term and/or that non-conventional oil production, in the meantime, does not become economically viable on a large-scale.

For at least the coming decade, and conceivably well beyond that, the Middle East will generally continue to be perceived as the oil supply region of the last resort. Unless, in the meantime, the upward pressures on the international oil market, achieved through restraint on trade via the OPEC quota system and via the vagaries of the trading system, are eliminated and replaced by long-run-supply-price-related prices, either in a fully competitive system or in a fully-ordered one.

Neither of these rational alternatives for the future of oil are, however, perceived as very likely – or, even very desirable, given governmental, corporate and public preferences for, first, less oil-intensive economies (oil currently supplies only 40% of total global energy supply, compared with 53% in 1975); and, second, for oil from sources other than the Middle East.

The latter development is reflected in a continuing and accelerating intra-regional trade in oil – as shown in Table II-13.2: from only 402 million tons in 1975 to 737 million tons in 1999. In 1975, such trade amounted to only 26% of total international trade and to 43.7% of Middle East exports. In 1999, the comparable percentages were 36.4% and 81.1% respectively.

Over the past 50 years Middle East oil enjoyed only a short period of acceptability – in the context of a rapidly rising global demand for energy, in general, and for oil, in particular. By the late 1950s its previously acceptable role was already being questioned – under the impact of changing power relationships within the oil industry itself and given oil's impact on broader geo-political and geo-economic issues. This evolving situation became even more fraught in the context of repeated supply interruptions and with the Middle East producers' use of oil as a weapon in support of their political and strategic objectives.

The next deterioration came when low-cost Middle East oil came to be priced at levels which bore no relation to costs and which not only made alternative oil supplies but also alternative sources of energy (notably gas) relatively much more attractive, so enhancing competition for Middle East oil.

Thus, the Middle East has not recovered from the perception – and the reality – of its lack of attraction as a supply source. Indeed, had oil production of the Soviet Union/FSU not declined from year to year from 1987 to 1996 (from 650 to 350 million tons), then competition for markets by the Middle East suppliers would have been even tougher than it has been – particularly after the end of the 'Cold War', when previous western perceptions of security and other dangers of reliance on Soviet oil were eliminated. In other words, though the Middle East is still replete with reserves and with the possibility of its oil being a non-depleting resource (Mahmoud and Beck, 1995), it has basically failed to establish an international perception of its being a reliable supplier on a continuing basis.

This central element in the *real politiek* of world oil is, however, tempered by widely-held conventional views of future global oil supply/demand relationships, *viz*. the inevitability of a world which cannot get by without a dramatic re-expansion of Middle East oil production and exports (IEA, 2000). In this context, the perceived international fears for dependence on Middle East oil are complemented by other fears arising from the hypothesised absence of any alternative to such dependence. The former fears are, as shown in this paper, justified by the past 25 years' experiences and by the continuing absence of evidence to indicate any near-future radical changes in the fundamentals of the situation. Not even the re-entry of the international oil companies to the arena would instantly alter the perception of the unreliability of supplies on a continuing basis at prices related to costs rather than to politics.

The latter fears, arising from the hypothesised inevitability of dependence on the Middle East, are, however, much less soundly based. They emerge from perceived supply prospects and demand developments which are hardly credible. On the demand side they predict a 2+% annual growth rate compared with an average 0.95% annual rate since 1975; while on the supply side, they grossly understate the potential from outside the Middle East. They omit even the most basic consideration of the industry's recent most rapid technological developments ever in the exploration for, and the exploitation of, oil reserves. Such expertise, having been initiated in a few limited areas (both onshore and offshore the United States, offshore Brazil and Angola and in the North Sea), is now diffusing around most of the world with oil potential; except, ironically, in the Middle East, where the upstream industry generally has been starved of investments and expertise. Under such conditions, any forecast of the inevitability of the world's future dependence on Middle East oil has a low probability of being correct.

## Bibliography

M. A. Adelman, *The World Petroleum Market*, J. Hopkins University Press, Baltimore, 1972.

M. A. Adelman, *The Economics of Oil Supply*, The MIT Press, Cambridge, Mass, 1993.

M. A. Adelman, *The Genie out of the Bottle, World Oil since 1970*, The MIT Press, Mass, 1995.

D. Hirst, *Oil and Public Opinion in the Middle East*, Faber and Faber, London, 1966.

C. Issawi and M. Yeganeh, *The Economics of Middle Eastern Oil*, Praeger, New York, 1962.

E. Kanovsky, *The Diminishing Importance of Middle East Oil*, Holmes and Meier, New York, 1982.

S. H. Longrigg, *Oil in the Middle East*, Oxford University Press, Oxford, 3rd Edition, 1968.

H. Lubell, *Middle East Oil Crises and W. Europe's Energy Supplies*, Elsevier Press, New York, 1971.

International Energy Agency, *World Energy Outlook to 2020*, OECD, Paris, 2000.

G. Lenczowski, *Oil and State in the Middle East*, Cornell University Press, Ithaca, New York, 1960.

R. F. Mahmoud and J. N. Beck, "Why the Middle East Fields may produce Oil forever", *Offshore*, April 1995, pp.56–62.

Ø. Noreng, *Oil and Islam*, Wiley and Sons, Chichester, 1998.

P. R. Odell, "The Significance of Oil, *Journal of Contemporary History*, Vol.3, No.3, 1968, pp.93–110.

P. R. Odell, *Oil and World Power*, Penguin Books, London, 8th Edition, 1986.

P. R. Odell, "The Global Oil Industry – Middle East Domination or Regionalization, *Regional Studies*, Vol.31, No.3, 1997.

P. R. Odell, *Fossil Fuel Resources in the 21st Century*, Financial Times Energy, London, 1999.

E.T. Penrose, *The Large International Firm in Developing Countries*, Allen and Unwin, London, 1968.

B. Shwadran, *The Middle East, Oil and the Great Powers*, Wiley, New York, 3rd Edition, 1979.

P. Stevens, *Joint Ventures in Middle East Oil*, 1957–1975, Graham and Trotman, London, 1976.

G. W. Stocking, *Middle East Oil: a Study in Political and Economic Controversy*, Allen Lane, London, 1971.

# Index

## A

# C

# D

# E

# I

# M

# O

# P

# Q

# R

# S

# T

# U